INDUSTRIAL ROBOTICS
Technology, Programming, and Applications

CAD/CAM, Robotics, and Computer Vision

Consulting Editor

Herbert Freeman, *Rutgers University*

Fu, Gonzalez, and Lee: *Robotics: Control, Sensing, Vision, and Intelligence*
Groover, Weiss, Nagel, and Odrey: *Industrial Robotics: Technology, Programming, and Applications*

INDUSTRIAL ROBOTICS
Technology, Programming, and Applications

Mikell P. Groover

Professor of Industrial Engineering
Lehigh University

Mitchell Weiss

ProgramMation Inc.
Cofounder of United States Robots, Inc.

Roger N. Nagel

Professor of Computer Science
and Electrical Engineering
Lehigh University

Nicholas G. Odrey

Associate Professor of Industrial Engineering
Lehigh University

McGraw-Hill Book Company

New York St. Louis San Francisco Auckland Bogotá Hamburg
Johannesburg London Madrid Mexico Montreal New Delhi
Panama Paris São Paulo Singapore Sydney Tokyo Toronto

This book was set in Times Roman by Scanway Graphics International Ltd.
The editor was Anne Murphy;
the production supervisor was Diane Renda;
the cover was designed by Mark Wieboldt.
Project supervision was done by Publishing Synthesis, Ltd.
R.R. Donnelley & Sons Company was printer and binder.

INDUSTRIAL ROBOTICS
Technology, Programming, and Applications

34567890 DOCDOC 89876

ISBN 0-07-024989-X

Library of Congress Cataloging-in-Publication Data

Groover, Mikell P., 1939–
   Industrial robotics.

   1. Robots, Industrial. I. Title.
TS191.8.G76   1986   629.8'92   85-18453
ISBN 0-07-024989-X

CONTENTS

Part 3 Robot Programming and Languages

Part 4 Applications Engineering for Manufacturing

Part 6 Implementation Principles and Issues

Part 7 Social Issues and the Future of Robotics

FOREWORD

In 1946 when I was a senior at Columbia University one course I would have revelled in is the one whose textbook might have been *Industrial Robotics: Technology, Programming, and Applications*. But, of course, in 1946 such a course would have been impossible. So much of the technology was not at hand and there was hardly any motivation, other than sheer fun, to build robots. Fun had already prompted the ingenious automatons that appear herein as background history.

Yet the beast was stirring. My friend, Isaac Asimov, was busy at Boston College telling us with prophetic science fiction how it could be; nay, how it should be. None of that doomsday view owing to Capek.

Surely Asimov left a subliminal message with me as I put my Columbia training to industrial control problems. Servo theory was born in World War II, an esoteric discipline called boolean algebra became digital logic and the transistor was invented after I graduated.

The story of my fortuitous association with inventor Devol is recounted herein. Collectively it all welled up, as Victor Hugo would have had it, "an idea whose time had come."

Devol and I started in 1956. We had a first robot in the field in 1961 when the all-in cost of an automotive worker was $3.50/hour. Ever since the cost of labor has increased to the current level of $21.00/hour in the automotive industry. Meanwhile, the cost of manufacturing a robot has tumbled from $60,000 in 1961 dollars to $25,000 in 1984 dollars. And a robot's capabilities today are so vastly superior to those of the first lumbering machines!

Enter *Industrial Robotics: Technology, Programming, and Applications*. Here is the whole spectrum of the state of the art. My mind boggles to think how an engineering senior armed with this background would have been greeted by my original employer's personnel office. Probably he or she would have been burned at the stake!

The authors bring it all together from design to use. And there is no shirking from peripheral activities such as CAD-CAM and Artificial Intelligence.

A would-be robot designer will learn the basic criteria. He or she will not have to "reinvent the wheel." His or her counterpart on the factory floor will be able to pick intelligently from the range of robots on the market.

My only problem in recommending this book is a selfish one. Since my colleagues and I sold Unimation, Inc., I have become a consultant. Will my clients still pay me dearly if they can read all about it in this exhaustive tutorial treatise?

Joseph F. Engelberger

PREFACE

This book is intended to provide a comprehensive survey of the technical topics related to industrial robotics. The field of robotics is emerging to become one of the important automation areas for the 1980s and 1990s. Engineers, technicians, and managers must be educated and trained in order to realize the full potential of this technology. It is our hope that this book might help to satisfy the need for text materials to develop these technically educated people.

Our book is designed principally as a text for use in undergraduate and first year graduate engineering programs. It should be suitable for courses in several departments, including mechanical, industrial, manufacturing, and electrical engineering. The book includes mechanical joint-link analysis, control systems, sensors, machine vision, end effector design, and other topics of interest to these engineering disciplines. The text would also be appropriate for courses in computer science since a substantial portion of the book is devoted to robot programming. We have also designed the book for industrial training courses, and it contains much material that is relevant to those who must install robot systems. In short, it is a book on the technology, programming, and applications of industrial robots that should serve the student of robotics making the transition from the classroom and laboratory environment of academia into the applied and practical world of industry.

We began developing the outline for this text in 1981. The contract with McGraw-Hill to write the book was signed in summer 1982. A great deal has happened in the field of robotics since then. The robotics industry has changed dramatically from one dominated by small companies to one consisting of a significant number of large corporations. We are beginning to see the fallout of the weaker companies in the industry. The technology has also developed dramatically during these several years. Computer control has become pervasive; machine vision and other sensors have captured much of the spotlight in robotics; and other advances have made robots a more sophisticated yet easier to use technology.

Also during the last several years we (the authors) have learned a great deal about the field of robotics. We have taught a course at Lehigh in the subject several times; we have developed new text materials and problem sets suitable for such a course; we have designed new robot systems; and we have developed a wide variety of industrial robot applications. The reader of this book is the beneficiary of these developments.

Something else that has happened since 1982 is that Lehigh University has hired two new faculty members whose expertise includes robotics: Roger Nagel in Fall 1982 and Nick Odrey in Fall 1983. Accordingly, we have seen our way to invite them to add their expertise to this book. Roger's education and interests are in computer science, and his professional experience includes research at the National Bureau of Standards in robot programming and machine vision. He was also Director of Automation at International Harvester where he managed projects in robotics before coming to Lehigh. Nick is an aerospace engineer turned industrial engineer. His background is heavily oriented toward the mathematical analysis of control systems and mechanical linkages (such as a robot's mechanical manipulator). Combining their knowledge with that of the two original coauthors we have a team whose expertise in robotics is substantial. The coauthors' educational backgrounds and affiliations include mechanical engineering, industrial engineering, computer science and electrical engineering. Their professional backgrounds include both academe and industry. We believe that the breadth and depth of knowledge and experience of this team has permitted us to provide a more complete and comprehensive coverage of industrial robotics than exists in any other available text on this subject.

The book contains 20 chapters, many of them technical with engineering problem sets at the end. Even the most ambitious and work-oriented instructor will have difficulty in packing all twenty chapters into a single semester. Accordingly, what must be done is to cover the chapters that are most appropriate for the particular course being offered, and send the students on their way with the hope that they will read the other chapters if the need to do so subsequently arises in their work in robotics.

In the *Instructor's and Solutions Manual* for the book (available from McGraw-Hill Book Co.), we provide suggestions about how the book might best be used to complement a course in the following disciplines: mechanical engineering, industrial or manufacturing engineering, computer science, and electrical engineering. We also recommend an outline of chapters that might be used in industrial training courses whose emphasis is on applications.

ACKNOWLEDGMENTS

There are many people and organizations to be acknowledged for their contributions and assistance in publishing this book. Our fear is that we may overlook some whose contributions were significant enough to merit inclusion. To those we have overlooked, if there are any, we apologize in advance. For their technical input and/or review of portions of the manuscript, we are indebted to the following: Robert Alexander of the Lord Corp; Jack Basiago, graduate student in Manufacturing Systems Engineering (MSE) at Lehigh; Geneen Budreau of the Lord Corp; Mark Dane of Cincinnati Milacron; Scott Dickenson, a student of robotics who questioned things; Joseph F. Engelberger, for providing us with a very nice Foreword to our book; Vernon E. Estes of the General Electric Company who, in addition to providing material for the book, was helpful during initial startup of our Robotics Laboratory at Lehigh; David Fitzpatrick of ORSI; Ray Floyd of IBM Corporation; Joseph Fromme of Cincinnati Milacron; David Hanan, MSE student; Don Hillman, Computer Science Professor at Lehigh and specialist in artificial intelligence;

Joan Juzwiak, advertising specialist for GMF Robotics; John Keefe, loyal graduate student and project engineer for special assignments; Art Kendall of IBM Corporation, good friend and advocate for Lehigh's MSE Program; Peter Kerstens, MSE student; Dana Morgan, MSE student who stimulated our thinking on a joint classification scheme; Paul J. Nelson of the IBM Corporation; John Ochs, Director of Lehigh's Computer-Aided Design Laboratory; Jim Reid, on leave from Cincinnati Milacron, currently a Lehigh MSE student; Chester A. Sadlow of Westinghouse (Unimation Inc.); Armand Small of Grumman Aerospace; Ramadan Taher, IE graduate student; Gordon Vanderbrug of Automatix; West Vogel, a graduate student in IE who prompted our thinking on robot economic analysis.

In addition, we acknowledge with gratitude the reviews of our academic peers whose comments were helpful in shaping the final version of the text: Rashpal S. Ahluwalia (Ohio State University), Stephen J. Derby (Rensselaer Polytechnic Institute), Steven Dickerson (Georgia Institute of Technology), Lyman L. Francis (University of Missouri, Rolla), Herbert Freeman (Rutgers University), Ernest L. Hall (University of Cincinnati), R. T. Johnson (University of Missouri, Rolla), Donald J. McAleece and Edward E. Messal (Indiana University and Purdue University), Daniel Metz (University of Illinois), Wolfgang Sauer (University of Massachusetts), Holger J. Sommer (Pennsylvania State University), Allen Tucker (Colgate University), Richard A. Wysk (Pennsylvania State University).

We are also indebted to B. J. Clarke, Rodger Klas, and Anne Murphy of McGraw-Hill Book Co. for their wisdom and perception in selecting us to do the book, and their patience and tolerance with us during manuscript preparation.

For their help in preparing our solutions manual for the book we would like to thank Cemal Doyden, IE graduate student; Thomas Grycan, IE undergraduate; and W. Scott Sendel, IE undergraduate. We must also acknowledge Fern Sotzing, for her secretarial assistance and friendly disposition during manuscript preparation.

Finally, we would like to express our appreciation to our wives, respectively, Bonnie, Nancy, Arlene, and Sandy, for their understanding and encouragement during the many hours that their husbands spent on the book.

Mikell P. Groover
Mitchell Weiss

ABOUT THE AUTHORS

MIKELL P. GROOVER received his B.A. in Applied Science, B.S. in Mechanical Engineering, and M.S. and Ph.D. in Industrial Engineering from Lehigh University. He is Professor of Industrial Engineering and Director of the Manufacturing Technology Laboratory at Lehigh. He is author and coauthor of two previous books on automation and CAD/CAM, respectively (published by Prentice-Hall). His areas of specialization include manufacturing technology, automation, and robotics.

MITCHELL WEISS received his B.S. in Mechanical Engineering from the Massachusetts Institute of Technology. He was employed as the applications engineer for the PUMA robot by Unimation, Inc., prior to cofounding United States Robots in 1980. He was involved in the design of robots for the company. He has started another company, ProgramMation, and is currently its President.

ROGER N. NAGEL received his B.S. from Stevens Institute of Technology and his M.S. and Ph.D. in Computer Science from the University of Maryland. His professional experience includes the National Bureau of Standards and International Harvester, Inc., as Corporate Director of Automation Technology. He is currently Director of the Institute for Robotics at Lehigh University, and Professor of Computer Science and Electrical Engineering.

NICHOLAS G. ODREY is currently the Director of the Robotics Laboratory within the Institute for Robotics at Lehigh University and is an Associate Professor of Industrial Engineering. His academic background includes a B.S. and M.S. in Aerospace Engineering. After considerable experience in the aerospace industry, he returned for his Ph.D. in Industrial Engineering with specialization in Manufacturing Systems at the Pennsylvania State University. Prior to joining Lehigh, he was associated with the University of Rhode Island, West Virginia University, the National Bureau of Standards, and was a faculty fellow with the U.S. Air Force ICAM program.

PART
ONE

FUNDAMENTALS OF ROBOTICS

This first part of our book is intended to introduce the subject of industrial robotics, both its social significance and its technological importance. Robotics is a prominent component of manufacturing automation which will affect human labor at all levels, from unskilled workers to professional engineers and managers of production. Future robots may find applications outside of the factory in banks, restaurants, and even homes. It is possible, perhaps likely, that robotics will become a field, like today's computer technology, which is pervasive throughout our society. Our book undertakes the ambitious objective of providing technical literacy in this fascinating field. Part One introduces the fundamentals.

Part One contains two chapters. Chapter One defines the term robot in the context of industrial automation. It also provides a short history of the development of the technology, including a section on how robotics has been perceived by the public in science fiction stories.

Chapter Two entitled Fundamentals of Robot Technology, Programming, and Applications can almost be considered a summary of the entire book. We survey each of these three areas. Each is interrelated to the others: robotics technology is controlled by means of programming, and the ability to program a robot is dependent on its level of technology. Successful implementation of robotics in useful applications is obviously a function of the technology and programming. It is important for the reader to have an appreciation of all three areas in order to understand the technical details of each separate area. Chapter Two attempts to generate this appreciation of the three areas.

ONE

INTRODUCTION

The field of robotics has its origins in science fiction. The term robot was derived from the English translation of a fantasy play written in Czechoslovakia around 1920. It took another 40 years before the modern technology of industrial robotics began. Today, robots are highly automated mechanical manipulators controlled by computers.

In this chapter, we survey some of the science fiction stories about robots, and we trace the historical development of robotics technology. Let us begin our chapter by defining the term robotics and establishing its place in relation to other types of industrial automation.

1-1 AUTOMATION AND ROBOTICS

Automation and robotics are two closely related technologies. In an industrial context, we can define automation as a technology that is concerned with the use of mechanical, electronic, and computer-based systems in the operation and control of production. Examples of this technology include transfer lines, mechanized assembly machines, feedback control systems (applied to industrial processes), numerically controlled machine tools, and robots. Accordingly, robotics is a form of industrial automation.

There are three broad classes of industrial automation: fixed automation, programmable automation, and flexible automation. Fixed automation is used when the volume of production is very high and it is therefore appropriate to design specialized equipment to process the product (or a component of a product) very efficiently and at high production rates. A good example of fixed automation can be found in the automobile industry, where highly integrated transfer lines consisting of several dozen workstations are used to perform

3

machining operations on engine and transmission components. The economics of fixed automation are such that the cost of the special equipment can be divided over a large number of units, and the resulting unit costs are low relative to alternative methods of production. The risk encountered with fixed automation is this; since the initial investment cost is high, if the volume of production turns out to be lower than anticipated, then the unit costs become greater than anticipated. Another problem with fixed automation is that the equipment is specially designed to produce the one product, and after that product's life cycle is finished, the equipment is likely to become obsolete. For products with short life cycles, the use of fixed automation represents a big gamble.

Programmable automation is used when the volume of production is relatively low and there are a variety of products to be made. In this case, the production equipment is designed to be adaptable to variations in product configuration. This adaptability feature is accomplished by operating the equipment under the control of a "program" of instructions which has been prepared especially for the given product. The program is read into the production equipment, and the equipment performs the particular sequence of processing (or assembly) operations to make that product. In terms of economics, the cost of the programmable equipment can be spread over a large number of products even though the products are different. Because of the programming feature, and the resulting adaptability of the equipment, many different and unique products can be made economically in small batches.

The relationship of the first two types of automation, as a function of product variety and production volume, is illustrated in Fig. 1-1. There is a third category between fixed automation and programmable automation, which is called "flexible automation." Other terms used for flexible automation include "flexible manufacturing systems," (or FMS) and "computer-integrated manufacturing systems." The concept of flexible automation has only developed into practice within the past 15 or 20 years. Experience thus far with this type of automation suggests that it is most suitable for the midvolume production range, as shown in Fig. 1-1. As indicated by its position relative to the other two types, flexible systems possess some of the features of both fixed automation and programmable automation. It must be programmed for different product configurations, but the variety of configurations is usually more limited than for programmable automation, which allows a certain amount of integration to occur in the system. Flexible automated systems typically consist of a series of workstations that are interconnected by a materials-handling and storage system. A central computer is used to control the various activities that occur in the system, routing the various parts to the appropriate stations and controlling the programmed operations at the different stations.

One of the features that distinguishes programmable automation from flexible automation is that with programmable automation, the products are made in batches. When one batch is completed, the equipment is reprogram-

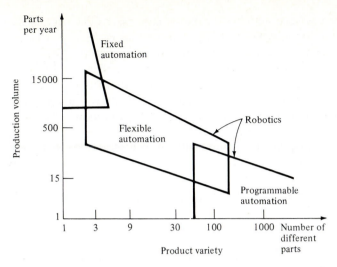

Figure 1-1 Relationship of fixed automation, programmable automation, and flexible automation as a function of production volume and product variety.

med to process the next batch. With flexible automation, different products can be made at the same time on the same manufacturing system. This feature allows a level of versatility that is not available in pure programmable automation, as we have defined it. This means that products can be produced on a flexible system in batches if that is desirable, or several different product styles can be mixed on the system. The computational power of the control computer is what makes this versatility possible.

Of the three types of automation, robotics coincides most closely with programmable automation. An industrial robot is a general-purpose, programmable machine which possesses certain anthropomorphic, or humanlike, characteristics. The most typical humanlike characteristic of present-day robots is their movable arms. The robot can be programmed to move its arm through a sequence of motions in order to perform some useful task. It will repeat that motion pattern over and over until reprogrammed to perform some other task. Hence, the programming feature allows robots to be used for a variety of different industrial operations, many of which involve the robot working together with other pieces of automated or semiautomated equipment. These operations include machine loading and unloading, spot welding, and spray painting.

The "official" definition of an industrial robot is provided by the Robotics Industries Association (RIA), formerly the Robotics Institute of America (RIA):

> An industrial robot is a reprogrammable, multifunctional manipulator designed to move materials, parts, tools, or special devices through variable programmed motions for the performance of a variety of tasks.

This definition reinforces our conclusion that industrial robots should be classified as a form of programmable automation.

While the robots themselves are examples of programmable automation, they are sometimes used in flexible automation and even fixed automation systems. These systems consist of several machines and/or robots working together, and are typically controlled by a computer or a programmable controller. A production line that performs spot welds on automobile bodies is an example of this kind of system. The welding line might consist of two dozen robots or more, and is capable of accomplishing hundreds of separate spot welds on two or three different body styles (e.g., sedans, coupes, and station wagons). The robot programs are contained in the computer or programmable controller and are downloaded to each robot for the particular automobile body that is to be welded at each station. Owing to this feature, such a line might appropriately be considered a high-production flexible automation system.

Today the human analogy of an industrial robot is very limited. Robots do not look like humans, and they do not behave like humans. Instead, they are one-armed machines which almost always operate from a fixed location on the factory floor. Future robots are likely to have a greater number of attributes similar to the attributes of humans. They are likely to have greater sensor capabilities, more intelligence, a higher level of manual dexterity, and a limited degree of mobility. There is no denying that the technology of robotics is moving in a direction to provide these machines with more and more capabilities like those of humans.

1-2 ROBOTICS IN SCIENCE FICTION

Notwithstanding the limitations of current-day robotic machines, the popular concept of a robot is that it looks and acts like a human being. This humanoid concept has been inspired and encouraged by a number of science fiction stories.

Certainly one of the first works of relevance to our discussion of robotics in science fiction was a novel by Mary Shelley, published in England in 1817. Titled *Frankenstein*, the story deals with the efforts of a scientist, Dr. Frankenstein, to create a humanoid monster which then proceeds to raise havoc in the local community. The story has been popularized in several versions over the years through the medium of motion pictures. The movie screen image of the Frankenstein monster gone astray from the plans of its well-intentioned creator has made a lasting impression on the minds of millions of people. This impression has carried over to robots where the word conjures up similar images of science and technology in danger of running amuck.

A Czechoslovakian play in the early 1920s by Karel Capek, called *Rossum's Universal Robots*, gave rise to the term robot. The Czech word "robota" means servitude or forced worker, and when translated into English, the translated word became robot. The story concerns a brilliant scientist named Rossum and his son who develop a chemical substance that is similar to protoplasm. They use the substance to manufacture robots. Their plan is that the robots will serve humankind obediently and do all physical labor. Rossum continues to make improvements in the design of the robots, eliminating unnecessary organs and other parts, and finally develops a "perfect" being. The plot takes a sour turn when the perfect robots begin to dislike their subservient role and proceed to rebel against their masters, killing all human life.

Among science fiction writers, Isaac Asimov has contributed a number of stories about robots, starting in 1939, and indeed is credited with coining the term "robotics." The picture of a robot that appears in his work is that of a well-designed, fail-safe machine that performs according to three principles. These principles were called the Three Laws of Robotics by Asimov, and they are:

1. A robot may not injure a human being or, through inaction, allow a human to be harmed.
2. A robot must obey orders given by humans except when that conflicts with the First Law.

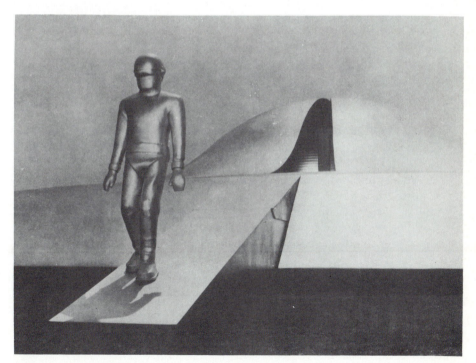

Figure 1-2 The robot Gort in the film *The Day the Earth Stood Still.* (Courtesy of Twentieth-Century-Fox.)

3. A robot must protect its own existence unless that conflicts with the First or Second Laws.

   A number of movies and television shows have added to the lore of robotics, some picturing robots as friendly servants and companions, others showing them in different ways. The 1951 movie *The Day the Earth Stood Still* was about a mission from a distant planet sent to earth in a flying saucer to try to establish the basis for peace among the world's nations. The crew of the flying saucer consisted of only two members, a humanlike being and an omniscient, omnipotent, indestructible robot named Gort. The robot was a "universal" (excuse the pun) peacekeeper, and when a planet got out of line, punishment was swift and final. The mission to earth was not a complete success (obviously), but it demonstrated the terrible destructive power of future weapons. Figure 1-2 shows a photo of Gort from the film.

Figure 1-3 The friendly robots R2D2 and C3PO from *Star Wars*. (Courtesy of LUCASFILM LTD., © Lucasfilm Ltd. (LFL) 1977. All rights reserved.)

The 1968 movie *2001: A Space Odyssey* contained not a mechanical robot but a highly intelligent, talking computer named HAL. The job of the computer was to monitor and control the systems on-board the spaceship on its way to the planet Jupiter, and to be a friend and companion to the spaceship crew. During the voyage, one of HAL's circuits fails and its personality goes bad. It begins killing off the members of the crew in order to protect itself and is only stopped in a final contest between itself and the one remaining crew member.

The *Star Wars* series (*Star Wars* in 1977, *The Empire Strikes Back* in 1980, and *The Return of the Jedi* in 1983) pictured robots as friendly, harmless machines. The robots, R2D2 and C3PO (Figure 1-3), are able to move around, they are intelligent, and they can communicate with their human masters. They do not play major roles in these movies except mostly as comic relief. However, to movie audiences, they stand as significant characters because they are so benevolent and because they represent the opportunities offered by robotics and other advanced technologies to be helpful and un-threatening to humans.

1-3 A BRIEF HISTORY OF ROBOTICS

Science fiction has no doubt contributed to the development of robotics, by planting ideas in the minds of young people who might embark on careers in robotics, and by creating awareness among the public about this technology. We should also identify certain technological developments over the years that have contributed to the substance of robotics. Table 1-1 presents a chrono-logical listing which summarizes the historical developments in the technology of robotics.

Some of the early developments in the field of automata deserve mention although not all of them deal directly with robotics. In the seventeenth and eighteenth centuries, there were a number of ingenious mechanical devices that had some of the features of robots. Jacques de Vaucanson built several human-sized musicians in the mid-1700s. Essentially these were mechanical robots designed for a specific purpose: entertainment. In 1805, Henri Mail-lardet constructed a mechanical doll which was capable of drawing pictures. A series of cams were used as the "program" to guide the device in the process of writing and drawing. Maillardet's writing doll is on display in the Franklin Institute in Philadelphia, Pennsylvania. These mechanical creations of human form must be regarded as isolated inventions reflecting the genius of men who were well ahead of their time. There were other mechanical inventions during the industrial revolution, created by minds of equal genius, many of which were directed at the business of textile production. These included Har-greaves' spinning jenny (1770), Crompton's mule spinner (1779), Cartwright's power loom (1785), the Jacquard loom (1801), and others.

In more recent times, numerical control and telecherics are two important

Table 1-1 Chronology of developments related to robotics technology, including significant robot applications

Date	Development
mid-1700s	J. de Vaucanson built several human-sized mechanical dolls that played music.
1801	J. Jacquard invented the Jacquard loom, a programmable machine for weaving threads or yarn into cloth.
1805	H. Maillardet constructed a mechanical doll capable of drawing pictures.
1946	American inventor G. C. Devol developed a controller device that could record electrical signals magnetically and play them back to operate a mechanical machine. U.S. patent issued in 1952.
1951	Development work on teleoperators (remote-control manipulators) for handling radioactive materials. Related U.S. patents issued to Goertz (1954) and Bergsland (1958).
1952	Prototype Numerical Control machine demonstrated at the Massachusetts Institute of Technology after several years of development. Part programming language called APT (Automatically Programmed Tooling) subsequently developed and released in 1961.
1954	British inventor C. W. Kenward applied for patent for robot design. British patent issued in 1957.
1954	G. C. Devol develops designs for "programmed article transfer." U.S. patent issued for design in 1961.
1959	First commercial robot introduced by Planet Corporation. It was controlled by limit switches and cams.
1960	First "Unimate" robot introduced, based on Devol's "programmed article transfer." It used numerical control principles for manipulator control and was a hydraulic drive robot.
1961	Unimate robot installed at Ford Motor Company for tending a die casting machine.
1966	Trallfa, a Norwegian firm, built and installed a spray painting robot.
1968	A mobile robot named "Shakey" developed at SRI (Stanford Research Institute). It was equipped with a variety of sensors, including a vision camera and touch sensors, and it can move about the floor.
1971	The "Stanford Arm," a small electrically powered robot arm, developed at Stanford University.
1973	First computer-type robot programming language developed at SRI for research called WAVE. Followed by the language AL in 1974. The two languages were subsequently developed into the commercial VAL language for Unimation by Victor Scheinman and Bruce Simano.
1974	ASEA introduced the all-electric drive IRb6 robot.
1974	Kawasaki, under Unimation license, installed arc-welding operation for motorcycle frames.
1974	Cincinnati Milacron introduced the T^3 robot with computer control.
1975	Olivetti "Sigma" robot used in assembly operation—one of the very first assembly applications of robotics.

Table 1-1 (*cont.*)

Date	Development
1976	Remote Center Compliance (RCC) device for part insertion in assembly developed at Charles Stark Draper Labs in United States.
1978	PUMA (Programmable Universal Machine for Assembly) robot introduced for assembly by Unimation, based on designs from a General Motors study.
1978	Cincinnati Milacron T^3 robot adapted and programmed to perform drilling and routing operations on aircraft components, under Air Force ICAM (Integrated Computer-Aided Manufacturing) sponsorship.
1979	Development of SCARA type robot (Selective Compliance Arm for Robotic Assembly) at Yamanashi University in Japan for assembly. Several commercial SCARA robots introduced around 1981.
1980	Bin-picking robotic system demonstrated at University of Rhode Island. Using machine vision, the system was capable of picking parts in random orientations and positions out of a bin.
1981	A "direct-drive robot" developed at Carnegie-Mellon University. It used electric motors located at the manipulator joints without the usual mechanical transmission linkages used on most robots.
1982	IBM introduces the RS-1 robot for assembly, based on several years of in-house development. It is a box-frame robot, using an arm consisting of three orthogonal slides. The robot language AML, developed by IBM, also introduced to program the RS-1.
1983	Report issued on research at Westinghouse Corp. under National Science Foundation sponsorship on "adaptable-programmable assembly system" (APAS), a pilot project for a flexible automated assembly line using robots.
1984	Several off-line programming systems demonstrated at the Robots 8 show. Typical operation of these systems allowed the robot program to be developed using interactive graphics on a personal computer and then downloaded to the robot.

technologies in the development of robotics. Numerical control (NC) was developed for machine tools in the late 1940s and early 1950s. As its name suggests, numerical control involves the control of the actions of a machine tool by means of numbers. It is based on the original work of John Parsons who conceived of using punched cards containing position data to control the axes of a machine tool. He demonstrated his concept to the United States Air Force, which proceeded to support a research and development project at the Massachusetts Institute of Technology. The MIT project used a three-axis milling machine to demonstrate the prototype for NC in 1952. Subsequent work at MIT led to the development of APT (Automatically Programmed Tooling), a part programming language to accomplish the programming of the NC machine tool. It is interesting to note that the Jacquard loom and the player piano, developed around 1876, can be considered to be precursors of the modern NC machine tool. Both operated using a form of punched paper tape as a program to control the actions of the respective machines.

The field of telecherics deals with the use of a remote manipulator controlled by a human being. Sometimes called a teleoperator, the remote manipulator is a mechanical device which translates the motions of the human operator into corresponding motions at a remote location. A common use of a teleoperator is in the handling of dangerous substances, such as radioactive materials. The human can remain in a safe location; yet by peering through a leaded glass window or by viewing on closed-circuit television, the operator can guide the movements of the remote arm. Early telecheric devices were entirely mechanical, but more modern systems use a combination of mechanical systems and electronic feedback control. Work on teleoperator designs for handling radioactive materials dates back to the 1940s. Telecheric devices were used by the Atomic Energy Commission starting around the same time.

It is the combination of numerical control and telecherics that forms the basis for the modern robot. The robot is a mechanical manipulator whose motions are controlled by programming techniques very similar to those used in numerical control. There are two individuals who must be credited with recognizing the confluence of these two technologies and the potential it might offer in industrial applications. The first was a British inventor named Cyril Walter Kenward who applied for a British patent for a robotic device in March 1954. This patent was issued in 1957. The sketch for this device is shown in Fig. 1-4.

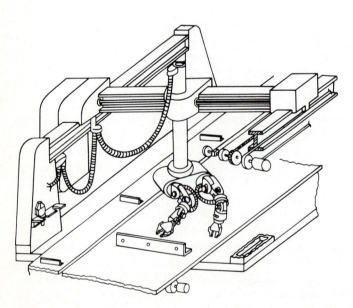

Figure 1-4 Sketch of the robotic device for which Cyril Walter Kenward was issued a British patent in 1957.

The second person who must be mentioned in this context is George C. Devol, the American inventor, who must be credited with two inventions that led to the development of modern day robots. The first was a device for recording electrical signals magnetically and playing them back to control a machine. The device is dated around 1946 and the U.S. patent for it was issued in 1952. The second invention was titled "Programmed Article Transfer" and the U.S. patent for this device was issued in 1961. The description of the device in the June 13, 1961 edition of the *U.S. Patent Record* is presented in Fig. 1-5, and the sketch based on the diagram which accompanied the description is shown in Fig. 1-6. Although Devol's patent followed Kenward's by several years, it was Devol's work that established the foundation for the modern industrial robot. What made Devol's invention into an industry in the United States rather than in the United Kingdom was the presence of a catalyst in the person of Joseph Engelberger.

Joseph F. Engelberger graduated from Columbia University with a graduate degree in physics in 1949. As a student, he had read with fascination several of Asimov's novels. By the mid-1950s he was the chief engineer for an aerospace division of a company located in Stamford, Connecticut. The division was in the business of making controls for jet engines. Hence, by the

2,988,237
PROGRAMMED ARTICLE TRANSFER
George C. Devol, Jr., Brookside Drive, Greenwich, Conn.
Filed Dec. 10, 1954, Ser. No. 474,574
28 Claims. (Cl. 214—11)

1. Apparatus having automatic control means, including a mechanical output device and power operating means therefor, position representing means coupled to said mechanical output device for conjoint operation therewith, said position representing means including an assembly of separate sensing units each having its own individual output and combinational code means sensed by and relatively movable with respect to said sensing units through a series of positions corresponding to positions of said device, said combinational code means including a uniquely identifying combination of control portions of different kinds opposite the respective sensing units in each said position such that the combination of control portions at each said position is different from the combinations of control portions in all the others of said series of positions, a program-controller having a series of recorded combinational code position symbols duplicating selected combinations in said series of positions, said program-controller having as many position-symbol sensing elements for sensing said combinational code position symbols as there are sensing units in said position representing means, a series of individual coincidence detectors each having its respective direct signal coupling to a corresponding one of said sensing units and to a corresponding one of said sensing elements, said power operating means having control means responsive to said coincidence detectors jointly.

Figure 1-5 Description of George C. Devol's "Programmed Article Transfer" as it appeared in the *U.S. Patent Record* dated June 13, 1961.

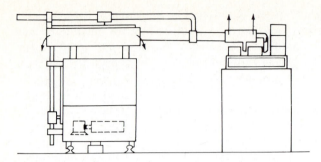

Figure 1-6 Sketch of Devol's "Programmed Article Transfer" similar to the diagram that accompanied the patent description.

time a chance meeting took place in 1956, Engelberger was predisposed by education, avocation, and occupation toward the notion of robotics. As fate would have it, Joseph Engelberger met George Devol at a cocktail party held in Fairfield, Connecticut. During the conversation, Devol told Engelberger about his invention of the programmed article transfer device, and the two subsequently began discussing the possibility of commercializing the invention. Through the financial backing of the Consolidated Diesel Electric Company (now Condec Corp.), Engelberger and Devol started to develop plans and prototypes for the universal helper, or "Unimate." In 1962, the Unimation Company was founded as a joint venture between Consolidated Diesel Electric and the Pullman Corporation. Engelberger became president of the company and has promoted the development and the application of robotics ever since. Figure 1-7 shows Engelberger and Devol being served by a Unimate.

The first recorded installation of a Unimate robot was at the Ford Motor Company for unloading a die-casting machine. (It is ironic to note that although Ford was one of the very first companies to use a robot, it refused for many years to recognize the word robot, preferring instead to use the term "universal transfer device" or UTD.) More applications followed, slowly at first, using robots not only from Unimation, but also from a number of other companies in the United States, Europe, and Japan. Some of the more significant robot installations are included in Table 1-1.

There were many other worthwhile contributions to the field of robotics, although space limits our including all of them. It is appropriate to note some of the pioneering work at Stanford University and Stanford Research Institute on computer-oriented robot languages. In 1973, the experimental language called WAVE was developed. This was followed by the development of the AL language in 1974, another language designed for research. The first commercial robot language was VAL, developed by Victor Scheinman and Bruce Simano for Unimation, Inc. The language was first used to program Unimation's PUMA robot, a relatively small jointed-arm robot whose design

Figure 1-7 Joseph Engelberger (left) and George Devol being served drinks by a Unimate robot. (Photo courtesy of Joseph F. Engelberger.)

was based on studies of assembly automation that had been done by General Motors. PUMA stands for Programmable Universal Machine for Assembly.

The Stanford work on robot languages, and much of the subsequent work that has been done in robotics, is largely based on developments in computer technology. Although computers were certainly available at the birth of the robotics industry, it was not until the mid to late 1970s that the economics were right for the use of a small computer as the robot controller. Today, nearly all robots introduced into the market use computer controls. Indeed, the field of robotics is often considered to be a combination of machine tool technology and computer science.

1-4 THE ROBOTICS MARKET AND THE FUTURE PROSPECTS

Annual sales for industrial robots have been growing in the United States at the rate of about 25 percent per year. Figure 1-8 presents the current statistics and projections for annual sales of industrial robots and the resulting number

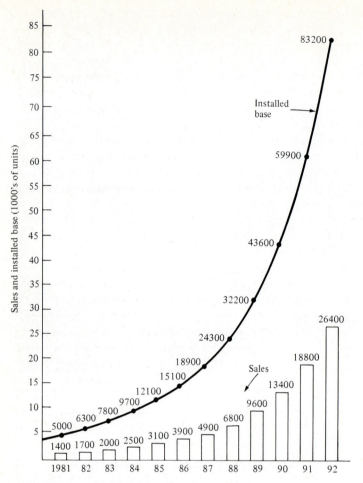

Figure 1-8 Actual and projected annual sales and the resulting number of installations of industrial robots in the United States through the early 1990s.

of robot installations. Past and present sales and installation statistics are based on our best estimates at time of writing. The sales projections show a continued average annual growth rate of 25 percent through 1987. Around 1987, we expect the growth rate to increase in the United States due to several factors. First, more people in industry are becoming aware of the technology and aware of its potential for useful applications. Second, the technology of robotics will improve over the next few years in such a way as to make robots more user friendly, easier to interface to other hardware and software, and easier to install. Third, as the market grows, we expect economies of scale in the production of robots to effect a reduction in the unit price, making robot application projects easier to justify. Fourth, the robotics market is expected to expand beyond the large corporation, which has been the traditional customer for this technology, and into the medium-sized and smaller companies. This will result in a substantial increase in the customer base for industrial robots. It cannot be determined whether these factors will all converge in the year 1988 to create a sudden surge in demand. However, for the purposes of the projection, we are using 1988 as the year when the sales growth rate will increase, and we are using 40 percent as our estimate of that new growth rate.

The projected installed base in Fig. 1-8 represents the accumulation of these annual sales, adjusted for obsolete robots that have been discarded. We believe it is reasonable to assume that installed robots will become worn out and/or technologically obsolete after an average seven-year service life. Advances in the technology and reductions in pricing will make new units relatively attractive compared to old units in service.

Robotics is a technology with a future, and it is a technology for the future. If present trends continue, and if some of the laboratory research currently underway is ultimately converted into practicable technology, robots of the future will be mobile units with one or more arms, multiple sensor capabilities, and the computational and data processing power of today's mainframe computers. They will be able to respond to human voice command. They will be able to receive general instructions and will translate those instructions using artificial intelligence into a specific set of actions required to carry them out. They will be able to see, hear, feel, apply a precisely measured force to an object, and move under their own power. In short, future robots will have many of the attributes of human beings. It is hard to imagine that robots will ever replace humans in the sense of Karel Capek's play, "Rossum's Universal Robots." On the contrary, robotics is a technology that can be harnessed solely for the benefit of humankind. However, like other technologies, there are potential dangers involved, and safeguards must be instituted to prevent its harmful use.

Getting from the present to the future will require much work in mechanical engineering, electrical engineering, computer science, industrial engineering, materials technology, manufacturing systems engineering, and the social sciences. The purpose of this book is to explore and examine these areas which constitute the technology, programming, and application of industrial robotics.

1-5 ORGANIZATION OF THIS BOOK

The text for this book is organized into seven parts. Part One is introductory. This first chapter provides the motivation and rationale for learning about robotics. Chapter Two presents an overview of robot technology and programming. This is important because in the flow of the book, we will be discussing technical topics which must be placed into context relative to other topics. Chapter Two provides the survey of the entire field necessary to establish those relationships.

Part Two examines the technical topics that relate to the robot and the peripheral hardware used with the robot. Chapter Three discusses the mechanical components of the robot and the control systems used to control the joints of the arm and wrist. Chapter Four presents some of the mathematical analysis that is used in robotics to study the motion of the manipulator. Chapter Five covers end effectors, the mechanical hands and other devices that are attached to the robot arm to perform useful work. Chapters Six and Seven are concerned with the sensors that are used in robotics, including machine vision, an important technology which is likely to find many applications in robotics work in the future.

Part Three deals with robot programming. Chapter Eight is concerned with the fundamentals of how to program robots and what the requirements for robot programming are. Chapter Nine and its appendixes cover some of the robot textual languages that are in common use. Chapter Ten presents a survey of artificial intelligence and its relationship to robotics. We anticipate that robots of the future will possess far greater intelligence and reasoning power than current-day robots, and the field of artificial intelligence will provide the methodologies for these capabilities.

Part Four is concerned with applications engineering. What are the engineering and economic problems that must be addressed in installing robots, and what are some of the applications of robots today? Chapter 11 describes work cell design and control—how to use the technology and programming of robotics in industrial applications. Chapter Twelve presents the methods that should be used to justify a robot investment. Part Five presents a survey of how robotics is used in industry. Chapters Thirteen through Fifteen discuss the various types of robot applications in manufacturing today.

Closely related to applications engineering are the implementation issues associated with the introduction of robotics into the factory. Part Six surveys some of these issues. Chapter Sixteen proposes a seven-step approach for the implementation of robotics in the firm, from initial familiarization with the technology to installation of the robot work cell. Chapter Seventeen discusses some of the additional problem areas that must be confronted during implementation. These areas include safety, training, maintenance, and quality control.

Part Seven deals with social issues and the future of robotics. Chapter Eighteen explores the possible social impact of robotics, giving particular attention to the problems confronting labor. In chapters Nineteen and Twenty, we speculate about the following questions. What will the technology of robotics be like in the future? And what kinds of applications will robots be performing in the future?

REFERENCES

1. J. S. Albus, *Brains, Behavior, and Robotics*, BYTE Books (subsidiary of McGraw-Hill), Peterborough, NH, 1981, chap. 8.

2. R. U. Ayres and S. M. Miller, "Industrial Robots on the Line," *Technology Review*, 34–47 (May/June 1982).
3. R. U. Ayres and S. M. Miller, *Robotics—Applications and Social Implications*, Ballinger, Cambridge, MA, 1983, chaps. 1 and 2.
4. R. C. Dorf, *Robotics and Automated Manufacturing*, Reston, Reston, VA, 1983, chap. 3.
5. J. F. Engelberger, *Robotics in Practice*, AMACOM (Division of American Management Association), 1980, author's preface.
6. R. M. Glorioso and F. C. C. Osorio, *Engineering Intelligent Systems*, Digital Press, Digital Equipment Corp., 1980, chap. 15.
7. M. P. Groover, *Automation, Production Systems, and Computer-Aided Manufacturing*, Prentice-Hall, Englewood Cliffs, NJ, 1980, chap. 1.
8. "Joseph F. Engelberger—The American Machinest Award," *American Machinist*, 68–75 (December 1982).
9. B. Rooks, "The Cocktail Party That Gave Birth to the Robot," *Decade of Robotics*, IFS Publications, Bedford, England, 1983.

CHAPTER
TWO

FUNDAMENTALS OF ROBOT TECHNOLOGY, PROGRAMMING, AND APPLICATIONS

Robotics is an applied engineering science that has been referred to as a combination of machine tool technology and computer science. It includes such seemingly diverse fields as machine design, control theory, microelectronics, computer programming, artificial intelligence, human factors, and production theory. Research and development are proceeding in all of these areas to improve the way robots work and think. It is likely that the research efforts will result in future robots that will make today's machines seem quite primitive. Advancements in technology will enlarge the scope of the industrial applications of robots.

Our problem in this chapter is to define the basic technology, programming, and applications of current day industrial robotics. The technical fields listed above are highly interdependent in the manner in which they are used in robotics. In order to appreciate robotics technology and programming, one must be aware of the way robots are applied in industry. In order to understand the use of sensors in robotics, one must be familiar with the way robots are programmed. To comprehend the use of an end effector, one must know that a fundamental function of a robot is to handle parts and tools. In this chapter, therefore, we provide that survey of the entire field of robotics to establish the necessary framework for the reader to relate the various topics in the chapters that follow.

To describe the technology of a robot, we must define a variety of technical features about the way the robot is constructed and the way it operates. Robots work with sensors, tools, and grippers, and these terms must be defined. The programming of robots is accomplished in several ways.

20

Although we discuss this subject in considerable detail later in the book, a concise description is presented in this chapter. Finally, robots are used to perform work in industry, and we provide a survey of these industrial applications. To survey these various topics, this chapter is organized into the following sections:

Robot anatomy
Work volume
Drive systems
Control systems and dynamic performance
Precision of movement
End effectors
Sensors
Robot programming and work cell control
Applications

For many of these topics, it is appropriate to delve much deeper into the subject, well beyond the basic introduction intended by this chapter. We discuss these topics in greater depth in subsequent chapters of the book.

2-1 ROBOT ANATOMY

Robot anatomy is concerned with the physical construction of the body, arm, and wrist of the machine. Most robots used in plants today are mounted on a base which is fastened to the floor. The body is attached to the base and the arm assembly is attached to the body. At the end of the arm is the wrist. The wrist consists of a number of components that allow it to be oriented in a variety of positions. Relative movements between the various components of the body, arm, and wrist are provided by a series of joints. These joint movements usually involve either rotating or sliding motions, which we will describe later in this section. The body, arm, and wrist assembly is sometimes called the manipulator.

Attached to the robot's wrist is a hand. The technical name for the hand is "end effector" and we will discuss end effectors later in this chapter and in much greater detail in a later chapter. The end effector is not considered as part of the robot's anatomy. The arm and body joints of the manipulator are used to position the end effector, and the wrist joints of the manipulator are used to orient the end effector.

Four Common Robot Configurations

Industrial robots are available in a wide variety of sizes, shapes, and physical configurations. The vast majority of today's commercially available robots possess one of four basic configurations:

1. Polar configuration
2. Cylindrical configuration
3. Cartesian coordinate configuration
4. Jointed-arm configuration

The four basic configurations are illustrated in the schematic diagrams of Fig. 2-1.

The polar configuration is pictured in part (*a*) of Fig. 2-1. It uses a telescoping arm that can be raised or lowered about a horizontal pivot. The pivot is mounted on a rotating base. These various joints provide the robot

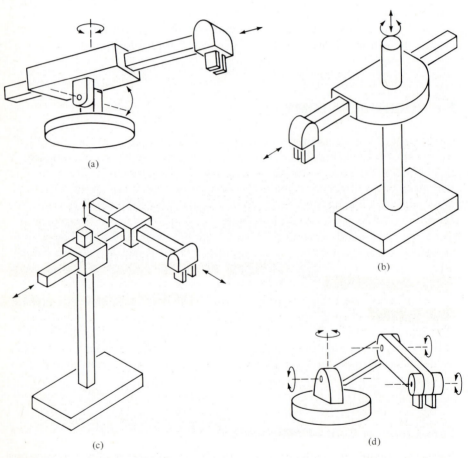

(a)

(b)

(c)

(d)

Figure 2-1 The four basic robot anatomies: (a) polar, (b) cylindrical, (c) cartesian, and (d) jointed-arm. (Reprinted from Reference [7].)

with the capability to move its arm within a spherical space, and hence the name "spherical coordinate" robot is sometimes applied to this type. A number of commercial robots possess the polar configuration. These include the familiar Unimate 2000 series, pictured in Fig. 2-2. Another robot which is much smaller than the Unimate is the MAKER 110, made by United States Robots, and illustrated in Fig. 2-3.

The cylindrical configuration, as shown in Fig. 2-1(b), uses a vertical column and a slide that can be moved up or down along the column. The robot arm is attached to the slide so that it can be moved radially with respect to the column. By rotating the column, the robot is capable of achieving a work space that approximates a cylinder. An example of the cylindrical configuration is pictured in Fig. 2-4.

The cartesian coordinate robot, illustrated in part (c) of Fig. 2-1, uses three perpendicular slides to construct the x, y, and z axes. Other names are sometimes applied to this configuration, including xyz robot and rectilinear robot. By moving the three slides relative to one another, the robot is capable

Figure 2-2 Unimate 2000—polar configuration. Here, the Unimate performs a machine loading and unloading operation. The 2000 series robots have provided many years of service. (Photo courtesy of Unimation, Inc.)

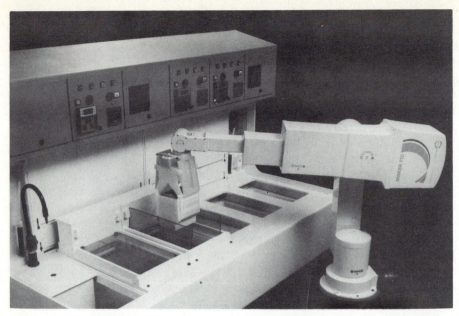

Figure 2-3 The MAKER 110—polar configuration. The MAKER performs a semiconductor wafer-etching application in the electronics industry. (Photo courtesy of United States Robots.)

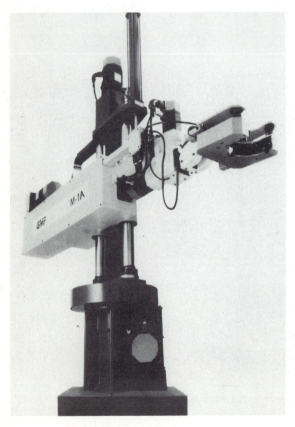

Figure 2-4 GMF Model M-1A—cylindrical configuration. (Photo courtesy of GMF Robotics.)

of operating within a rectangular work envelope. An example of this configuration is the IBM RS-1 robot (currently called the Model 7565), pictured in Fig. 2-5. The RS-1, because of its appearance and construction, is occasionally referred to as a "box" configuration. "Gantry" robot is another name used for cartesian robots that are generally large and possess the appearance of a gantry-type crane. An example is shown in Fig. 2-6.

The jointed-arm robot is pictured in Fig. 2-1(*d*). Its configuration is similar to that of the human arm. It consists of two straight components, corresponding to the human forearm and upper arm, mounted on a vertical pedestal. These components are connected by two rotary joints corresponding to the shoulder and elbow. A wrist is attached to the end of the forearm, thus providing several additional joints. Several commercially available robots possess the jointed-arm configuration, including the Cincinnati Milacron T3 (Model 776) robot, illustrated in Fig. 2-7. A special version of the jointed arm robot is the SCARA, whose shoulder and elbow joints rotate about vertical axes. SCARA stands for Selective Compliance Assembly Robot Arm, and this configuration provides substantial rigidity for the robot in the vertical direction, but compliance in the horizontal plane. This makes it ideal for many assembly tasks. A SCARA robot is pictured in Fig. 2-8.

There are relative advantages and disadvantages to the four basic robot anatomies simply because of their geometries. In terms of repeatability of

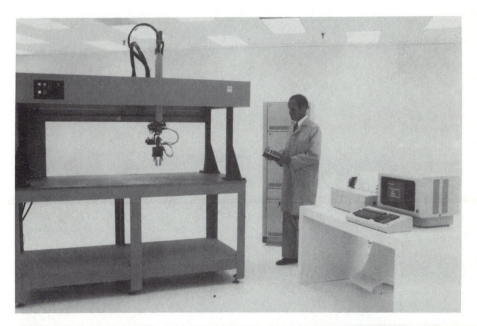

Figure 2-5 The RS-1 (Model 7565)—cartesian coordinate robot. (Photo courtesy of IBM Corporation.)

Figure 2-6 Cincinnati Milacron T3-800 series robot—gantry type robot, a large cartesian coordinate configuration. (Photo courtesy of Cincinnati Milacron.)

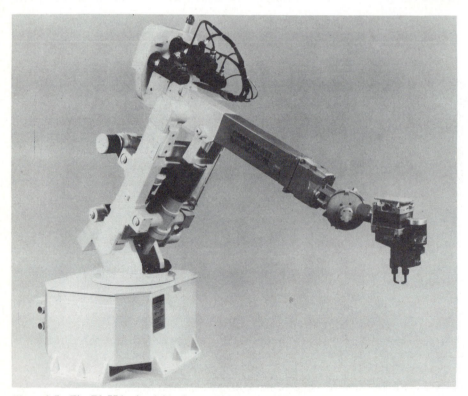

Figure 2-7 The T3-776 robot jointed-arm configuration. (Photo courtesy of Cincinnati Milacron.)

Figure 2-8 SCARA (Selective Compliance Assembly Robotic Arm) robot. (Photo courtesy of United States Robots.)

motion (the capability to move to a taught point in space with minimum error), the box-frame cartesian robot probably possesses the advantage because of its inherently rigid structure. (We will define repeatability and other related terms in Sec. 2-5.) In terms of reach (the ability of the robot to extend its arm significantly beyond its base), the polar and jointed arm configurations have the advantage. The lift capacity of the robot is important in many applications. The cylindrical configuration and the gantry *xyz* robot can be designed for high-rigidity and load-carrying capacity. For machine-loading applications, the ability of the robot to reach into a small opening without interference with the sides of the opening is important. The polar configuration and the cylindrical configuration possess a natural geometric advantage in terms of this capability.

Robot Motions

Industrial robots are designed to perform productive work. The work is accomplished by enabling the robot to move its body, arm, and wrist through a series of motions and positions. Attached to the wrist is the end effector which is used by the robot to perform a specific work task. The robot's movements can be divided into two general categories: arm and body motions, and wrist

motions. The individual joint motions associated with these two categories are sometimes referred to by the term "degrees of freedom," and a typical industrial robot is equipped with 4 to 6 degrees of freedom.

The robot's motions are accomplished by means of powered joints. Three joints are normally associated with the action of the arm and body, and two or three joints are generally used to actuate the wrist. Connecting the various manipulator joints together are rigid members that are called links. In any link–joint–link chain, we shall call the link that is closest to the base in the chain the input link. The output link is the one that moves with respect to the input link.

The joints used in the design of industrial robots typically involve a relative motion of the adjoining links that is either linear or rotational. Linear joints involve a sliding or translational motion of the connecting links. This motion can be achieved in a number of ways (e.g., by a piston, a telescoping mechanism, and relative motion along a linear track or rail). Our concern here is not with the mechanical details of the joint, but rather with the relative motion of the adjacent links. We shall refer to the linear joint as a type L joint (L for Linear). The table of Fig. 2-9 illustrates the linear joint. The term prismatic joint is sometimes used in the literature in place of linear joint.

There are at least three types of rotating joint that can be distinguished in robot manipulators. The three types are illustrated in Fig. 2-9. We shall refer to the first as a type R joint (R for Rotational). In the type R joint the axis of rotation is perpendicular to the axes of the two connecting links. The second type of rotating joint involves a twisting motion between the input and output links. The axis of rotation of the twisting joint is parallel to the axes of both links. We shall call this a type T joint (T for Twisting). The third type of rotating joint is a revolving joint in which the input link is parallel to the axis of rotation and the output link is perpendicular to the axis of rotation. In essence, the output link revolves about the input link, as if it were in orbit. This joint will be designated as a type V joint (V for re$V$olving).

The arm and body joints are designed to enable the robot to move its end effector to a desired position within the limits of the robot's size and joint movements. For robots of polar, cylindrical, or jointed-arm configuration, the 3 degrees of freedom associated with the arm and body motions are:

1. Vertical traverse—This is the capability to move the wrist up or down to provide the desired vertical attitude.
2. Radial traverse—This involves the extension or retraction (in or out movement) of the arm from the vertical center of the robot.
3. Rotational traverse—This is the rotation of the arm about the vertical axis.

The degrees of freedom associated with the arm and body of the robot are shown in Fig. 2-10 for a polar configuration robot. Similar degrees of freedom are associated with the cylindrical configuration and jointed-arm robot. For a cartesian coordinate robot, the three degrees of freedom are vertical move-

Type	Name	Illustration
L	Linear	
R	Rotational	
T	Twisting	
V	Revolving	

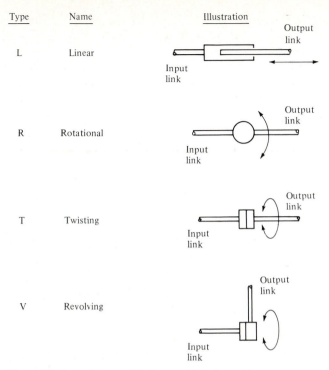

Figure 2-9 Several types of joints used in robots: (a) rotational joint with rotation along an axis perpendicular to arm member axes, (b) rotational joint with twisting action, (c) linear motion joint, usually achieved by a sliding action.

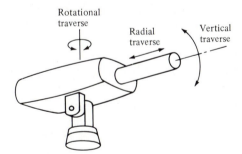

Figure 2-10 Three degrees of freedom associated with arm and body of a polar coordinate robot.

ment (z-axis motion), in-and-out movement (y-axis motion), and right-or-left movement (x-axis motion). These are achieved by corresponding movements of the three orthogonal slides of the robot arm.

The wrist movement is designed to enable the robot to orient the end effector properly with respect to the task to be performed. For example, the hand must be oriented at the appropriate angle with respect to the workpiece

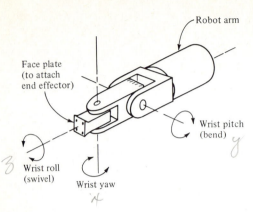

Figure 2-11 Three degrees of freedom associated with the robot wrist.

in order to grasp it. To solve this orientation problem, the wrist is normally provided with up to 3 degrees of freedom (the following is a typical configuration):

1. Wrist roll—Also called wrist swivel, this involves rotation of the wrist mechanism about the arm axis.
2. Wrist pitch—Given that the wrist roll is in its center position, the pitch would involve the up or down rotation of the wrist. Wrist pitch is also sometimes called wrist bend.
3. Wrist yaw—Again, given that the wrist swivel is in the center position of its range, wrist yaw would involve the right or left rotation of the wrist.

These degrees of freedom for the wrist are illustrated in Fig. 2-11. The reason for specifying that the wrist roll be in its center position in the definitions of pitch and yaw is because rotation of the wrist about the arm axis will alter the orientation of the pitch and yaw movements.

Joint Notation Scheme

The physical configuration of the robot manipulator can be described by means of a joint notation scheme, using the joint types defined earlier in this section (L, R, T, and V). Considering the arm and body joints first, the letters can be used to designate the particular robot configuration starting with the joint closest to the base and proceeding to the joint that connects to the wrist. Accordingly, a jointed-arm robot (excluding the wrist assembly) would have three rotational joints and would be designated as either TRR or VVR. Typical notations for the four basic configurations are summarized in Table 2-1.

The joint notation scheme permits the designation of more or less than the three joints typical of the basic configurations indicated in the table. It can also be used to explore other possibilities for configuring robots, beyond the four basic types.

Table 2-1 Notation scheme for designating robot configurations

Robot configuration (arm and body)	Symbol
Polar configuration	*TRL*
Cylindrical configuration	*TLL, LTL, LVL*
Cartesian coordinate robot	*LLL*
Jointed arm configuration	*TRR, VVR*

Robot configuration (wrist)	Symbol
Two-axis wrist (typical)	:*RT*
Three-axis wrist (typical)	:*TRT*

The notation system can be expanded to include wrist motions by designating the two or three (or more) types of wrist joint. The notation starts with the joint closest to the arm interface, and proceeds to the mounting plate for the end effector. Wrist joints are predominantly rotating joints of type R and T. Hence, a typical wrist mechanism with three rotational joints would be indicated by *TRR* (Fig. 2-11). This notation is simply added to the notation for the arm and body configuration. For example, a polar coordinate robot with a three-axis wrist might be designated as *TRL:TRT*.

The scheme can also provide for the possibility of robots that move on a track in the floor or along an overhead rail system in the factory. As an illustration, a *TRL:TRT* robot fastened to a platform on wheels that can be driven along a track between several machine tools would be designated by the following notation: *L-TRL:TRT*. In this case, even though the wheels of the platform rotate, the motion of the robot is linear.

2-2 WORK VOLUME

Work volume is the term that refers to the space within which the robot can manipulate its wrist end. The convention of using the wrist end to define the robot's work volume is adopted to avoid the complication of different sizes of end effectors that might be attached to the robot's wrist. The end effector is an addition to the basic robot and should not be counted as part of the robot's working space. A long end effector mounted on the wrist would add significantly to the extension of the robot compared to a smaller end effector. Also, the end effector attached to the wrist might not be capable of reaching certain points within the robot's normal work volume because of the particular combination of joint limits of the arm.

The work volume is determined by the following physical characteristics

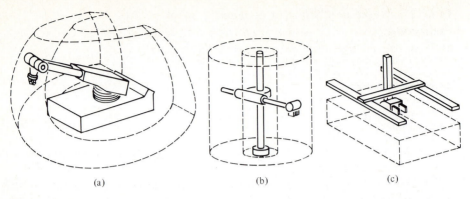

Figure 2-12 Work volumes for various robot anatomies: (a) polar, (b) cylindrical, and (c) cartesian. (Reprinted from Reference [7].)

of the robot:

The robot's physical configuration
The sizes of the body, arm, and wrist components
The limits of the robot's joint movements

The influence of the physical configuration on the shape of the work volume is illustrated in Fig. 2-12. A polar coordinate robot has a work volume that is a partial sphere, a cylindrical coordinate robot has a cylindrical work envelope, a cartesian coordinate robot has a rectangularly shaped work space, and a jointed-arm robot approximates a work volume that is spherical. The size of each work volume shape is influenced by the dimensions of the arm components and by the limits of its joint movements. Using the cylindrical configuration as an example, limits on the rotation of the column about the base would determine what portion of a complete cylinder the robot could reach with its wrist end.

2-3 ROBOT DRIVE SYSTEMS

The robot's capacity to move its body, arm, and wrist is provided by the drive system used to power the robot. The drive system determines the speed of the arm movements, the strength of the robot, and its dynamic performance. To some extent, the drive system determines the kinds of applications that the robot can accomplish. In this and the following sections, we will discuss some of these technical features.

Types of Drive Systems

Commercially available industrial robots are powered by one of three types of drive systems. These three systems are:

1. Hydraulic drive
2. Electric drive
3. Pneumàtic drive

Hydraulic drive and electric drive are the two main types of drives used on more sophisticated robots.

Hydraulic drive is generally associated with larger robots, such as the Unimate 2000 series (Fig. 2-2). The usual advantages of the hydraulic drive system are that it provides the robot with greater speed and strength. The disadvantages of the hydraulic drive system are that it typically adds to the floor space required by the robot, and that a hydraulic system is inclined to leak oil which is a nuisance. Hydraulic drive systems can be designed to actuate either rotational joints or linear joints. Rotary vane actuators can be utilized to provide rotary motion, and hydraulic pistons can be used to accomplish linear motion.

Electric drive systems do not generally provide as much speed or power as hydraulic systems. However, the accuracy and repeatability of electric drive robots are usually better. Consequently, electric robots tend to be smaller, requiring less floor space, and their applications tend toward more precise work such as assembly. The MAKER 110 (Fig. 2-3) is an example of an electric drive robot that is consistent with these tendencies. Electric drive robots are actuated by dc stepping motors or dc servomotors. These motors are ideally suited to the actuation of rotational joints through appropriate drive train and gear systems. Electric motors can also be used to actuate linear joints (e.g., telescoping arms) by means of pulley systems or other translational mechanisms.

The economics of the two types of drive systems are also a factor in the decision to utilize hydraulic drive on large robots and electric drive on smaller robots. It turns out that the cost of an electric motor is much more proportional to its size, whereas the cost of a hydraulic drive system is somewhat less dependent on its size. These relationships are displayed conceptually in Fig. 2-13. As the illustration suggests, there is a hypothetical break-even point, below which it is advantageous to use electric drive and above which it is appropriate to use hydraulic drive. Having explained these factors, it should be noted that there is a trend in the design of industrial robots toward all electric drives, and away from hydraulic robots because of the disadvantages discussed above.

Pneumatic drive is generally reserved for smaller robots that possess fewer degrees of freedom (two- to four-joint motions). These robots are often limited to simple "pick-and-place" operations with fast cycles. Pneumatic power can

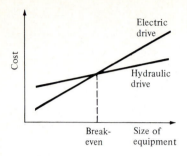

Figure 2-13 Cost vs. size for electric drive and hydraulic drive.

be readily adapted to the actuation of piston devices to provide translational movement of sliding joints. It can also be used to operate rotary actuators for rotational joints.

Speed of Motion

The speed capabilities of current industrial robots range up to a maximum of about 1.7 m/s (about 5 ft/sec). This speed would be measured at the wrist. Accordingly, the highest speeds can be obtained by large robots with the arm extended to its maximum distance from the vertical axis of the robot. As mentioned previously, hydraulic robots tend to be faster than electric drive robots.

The speed, of course, determines how quickly the robot can accomplish a given work cycle. It is generally desirable in production to minimize the cycle time of a given task. Nearly all robots have some means by which adjustments in the speed can be made. Determination of the most desirable speed, in addition to merely attempting to minimize the production cycle time, would also depend on other factors, such as:

The accuracy with which the wrist (end effector) must be positioned
The weight of the object being manipulated
The distances to be moved.

There is generally an inverse relationship between the accuracy and the speed of robot motions. As the required accuracy is increased, the robot needs more time to reduce the location errors in its various joints to achieve the desired final position. The weight of the object moved also influences the operational speed. Heavier objects mean greater inertia and momentum, and the robot must be operated more slowly to safely deal with these factors. The influence of the distance to be moved by the robot manipulator is illustrated in Fig. 2-14. Because of acceleration and deceleration problems, a robot is capable of traveling one long distance in less time than a sequence of short distances whose sum is equal to the long distance. The short distances may not permit the robot to ever reach the programmed operating speed.

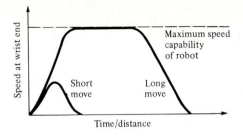

Figure 2-14 Influence of distance versus speed.

Load-Carrying Capacity

The size, configuration, construction, and drive system determine the load-carrying capacity of the robot. This load capacity should be specified under the condition that the robot's arm is in its weakest position. In the case of a polar, cylindrical, or jointed-arm configuration, this would mean that the robot arm is at maximum extension. Just as in the case of a human, it is more difficult to lift a heavy load with arms fully extended than when the arms are held in close to the body.

The rated weight-carrying capacities of industrial robots ranges from less than a pound for some of the small robots up to several thousand pounds for very large robots. An example is the Prab Versatran Model FC which has a rated load capacity of 2000 lb. The small assembly robots, such as the MAKER 110, have weight-carrying capabilities in the vicinity of 5 lb. The manufacturer's specification of this feature is the gross weight capacity. To use this specification, the user must consider the weight of the end effector. For example, if the rated load capacity of a given robot were 5 lb, and the end effector weighed 2 lb, then the net weight-carrying capacity of the robot would be only 3 lb.

2-4 CONTROL SYSTEMS AND DYNAMIC PERFORMANCE

In order to operate, a robot must have a means of controlling its drive system to properly regulate its motions. In this section we briefly describe the various types of control systems and the associated performance characteristics which are determined by the control system. A more thorough treatment of these topics is provided in Chaps. 3 and 4.

Four Types of Robot Controls

Commercially available industrial robots can be classified into four categories according to their control systems. The four categories are:

1. Limited-sequence robots
2. Playback robots with point-to-point control

3. Playback robots with continuous path control
4. Intelligent robots

Of the four categories, the limited-sequence robots represent the lowest level of control and the intelligent robots are the most sophisticated.

Limited-sequence robots do not use servo-control to indicate relative positions of the joints. Instead, they are controlled by setting limit switches and/or mechanical stops to establish the endpoints of travel for each of their joints. Establishing the positions and sequence of these stops involves a mechanical setup of the manipulator rather than robot programming in the usual sense of the term. With this method of control, the individual joints can only be moved to their extreme limits of travel. This has the effect of severely limiting the number of distinct points that can be specified in a program for these robots. The sequence in which the motion cycle is played out is defined by a pegboard or stepping switch or other sequencing device. This device, which constitutes the robot controller, signals each of the particular actuators to operate in the proper succession. There is generally no feedback associated with a limited sequence robot to indicate that the desired position has been achieved. Any of the three drive systems can be used with this type of control system; however, pneumatic drive seems to be the type most commonly employed. Applications for this type of robot generally involve simple motions, such as pick-and-place operations.

Playback robots use a more sophisticated control unit in which a series of positions or motions are "taught" to the robot, recorded into memory, and then repeated by the robot under its own control. The term "playback" is descriptive of this general mode of operation. The procedure of teaching and recording into memory is referred to as programming the robot. Playback robots usually have some form of servo-control (e.g., closed loop feedback system) to ensure that the positions achieved by the robot are the positions that have been taught.

Playback robots can be classified into two categories: point-to-point (PTP) robots and continuous-path (CP) robots. Point-to-point robots are capable of performing motion cycles that consist of a series of desired point locations and related actions. The robot is taught each point, and these points are recorded into the robot's control unit. During playback, the robot is controlled to move from one point to another in the proper sequence. Point-to-point robots do not control the path taken by the robot to get from one point to the next. If the programmer wants to exercise a limited amount of control over the path followed, this must be done by programming a series of points along the desired path. Control of the sequence of positions is quite adequate for many kinds of applications, including loading and unloading machines and spot welding.

Continuous-path robots are capable of performing motion cycles in which the path followed by the robot is controlled. This is usually accomplished by making the robot move through a series of closely spaced points which describe the desired path. The individual points are defined by the control unit rather than the programmer. Straight line motion is a common form of continuous-path control for industrial robots. The programmer specifies the starting point and the end point of the path, and the control unit calculates the sequence of individual points that permit the robot to follow a straight line trajectory. Some robots have the capability to follow a smooth, curved path that has been defined by a programmer who manually moves the arm through the desired motion cycle. To achieve continuous-path control to more than a limited extent requires that the controller unit be capable of storing a large number of individual point locations that define the compound curved path. Today this usually involves the use of a digital computer (a microprocessor is typically used as the central processing unit for the computer) as the robot controller. CP control is required for certain types of industrial applications such as spray coating and arc welding.

Intelligent robots constitute a growing class of industrial robot that possesses the capability not only to play back a programmed motion cycle but to also interact with its environment in a way that seems intelligent. Invariably, the controller unit consists of a digital computer or similar device (e.g., programmable controller). Intelligent robots can alter their programmed cycle in response to conditions that occur in the workplace. They can make logical decisions based on sensor data received from the operation. The robots in this class have the capacity to communicate during the work cycle with humans or computer-based systems. Intelligent robots are usually programmed using an English-like and symbolic language not unlike a computer programming language. Indeed, the kinds of applications that are performed by intelligent robots rely on the use of a high-level language to accomplish the complex and sophisticated activities that can be accomplished by these robots. Typical applications for intelligent robots are assembly tasks and arc-welding operations.

Speed of Response and Stability

Speed of response and stability are two important characteristics of dynamic performance related to control systems design. The speed of response refers to the capability of the robot to move to the next position in a short amount of time. This response time is obviously related to the robot's motion speed, a feature that we have already discussed. It is also a function of the control system. In robotics, stability is generally defined as a measure of the oscillations which occur in the arm during movement from one position to the next. A robot with good stability will exhibit little or no oscillations either during or at the termination of the arm movement. Poor stability would be indicated by a large amount of oscillation. It is generally desirable in control systems design for the system to have good stability and a fast response time. Unfortunately, these are competing objectives.

The stability of a robot can be controlled to a certain extent by incorporating damping elements into the robot's design. A high level of damping will increase the robot's stability (reduce its tendency toward oscillation).

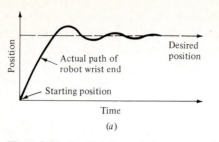 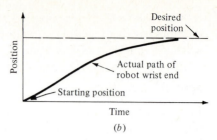

Figure 2-15 Concept of speed of response and stability in robotics: (a) low damping—fast response, (b) high damping—slow response.

The problem with high damping is that it reduces the speed of response. Accordingly, there is a compromise that must be struck between the stability of the robot and its ability to operate at high speeds.

The concept of stability and its relation to damping is illustrated in Fig. 2-15. In the two diagrams of the figure, the position of the robot's wrist is shown as a function of time for two cases: small damping and large damping. With low damping, the robot arm moves to the programmed position quickly, but exhibits considerable oscillation about the position. With a large amount of damping built into the system, the arm movement to the desired position is very sluggish but there is no oscillatory motion about the final position.

2-5 PRECISION OF MOVEMENT

The preceding discussion of response speed and stability is concerned with the dynamic performance of the robot. Another measure of performance is precision of the robot's movement. We will define precision as a function of three features:

1. Spatial resolution
2. Accuracy
3. Repeatability

These terms will be defined with the following assumptions. First, the definitions will apply at the robot's wrist end with no hand attached to the wrist. Second, the terms apply to the worst case conditions, the conditions under which the robot's precision will be at its worst. This generally means that the robot's arm is fully extended in the case of a jointed arm or polar configuration. Third, our definitions will be developed in the context of a point-to-point robot. That is, we will be concerned with the robot's capability to achieve a given position within its work volume. It is easier to define the various precision features in a static context rather than a dynamic context. It is considerably more difficult to define, and measure, the robot's capacity to achieve a defined motion path in space because it would be complicated by speed and other factors.

Spatial Resolution

The spatial resolution of a robot is the smallest increment of movement into which the robot can divide its work volume. Spatial resolution depends on two factors: the system's control resolution and the robot's mechanical inaccuracies. It is easiest to conceptualize these factors in terms of a robot with 1 degree of freedom.

The control resolution is determined by the robot's position control system and its feedback measurement system. It is the controller's ability to divide the total range of movement for the particular joint into individual increments that can be addressed in the controller. The increments are sometimes referred to as "addressable points." The ability to divide the joint range into increments depends on the bit storage capacity in the control memory. The number of separate, identifiable increments (addressable points) for a particular axis is given by

$$\text{Number of increments} = 2^n$$

where $n =$ the number of bits in the control memory.

For example, a robot with 8 bits of storage can divide the range into 256 discrete positions. The control resolution would be defined as the total motion range divided by the number of increments. We assume that the system designer will make all of the increments equal.

Example 2-1 Using our robot with 1 degree of freedom as an illustration, we will assume it has one sliding joint with a full range of 1.0 m (39.37 in.). The robot's control memory has a 12-bit storage capacity. The problem is to determine the control resolution for this axis of motion.

The number of control increments can be determined as follows:

$$\text{Number of increments} = 2^{12} = 4096$$

The total range of 1 m is divided into 4096 increments. Each position will be separated by

$$1 \text{ m}/4096 = 0.000244 \text{ m} \quad \text{or} \quad 0.244 \text{ mm}$$

The control resolution is 0.244 mm (0.0096 in.).

This example deals with only one joint. A robot with several degrees of freedom would have a control resolution for each joint of motion. To obtain the control resolution for the entire robot, component resolutions for each joint would have to be summed vectorially. The total control resolution would depend on the wrist motions as well as the arm and body motions. Since some of the joints are likely to be rotary while others are sliding, the robot's control resolution can be a complicated quantity to determine.

Mechanical inaccuracies in the robot's links and joint components and its feedback measurement system (if it is a servo-controlled robot) constitute the other factor that contributes to spatial resolution. Mechanical inaccuracies come from elastic deflection in the structural members, gear backlash, stretching of pulley cords, leakage of hydraulic fluids, and other imperfections in the mechanical system. These inaccuracies tend to be worse for larger robots simply because the errors are magnified by the larger components. The inaccuracies would also be influenced by such factors as the load being handled, the speed with which the arm is moving, the condition of maintenance of the robot, and other similar factors.

The spatial resolution of the robot is the control resolution degraded by these mechanical inaccuracies. Spatial resolution can be improved by increasing the bit capacity of the control memory. However, a point is reached where it provides little additional benefit to increase the bit capacity further because the mechanical inaccuracies of the system become the dominant component in the spatial resolution.

Accuracy

Accuracy refers to a robot's ability to position its wrist end at a desired target point within the work volume. The accuracy of a robot can be defined in terms of spatial resolution because the ability to achieve a given target point depends on how closely the robot can define the control increments for each of its joint motions. In the worst case, the desired point would lie in the middle between two adjacent control increments. Ignoring for the moment the mechanical inaccuracies which would reduce the robot's accuracy, we could initially define accuracy under this worst case assumption as one-half of the control resolution. This relationship is illustrated in Fig. 2-16. In fact, the mechanical inaccuracies would affect the ability to reach the target position. Accordingly, we define the robot's accuracy to be one-half of its spatial resolution as portrayed in Fig. 2-17.

Our definition of accuracy applies to the worst case, where the target point is directly between two control points. Our definition also implies that the accuracy is the same anywhere in the robot's work volume. In fact, the

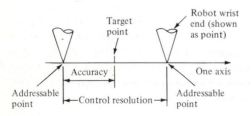

Figure 2-16 Illustration of accuracy and control resolution when mechanical inaccuracies are assumed to be zero.

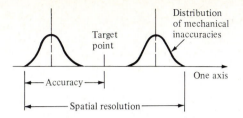

Figure 2-17 Illustration of accuracy and spatial resolution in which mechanical inaccuracies are represented by a statistical distribution.

accuracy of a robot is affected by several factors. First, the accuracy varies within the work volume, tending to be worse when the arm is in the outer range of its work volume and better when the arm is closer to its base. The reason for this is that the mechanical inaccuracies are magnified with the robot's arm fully extended. The term error map is used to characterize the level of accuracy possessed by the robot as a function of location in the work volume. Second, the accuracy is improved if the motion cycle is restricted to a limited work range. The mechanical errors will tend to be reduced when the robot is exercised through a restricted range of motions. The robot's ability to reach a particular reference point within the limited work space is sometimes called its local accuracy. When the accuracy is assessed within the robot's full work volume, the term global accuracy is used. A third factor influencing accuracy is the load being carried by the robot. Heavier workloads cause greater deflection of the mechanical links of the robot, resulting in lower accuracy.

Repeatability

Repeatability is concerned with the robot's ability to position its wrist or an end effector attached to its wrist at a point in space that had previously been taught to the robot. Repeatability and accuracy refer to two different aspects of the robot's precision. Accuracy relates to the robot's capacity to be programmed to achieve a given target point. The actual programmed point will probably be different from the target point due to limitations of control resolution. Repeatability refers to the robot's ability to return to the programmed point when commanded to do so.

These concepts are illustrated in Figure 2-18. The desired target point is denoted by the letter T. During the teach procedure, the robot is commanded to move to point T, but because of the limitations on its accuracy, the programmed position becomes point P. The distance between points T and P is a manifestation of the robot's accuracy in this case. Subsequently, the robot is instructed to return to the programmed point P; however, it does not return to the exact same position. Instead, it returns to position R. The difference between P and R is a result of limitations on the robot's repeatability. The robot will not always return to the same position R on subsequent repetitions of the motion cycle. Instead, it will form a cluster of points on both sides of the position P in Fig. 2-18.

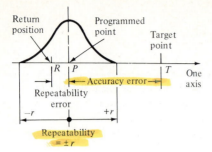

Figure 2-18 Illustration of repeatability and accuracy.

Repeatability errors form a random variable and constitute a statistical distribution as shown in the figure. It would be convenient if the repeatability errors formed a nice bell-shaped curve, suggesting a normally distributed random variable. What is closer to reality is that for each joint, the mechanical inaccuracies that are principally responsible for repeatability errors do not form the nice symmetric bell-shaped distribution shown in the figure. However, when the errors from several axes of motion are combined together, the resulting aggregate error is influenced by the central limit theorem in probability. This theorem states that the sums of random variables tend to form a normally distributed variable, even though the individuals come from a distribution other than the normal. Accordingly, we can infer that the repeatability error of a robot with five or six axes is approximately normal, even if the error due to each individual axis is not normal.

In three-dimensional space, the repeatability errors will surround the programmed point P, forming a distribution whose outer boundary can be conceptualized as a sphere. A robot manufacturer typically quotes the repeatability of its manipulator as the radius of the idealized sphere, usually expressing the specification as plus or minus a particular value. The size of the sphere will tend to be larger in the regions of the work volume that are further away from the center of the robot. It is likely that the shape of the sphere is not perfectly round, but instead is oblong in certain directions due to compliance of the robot arm.

Compliance

A feature of the robot that is related to our preceding discussion is compliance. The compliance of the robot manipulator refers to the displacement of the wrist end in response to a force or torque exerted against it. A high compliance means that the wrist is displaced a large amount by a relatively small force. The term "springy" is sometimes used to describe a robot with high compliance. A low compliance means that the manipulator is relatively stiff and is not displaced by a significant amount.

Robot manipulator compliance is a directional feature. That is, the compliance of the robot arm will be greater in certain directions than in other directions because of the mechanical construction of the arm.

Compliance is important because it reduces the robot's precision of movement under load. If the robot is handling a heavy load, the weight of the load will cause the robot arm to deflect. If the robot is pressing a tool against a workpart, the reaction force of the part may cause deflection of the manipulator. If the robot has been programmed under no-load conditions to position its end effector, and accuracy of position is important in the application, the robot's performance will be degraded because of compliance when it operates under loaded conditions.

2-6 END EFFECTORS

For industrial applications, the capabilities of the basic robot must be augmented by means of additional devices. We might refer to these devices as the robot's peripherals. They include the tooling which attaches to the robot's wrist and the sensor systems which allow the robot to interact with its environment. We provide a more comprehensive treatment of these robot technology areas in Chaps. 5, 6, and 7.

In robotics, the term end effector is used to describe the hand or tool that is attached to the wrist. The end effector represents the special tooling that permits the general-purpose robot to perform a particular application. This special tooling must usually be designed specifically for the application.

End effectors can be divided into two categories: grippers and tools. Grippers would be utilized to grasp an object, usually the workpart, and hold it during the robot work cycle. There are a variety of holding methods that can be used in addition to the obvious mechanical means of grasping the part between two or more fingers. These additional methods include the use of suction cups, magnets, hooks, and scoops. A tool would be used as an end effector in applications where the robot is required to perform some operation on the workpart. These applications include spot welding, arc welding, spray painting, and drilling. In each case, the particular tool is attached to the robot's wrist to accomplish the application.

2-7 ROBOTIC SENSORS

Sensors used as peripheral devices in robotics include both simple types such as limit switches and sophisticated types such as machine vision systems. Of course, sensors are also used as integral components of the robot's position feedback control system. We discuss this use in Chap. 3. Their function as peripheral devices in a robotic work cell is to permit the robot's activities to be coordinated with other activities in the cell. The sensors used in robotics include the following general categories:

1. Tactile sensors. These are sensors which respond to contact forces with another object. Some of these devices are capable of measuring the level of force involved.
2. Proximity and range sensors. A proximity sensor is a device that indicates when an object is close to another object but before contact has been made. When the distance between the objects can be sensed, the device is called a range sensor.
3. Miscellaneous types. The miscellaneous category includes the remaining kinds of sensors that are used in robotics. These include sensors for temperature, pressure, and other variables.
4. Machine vision. A machine vision system is capable of viewing the workspace and interpreting what it sees. These systems are used in robotics to perform inspection, parts recognition, and other similar tasks.

Sensors are an important component in work cell control and in safety monitoring systems.

2-8 ROBOT PROGRAMMING AND WORK CELL CONTROL

In its most basic form, a robot program can be defined as a path in space through which the manipulator is directed to move. This path also includes other actions such as controlling the end effector and receiving signals from sensors. The purpose of robot programming is to teach these actions to the robot.

There are various methods used for programming robots. The two basic categories of greatest commercial importance today are leadthrough programming and textual language programming. Chapters 8 and 9 describe these two methods, respectively.

Leadthrough programming consists of forcing the robot arm to move through the required motion sequence and recording the motions into the controller memory. Leadthrough methods are used to program playback robots. In the case of point-to-point playback robots, the usual procedure is to use a control box (called a teach pendant) to drive the robot joints to each of the desired points in the workspace, and record the points into memory for subsequent playback. The teach pendant is equipped with a series of switches and dials to control the robot's movements during the teach procedure. Owing to its ease and convenience and the wide range of applications suited to it, this leadthrough method is the most common programming method for playback-type robots.

Continuous-path playback robots also use leadthrough programming. For well-defined paths, such as moving along a straight line between two points, a teach pendant can be employed to teach the locations of the two points, and the robot controller then computes the trajectory to be followed in order to move along the straight line path. For more complex motions (e.g., those encountered in spray-painting operations), it is usually more convenient for the programmer to physically move the robot arm and end effector through the desired motion path and record the positions at closely spaced sampling intervals. Certain parameters of the motion cycle, such as the robot's speed, would be controlled independently when the job is set up to operate. Accordingly, the programmer does not need to be concerned with these aspects of the program. The programmer's principal concern is to make sure that the motion sequence is correct.

Textual programming methods use an English-like language to establish the logic and sequence of the work cycle. A computer terminal is used to input the program instructions into the controller but a teach pendant is also used to define the locations of the various points in the workspace. The robot programming language names the points as symbols in the program and these symbols are subsequently defined by showing the robot their locations. In addition to identifying points in the workspace, the robot languages permit the use of calculations, more detailed logic flow, and subroutines in the programs, and greater use of sensors and communications. Accordingly, the use of the textual languages corresponds largely to the so-called intelligent robots.

Some examples of the kinds of programming statements that would be found in the textual robot languages include the following sequence:

SPEED 35 IPS
MOVE P1
CLOSE 40 MM
WAIT 1 SEC
DEPART 60 MM

The series of commands tells the robot that its velocity at the wrist should be 35 in./sec in the motions which follow. The MOVE statement indicates that the robot is to move its gripper to point P1 and close to an opening of 40 mm. It is directed to wait 1.0 s before departing from P1 by a distance of 60 mm above the point.

A future enhancement of textual language programming will be to enter the program completely off-line, without the need for a teach pendant to define point locations in the program. The potential advantage of this method is that the programming can be accomplished without taking the robot out of production. All of the current methods of programming require the participation of the robot in order to perform the programming function. With off-line programming, the entire program can be entered into a computer for later downloading to the robot. Off-line programming would hasten the changeover from one robot work cycle to a new work cycle without a major time delay for reprogramming. Unfortunately, there are certain technical problems associated with off-line programming. These problems are mainly concerned with defining the spatial locations of the positions to be used in the work cycle, and that is why the teach pendant is required in today's textual robot languages.

In addition to the leadthrough and textual language programming methods, there is another form of programming for the low-technology-limited sequence robots. These robots are programmed by setting limit switches, mechanical stops, and other similar means to establish the endpoints of travel for each of the joints. This is sometimes called mechanical programming; it really involves more of a manual setup procedure rather than a programming method. The work cycles for these kinds of robots generally consist of a limited number of simple motions (e.g., pick-and-place applications), for which this manual programming method is adequate.

Work cell control deals with the problem of coordinating the robot to operate with other equipment in the work cell. A robot cell usually consists of not only the robot, but also conveyors, machine tools, inspection devices, and possibly human operators. Some of the activities in the robot work cell occur sequentially, while other activities occur simultaneously. A method of controlling and synchronizing these various activities is required, and that is the purpose of the work cell controller. Work cell control is accomplished either by the robot controller or a separate small computer or programmable controller. During operation, the controller communicates signals to the equipment in the cell and receives signals back from the equipment. These signals are sometimes called interlocks. By communicating back and forth with the different components of the work cell, the various activities in the cell are accomplished in the proper sequence.

2-9 ROBOT APPLICATIONS

Robots are employed in a wide assortment of applications in industry. Today most of the applications are in manufacturing to move materials, parts, and tools of various types. Future applications will include nonmanufacturing tasks, such as construction work, exploration of space, and medical care. At some time in the distant future, a household robot may become a mass produced item, perhaps as commonplace as the automobile is today.

For the present, most industrial applications of robots can be divided into the following three categories:

1. Material-handling and machine-loading and -unloading applications. In these applications the robot's function is to move materials or parts from one location in the work cell to some other location. The MAKER 110 in Figure 2-3 is shown performing a material-handling operation. Loading and/or unloading of a production machine is included within the scope of this material-handling activity. The Unimate 2000 in Fig. 2-2 is performing a machine load/unload operation for a machine tool.
2. Processing applications. This category includes spot welding, arc welding, spray painting, and other operations in which the function of the robot is to manipulate a tool to accomplish some manufacturing process in the work

cell. Spot welding represents a particularly important application in the processing category.

3. **Assembly and inspection.** These are two separate operations which we include together in this category. Robotic assembly is a field in which industry is showing great interest because of the economic potential. The SCARA robot shown in Fig. 2-8 is performing an assembly operation. Inspection robots would make use of sensors to gauge and measure quality characteristics of manufactured product.

We examine these applications of industrial robots, as well as the general problems associated with their installation, in more detail in later chapters of the book.

REFERENCES

1. S. Dickenson, *Report on Robot Joint Notation Scheme*, Course Report for IE 398, Lehigh University, 1984.
2. J. F. Engelberger, *Robotics in Practice*, AMACOM (American Management Association), New York, 1980, chaps. 1 and 2.
3. M. P. Groover, "Industrial Robots: A Primer on the Present Technology," *Industrial Engineering*, 54–61 (November 1980).
4. M. P. Groover and E. W. Zimmers, Jr., *CAD/CAM: Computer-Aided Design and Manufacturing*, Prentice-Hall, Englewood Cliffs, NJ, 1984, chap. 10.
5. R. N. Nagel, "Robots: Not Yet Smart Enough," *IEEE Spectrum*, 78–83 (May 1983).
6. L. V. Ottinger, "Robotics for the IE: Terminology, Types of Robots," *Industrial Engineering*, 28–35 (November 1981).
7. L. L. Toepperwein, M. T. Blackman, et al., "ICAM Robotics Application Guide," *Technical Report AFWAL-TR-80-4042*, vol. II, Materials Laboratory, Air Force Wright Aeronautical Laboratories, Ohio, April, 1980.

PROBLEMS

2-1 The notation scheme described in Sec. 2-1 provides a shorthand method of identifying robot configurations. For the following arm and body designations, describe the particular robot system, using sketches where possible to illustrate the robot.
   (a) LLR
   (b) RLR
   (c) LRR
   (d) LVR
   (e) LL-TRL

2-2 Sketch the two configurations for the jointed arm robot given in Table 2-1.

2-3 For the following robot wrist notations, describe the particular wrist configuration, using sketches similar to Fig. 2-11 to illustrate the wrist.
   (a) :TR
   (b) :RT
   (c) :TRT
Analyze the differences in the capability of the three wrist configurations to position and orient an end effector.

2-4 If your school or company operates a robotics laboratory, prepare a catalog of all of the robots in the lab in terms of their respective anatomy notations. That is, write the notation scheme codes for all the robots in the lab.

2-5 One of the axes of a robot is a telescoping arm with a total range of 0.50 m (slightly less than 20 in.). The robot's control memory has an 8-bit storage capacity for this axis. Determine the control resolution for the axis.

2-6 Solve Prob. 2-5 except that the robot has the following bit storage capacity in its control memory:
   (*a*) A 10-bit storage memory.
   (*b*) A 12-bit storage memory.

2-7 A large cartesian coordinate robot has one orthogonal slide with a total range of 30 in. One of the specifications on the robot is that it have a maximum control resolution of 0.010 in. on this particular axis. Determine the number of bits of storage capacity which the robot's control memory must possess to provide this level of precision.

2-8 One of the axes of a *RRL* robot is a sliding mechanism with a total of 0.70 m (about 27.5 in.). The robot's control memory has a 10-bit capacity. In addition, it has been observed that the mechanical inaccuracies associated with moving the arm to any given programmed point form a normally distributed random variable with the mean at the taught point and the standard deviation equal to 0.10 mm (about 0.004 in.). Assume that the standard deviation is isotropic (it is equal in all directions). By definition, three standard deviations include "all" of the mechanical errors in the arm movement. With these definitions and assumptions, determine the following:
   (*a*) The control resolution for this axis.
   (*b*) The spatial resolution for this axis.
   (*c*) The defined accuracy of the robot for this axis.
   (*d*) The repeatability of the robot.

2-9 The telescoping arm of a certain industrial robot obtains its vertical motion by rotating (type *R* joint) about a horizontal axis. The total range of rotation is 90°. The robot possesses a 10-bit storage capacity for this axis. When fully extended, the robot's telescoping arm measures 50 in. from the pivot point. When fully retracted the arm measures 30 in. from the pivot point.
   (*a*) Determine the robot's control resolution for this axis in degrees of rotation.
   (*b*) Determine the robot's control resolution on a linear scale in both the fully extended and fully retracted position.
   (*c*) Sketch the side view of the robot's work volume as determined by this pivoting axis.

2-10 The mechanism connecting the wrist assembly is a type *T* (twisting joint which can be rotated through eight full revolutions from one extreme position to the other. It is desired to have a control resolution of plus or minus 0.2 degrees of rotation (or better). What is the required bit storage capacity in order to achieve this resolution?

TWO

ROBOT TECHNOLOGY: THE ROBOT AND ITS PERIPHERALS

Part Two of the book is focused on the mechanical and electronic technology of industrial robots. This includes the technology of both the manipulator itself and the peripheral devices and systems that work with the robot.

The design of the robot manipulator represents a significant challenge to mechanical and electrical engineers. As suggested by our discussion in Chap. Two, the problem is to configure a physical system that is capable of positioning its end effector to within several thousandths of an inch of a desired target location, is relatively lightweight, possesses high-lift capacity, and moves at high speeds between positions in the workspace. Chapters Three and Four consider this design problem. Chapter Three examines the various components of the individual joints that make up the manipulator. Chapter Four treats the mathematical analysis of the arm position and motion. These mathematical methods are utilized to aid engineers in the design of robots, and by the robot control computer to calculate motion trajectories and frame transformations during the work cycle.

Chapter Five considers one of the important peripheral devices used with robots in industrial applications—their end effectors. The different types of end effector are discussed, along with the engineering considerations for their design. We concentrate in this chapter on the grippers used to handle workparts. Chapters Six and Seven examine another type of peripheral used in industrial robotics—sensors. Chapter Six discusses the variety of sensors commonly found in robot applications, with emphasis on touch and force sensors. Chapter Seven surveys the technology of machine vision. Sensors, combined with a sufficiently powerful computer brain to use them, provide robots with significant capabilities to perform useful work.

THREE

CONTROL SYSTEMS AND COMPONENTS

A robot is a mechanical system that must be controlled in order to accomplish a useful task. The task involves the movement of the manipulator arm, so the primary function of the robot control system is to position and orient the wrist (and end effector) with a specified speed and precision. This chapter and the following chapter deal with the problem of controlling the mechanical joints of an industrial robot. In the present chapter, we consider the basic theory of control systems and the components that are commonly used for applying this theory to a single joint of a robot manipulator. Chapter Four considers the mathematical analysis required to control a complete robot system involving a number of joints.

3-1 BASIC CONTROL SYSTEMS CONCEPTS AND MODELS

This and the following two sections will review some of the basic concepts and mathematical modeling techniques used to analyze control systems. The emphasis will be on mechanical systems since a robot manipulator falls within this class. Readers already familiar with linear feedback systems may want to skim quickly over these sections and go to Sec. 3-4.

When studying a mechanical system we are concerned with the response of the system to certain inputs. These inputs include commands to drive the system and disturbances from the environment. We can divide a system into five major components:

The input (or inputs) to the system
The controller and actuating devices

The plant (the mechanism or process being controlled)
The output (the controlled variable)
Feedback elements (sensors)

By analyzing the effects that the controller and the plant have on the inputs, we can predict what the outputs will be under certain conditions. In order to do this analysis we must be able to model the system mathematically.

Mathematical Models

Mathematical models are simpyy mathematical representations of real world systems. They are developed by applying the known rules of behavior for the elements in a system. Hooke's law for the operation of a spring is an example:

$$F = K_s x \qquad (3\text{-}1)$$

where F is the force applied to the spring, x is the displacement due to the application of the force, and K_s is the "spring constant." Using physical relationships of this sort we can develop models of more complex systems than just a spring. As an example, let us formulate the model for a familiar mechanical system: the spring–mass–damper system.

The system is illustrated in Fig. 3-1 and consists of a block with a certain mass suspended from a fixed wall by a dashpot. The mass is connected to a spring, the other end of which is given some displacement from its equilibrium position. We will use the following symbols to represent the various parameters of system behavior:

y = the displacement of the mass
M = the mass of the block
K_s = the spring constant
K_d = the damping coefficient of the dashpot
x = the displacement of the end of the spring

The operation of the system can be described as a sum of the forces on the mass. The force due to acceleration of the mass is

$$M \frac{d^2 y}{dt^2}$$

The force due to the dashpot is

$$K_d \frac{dy}{dt}$$

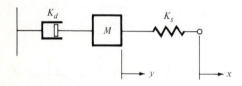

Figure 3-1 Schematic diagram of a spring-mass-damper system.

The force due to the spring is

$$K_s y - K_s x$$

Summing all of the forces we get

$$M\frac{d^2 y}{dt^2} + K_d \frac{dy}{dt} + K_s y = K_s x \tag{3-2}$$

In this system the input is x, representing the displacement of the end of the spring, and the system output is y, representing the displacement of the block. The system has been described by a second-order linear differential equation which relates the input and the output. This mathematical description of the system allows us to analyze its behavior. Before beginning the analysis, however, we will develop other useful tools for building the models.

Transfer Functions

Linear differential equations can be rewritten using the differential operator, s. The variable, s, is used to represent the mathematical operation of taking the derivative of a time-dependent variable with respect to time. Thus, functions which are variables of time [e.g., $x(t)$ and $y(t)$] become functions of the variable s [e.g., $X(s)$ and $Y(s)$]. By using s with Laplace transforms, linear differential equations can be converted to equivalent expressions which are functions of s. (It is assumed that the reader is familiar with the Laplace transform and its s operator for linear systems analysis.) Using s, Eq. (3-2) can be written as

$$Ms^2 Y(s) + K_d s Y(s) + K_s Y(s) = K_s X(s) \tag{3-3}$$

The transfer function relates the output of the system to an input. The spring–mass–damper transfer function can be derived by rewriting Eq. (3-3) as

$$\frac{Y(s)}{X(s)} = \frac{K_s}{Ms^2 + K_d s + K_s} \tag{3-4}$$

Block Diagrams

It is often useful to provide a schematic representation of the system in addition to the mathematical model. A common means of graphically representing the relationships among the components of the system is the block diagram. Block diagrams are constructed from four basic elements:

Function blocks
Signal arrows
Summing junctions
Takeoff points

The four components are illustrated in Fig. 3-2. A function block, shown in

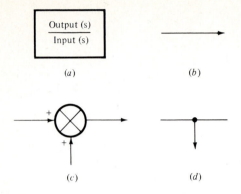

(a)   (b)

(c)   (d)

Figure 3.2 Block diagram elements. (a) function block, (b) signal arrow, (c) summing junction, (d) takeoff point.

Fig. 3-2(*a*), represents one of the components of the system and contains the transfer function for the component. The signal arrows, illustrated in Fig. 3-2(*b*), indicate the direction of the signals and variables in the diagram, and are used to connect function blocks and other system components. Summing junctions (also called summing points) permit two or more signals to be added (algebraically), as shown in Fig. 3-2(*c*). Takeoff points are pictured in part (*d*) of the figure and permit signals and variables to be shared among more than a single component.

By assembling these components it is possible to describe any linear system in the form of a block diagram. By convention, block diagrams are usually read from left to right, with inputs coming in from the left and outputs going to the right. Summing junctions may have any number of arrows entering, but only one leaving. Feedback loops generally run from right (the output side) to left (the input side).

> **Example 3-1** Draw the block diagram that corresponds to the spring–mass–damper system represented by Eqs. (3-2) and (3-3) in the text.
> The resulting block diagram is displayed in Fig. 3-3.

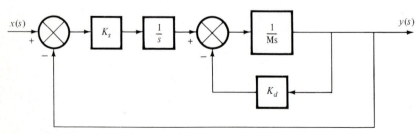

Figure 3-3 Block diagram for Example 3-1.

CONTROL SYSTEMS AND COMPONENTS **55**

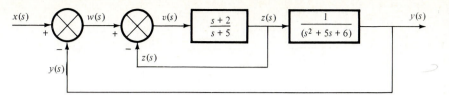

Figure 3-4 Block diagram for Example 3-2.

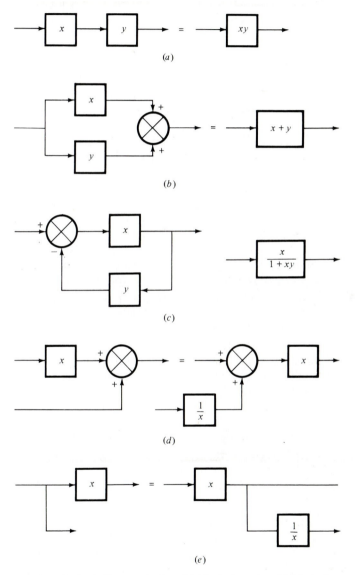

Figure 3-5 Block diagram algebra. (a) blocks in series, (b) blocks in parallel, (c) elimination of a feedback loop, (d) shifting a summing junction, (e) shifting a takeoff point.

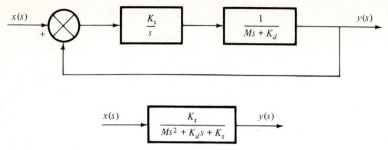

Figure 3-6 Block diagram reduction for Example 3-3.

Example 3-2 For the set of equations below, develop the block diagram which uses $X(s)$ as the system input and $Y(s)$ as the system output:

$$W(s) = X(s) - Y(s)$$

$$V(s) = W(s) - Z(s)$$

$$Z(s)(s+5) = V(s)(s+2)$$

$$Y(s)(s^2 + 5s + 6) = Z(s)$$

The resulting block diagram is shown in Fig. 3-4.

Complicated block diagrams may be simplified by using reduction techniques called block diagram algebra. For example, two function blocks in series may be combined into a single block. Figure 3-5 illustrates these reduction techniques. The usual procedure is to first combine all series blocks in the diagram into the corresponding single block; then any parallel blocks are combined; then the basic feedback loops are reduced to equivalent single blocks; finally, the summing points are shifted to the left and the takeoff points to the right. For very complex block diagrams, the above procedures may have to be repeated in order to simplify the diagram to a single block representing the system transfer function.

Example 3-3 Using block diagram algebra, reduce the block diagram of Fig. 3-3 (Example 3-1) to a single block.
Starting with Fig. 3-3, we reduce the diagram in two steps as shown in Fig. 3-6. The system transfer function is in the resulting block.

The Characteristic Equation

The characteristic equation for the spring–mass–damper system can be written as

$$Ms^2 + K_d s + K_s = 0 \qquad (3\text{-}5)$$

and the roots of the characteristic equation, Eq. (3-5), are given by

$$s_{1,2} = -\frac{K_d}{2M} \pm \frac{\sqrt{K_d^2 - 4MK_s}}{2M} \tag{3-6}$$

The performance of the system is dependent on the values of M, K_d, and K_s. One aspect of the system performance that can be determined by analyzing the roots of the characteristic equation is the "damping" of the system. Depending on the values of the parameters in the characteristic equation, the system may respond in one of four ways. The four responses classify the system into one of the following types:

1. Undamped system
2. Underdamped system
3. Critically damped system
4. Overdamped system

We will now briefly describe these four types of system response.

Undamped In order for the system to be undamped, the damping coefficient, K_d, must be equal to zero. In this case the roots of the characteristic equation are given by

$$s_{1,2} = \pm j\sqrt{\frac{K_s}{M}} \tag{3-7}$$

These are imaginary roots. Assuming a step input X to the system, the response can be described as

$$y = C_1 \sin(\omega_n t) + C_2 \cos(\omega_n t) + X \tag{3-8}$$

where $\omega_n = \sqrt{K_s/M}$ and is called the natural frequency of the system. The response represented by Eq. (3-8) is shown in Fig. 3-7(a) where it can be seen that the undamped response is oscillatory.

Underdamped When there is a small amount of damping in the system, that is, where

$$K_d^2 < 4MK_s$$

the roots may be rewritten as

$$s_{1,2} = -\frac{K_d}{2M} \pm j\frac{\sqrt{4MK_s - K_d^2}}{2M} \tag{3-9}$$

Substituting $a = Kd/2M$ and $\omega_d = \sqrt{(4MK_s - K_d^2)}/2M$, Eq. (3-9) may be written as

$$a_{1,2} = -a \pm j\omega_d \tag{3-10}$$

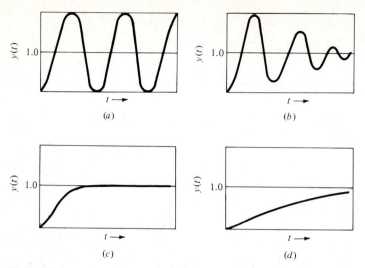

Figure 3-7 Response curves for the four possible cases in a second order linear system. (a) undamped, (b) underdamped, (c) critically damped, (d) overdamped.

The response of the system is described by

$$y = e^{-at}[C_1 \sin(\omega_d t) + C_2 \cos(\omega_d t)] + X \qquad (3\text{-}11)$$

which is similar to the previous response except that ω_d represents the damped natural frequency and the term e^{-at} results in a decaying amplitude envelope for the oscillations as represented in Fig. 3-7(b).

Critically Damped This situation occurs in the special case where

$$K_d^2 = 4MK_s$$

which results in the roots of the characteristic equation being

$$-\frac{K_d}{2M}$$

In this case the response of the system is given by

$$y = C_1 e^{-at} + C_2 t e^{-at} + X \qquad (3\text{-}12)$$

and is represented in Fig. 3-7(c). The critically damped response provides the fastest response without overshoot to the input of the four types.

Overdamped In the case of an overdamped system

$$K_d^2 > 4MK_s$$

and the roots are

$$s_{1,2} = -a \pm b \qquad (3\text{-}13)$$

where $b = \sqrt{K_d^2 - 4MK_s}/2M$. The response of the system is

$$y = C_1 e^{(-a+b)t} + C_2 e^{(-a-b)t} + X \qquad (3\text{-}14)$$

This is illustrated in Fig. 3-7(*d*). As with the critically damped system this response does not oscillate, but the time to reach the desired steady-state response is longer.

Although a robot manipulator is mechanically far more complex than the spring–mass–damper system we have analyzed here, it exhibits many of the same operating features. It has mass, it has stiffness that can be likened to the spring constant, and the joints possess damping. The resulting motions of the robot arm behave in a manner similar to the performance of a system described by a second-order linear differential equation with constant coefficients. The complications of the robot manipulator are: first, it represents a higher-order equation than second order; second, its degrees of freedom are greater than the single degree of freedom of the spring–mass–damper example, and these degrees of freedom interact to some extent; and third, its behavior includes features that are nonlinear.

Notwithstanding the complications listed above, the design of a robot must address the same kinds of performance characteristics discussed in our example. For instance, the robot stability problem discussed in Chap. 2 (Sec. 2-4) takes on additional meaning in the light of the discussion in this section. There are trade-offs that must be made between the response of the system and the stability of the system. In some cases it would be undesirable for the system to overshoot the target point and, therefore, it might be necessary to overdamp, with the result that the speed of response would be slower.

Figure 3-8 illustrates a general block diagram of the control system components for one joint of a robot manipulator. The input command is the defined position (and possibly speed) to which the joint is directed to move. The output variable is the actual position (and speed) of the joint. In nearly all robots today, most of the computational functions of the joint controller are carried out by a microprocessor as portrayed in Fig. 3-8.

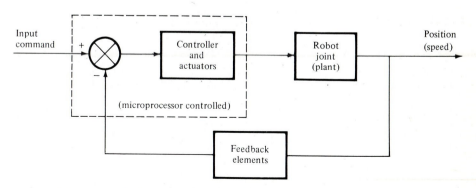

Figure 3-8 Typical block diagram configuration of a control system for a robot joint.

Up to this point we have focused on the modeling of physical systems and the ways in which a system may respond. Let us turn our attention to the methods of controlling the response of a system.

3-2 CONTROLLERS

As indicated in Fig. 3-8, the components of a control system include the controller and actuator. The purpose of the controller is to compare the actual output of the plant with the input command and to provide a control signal which will reduce the error to zero or as close to zero as possible. In this section we present the theoretical operation of these controllers and in the next section we discuss the physical devices which are used to implement this control in a robot joint.

The controller generally consists of a summing junction where the input and output signals are compared, a control device which determines the control action, and the necessary power amplifiers and associated hardware devices to accomplish the control action in the plant. The actuator is used in robotics to convert the control action into physical movement of the manipulator. The controller and actuator may be operated by pneumatic, hydraulic, mechanical, or electronic means, or combinations of these.

There are four basic control actions which are used singly or in combination to provide six common types of controller: on–off control, proportional control, derivative control and integral control. The six controller types are:

1. On–off
2. Proportional
3. Integral
4. Proportional-plus-integral (P-I)
5. Proportional-plus-derivative (P-D)
6. Proportional-plus-integral-plus-derivative (P-I-D)

Each one of these controller types is best suited to certain applications. The following subsections describe the operation of each type of controller.

On–Off Control

In the on–off controller the control element provides only two levels of control, full-on or full-off. An example of a common implementation of this type of controller is the household thermostat. If the error which is present at the controller is $e(t)$ and the control signal which is produced by the controller is $m(t)$, then the on–off controller is represented by

$$m(t) = M_1 \quad \text{for} \quad e(t) > 0$$
$$= M_2 \quad \text{for} \quad e(t) < 0 \tag{3-15}$$

In most on–off controllers either M_1 or M_2 is zero. The practical use of an on–off controller usually requires that the error must move through some range before switching actually takes place. This prevents the controller from oscillating at too high a frequency. This range is referred to as the differential gap.

Proportional Control

In cases where a smoother control action is required a proportional controller may be used. Proportional control provides a control signal that is proportional to the error. Essentially it acts as an amplifier with a gain K_p. Its action is represented by

$$m(t) = K_p e(t) \qquad\qquad (3\text{-}16)$$

Using the differential operator notation introduced earlier the transfer function would be

$$\frac{M(s)}{E(s)} = K_p \qquad\qquad (3\text{-}17)$$

Integral Control

In a controller employing an integral control action the control signal is changed at a rate proportional to the error signal. That is, if the error signal is large, the control signal increases rapidly; if it is small, the control signal increases slowly. This may be represented by

$$m(t) = K_i \int e(t)\, dt \qquad\qquad (3\text{-}18)$$

where K_i is the integrator gain. The corresponding transfer function is

$$\frac{M(s)}{E(s)} = K_i / s \qquad\qquad (3\text{-}19)$$

using $1/s$ as the operator for integration. If the error were to go to zero, the output of the controller would remain constant. This feature allows integral controllers to be used when there is some type of constant load on the system. Even if there is no error the controller would still maintain an output signal to counteract the load.

Proportional-plus-Integral Control

Sometimes it is necessary to combine control actions. A proportional controller is incapable of counteracting a load on the system without an error. An integral controller can provide zero error but usually provides slow response.

One way to overcome this is with the P-I controller. This is represented by

$$m(t) = K_p e(t) + \frac{K_p}{T_i} \int e(t) \, dt \qquad (3\text{-}20)$$

where T_i adjusts the integrator gain and K_p adjusts both the integrator and the proportional gain. The transfer function is

$$\frac{M(s)}{E(s)} = K_p \left(1 + \frac{1}{T_i s}\right) \qquad (3\text{-}21)$$

Proportional-plus-Derivative Control

Derivative control action provides a control signal proportional to the rate of change of the error signal. Since this would generate no output unless the error is changing, it is rarely used alone. The P-D controller is represented by

$$m(t) = K_p e(t) + K_p T_d \frac{de(t)}{dt} \qquad (3\text{-}22)$$

and the transfer function is

$$\frac{M(s)}{E(s)} = K_p (1 + T_d s) \qquad (3\text{-}23)$$

The effect of derivative control action is to anticipate changes in the error and provide a faster response to changes.

Proportional-plus-Integral-plus-Derivative Control

Three of the control actions can be combined to form the P-I-D controller. The P-I-D controller can be represented by

$$m(t) = K_p e(t) + \frac{K_p}{T_i} \int e(t) \, dt + K_p T_d \frac{de(t)}{dt} \qquad (3\text{-}24)$$

and the transfer function is

$$\frac{M(s)}{E(s)} = K_p \left(1 + \frac{1}{T_i s} + T_d s\right) \qquad (3\text{-}25)$$

P-I-D control is the most general control type and probably the most commonly used type of controller. It provides quick response, good control of system stability and low steady-state error. As indicated previously, the computations associated with any of the above controllers are typically performed by microcomputers in a modern robot controller.

3-3 CONTROL SYSTEM ANALYSIS

Analysis of a control system may be divided into two parts: transient response and steady-state response. The transient response of a system is the behavior of the system during the transition from some initial state to the final state. The steady-state response is the behavior of the system as time approaches infinity.

Transient Response of Second-Order Systems

Second-order linear systems are frequently used in control systems analysis, even when it is known that the particular system of interest may be of higher order. Second-order systems can often approximate complex physical systems with reasonable fidelity. Let us return to our transfer function for the second-order system derived in Sec. 3-1.

$$\frac{Y(s)}{X(s)} = \frac{K_s}{Ms^2 + K_d s + K_s} \tag{3-26}$$

The natural frequency of the system was represented by

$$\omega_n = \sqrt{\frac{K_s}{M}}$$

The damping ratio of a second-order system can be defined as

$$z = \frac{Kd/2M}{\omega_n}$$

If the damping ratio is equal to zero then the system will oscillate continuously, if $z < 1$ but greater than zero then the system is underdamped. If $z = 1$ then the system is critically damped and if $z > 1$ the system is overdamped. Figure 3-9 illustrates the transient response of a second-order system with different damping ratios to a unit step input. There are other parameters of interest in

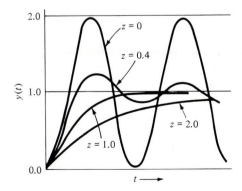

Figure 3-9 Transient responses for different damping ratios.

the transient response of a system. These are:

Delay time, t_d—Delay time is the time that it takes the system to reach one-half of the final value for the first time.

Rise time, t_r—Rise time is the time that it takes the system to go from 10 to 90 percent, 5 to 95 percent, or 0 to 100 percent of the final value.

Peak time, t_p—Peak time is the time that it takes the system to reach the maximum overshoot for the first time.

Maximum overshoot, M_p—Maximum overshoot is the maximum peak value measured from the steady state value in Fig. 3-10.

Settling time, t_s—Settling time is the time required for the system to stay within a range about the final value. This is usually within 2 and 5 percent.

Figure 3-10 illustrates these system parameters. In some cases, certain of the parameters are not relevant. In the case of a critically damped system there is no overshoot and hence M_p and t_p do not apply. In robotics it is sometimes critical that the system not be allowed to overshoot, while in other applications it may be necessary, for the sake of speed, to allow overshoot. The balancing of these parameters when designing the system is the responsibility of the controls engineer.

Included within the scope of the transient response is the question of whether the system will be stable for all inputs. System stability is interpreted to mean that the system output will not be driven toward a value of infinity in response to noninfinite input. Stability is assured if the transients gradually go toward zero as time increases. System instability occurs when the transient response increases with time.

The stability of any linear system can be determined if the characteristic equation of the system is known and if it is possible to factor the equation. Referring back to our example of a second-order linear system, let us relate

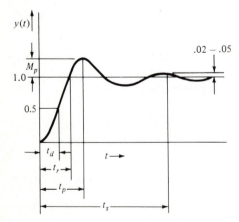

Figure 3-10 Transient response parameters.

our discussion of stability to the four system responses described in Sec. 3-1. In the case of the underdamped, critically damped, and overdamped systems, the transient responses gradually decrease with time as the output assumes some steady-state value. These systems are all stable. The common feature which makes each of these systems stable is the fact that their characteristic equations have roots that are negative real numbers or complex numbers with negative real parts. This is the requirement for stability. If the roots are negative real numbers or complex numbers with negative real parts, the transient response will always approach zero with time.

In the undamped case, the response continues to oscillate because the system possesses no damping. The roots of the characteristic equation are imaginary numbers with no real components. This case is said to be marginally stable. It represents the dividing line between system stability and instability.

Steady-State Response

The steady-state analysis of a control system is concerned with determining the response of the system after the transient response has disappeared. It is assumed that the system of interest is one which is stable. In steady-state analysis, the system designer wants to know whether the system will achieve the desired final value as time of operation increases. One approach to the problem would be to solve the differential equation of the system as it is subjected to some appropriate input. Depending on the degree of difficulty of the differential equation, this could prove to be a difficult approach. A more direct method is to make use of the final value theorem from control theory which uses the Laplace transform of the system output.

The final value theorem states that the final value of the function is given by

$$\lim_{t \to \infty} f(t) = \lim_{s \to 0} sF(s) \tag{3-27}$$

where $F(s)$ is the Laplace transform of the function $f(t)$. Implicit in the above statement of the final value theorem is that the limit of $f(t)$ exists as time approaches infinity.

Example 3-4 Suppose that in Eq. (3-2) the values of the constants are $M = 2$, $K_d = 6$, and $K_s = 5$. Further suppose that the input, x, to the system is a unit step function. Determine the steady-state response of the system according to the final value theorem.

The unit step response has a Laplace transform $= 1/s$. The transfer function for the system represented by Eq. (3-2) for the values given is

$$\frac{Y(s)}{X(s)} = \frac{5}{2s^2 + 6s + 5}$$

Substituting the unit step input, $1/s$, for $X(s)$ gives us the Laplace transform of the system response, $y(t)$. Expressed in the s domain, we have

$$Y(s) = \frac{1}{s}\frac{5}{2s^2 + 6s + 5}$$

Substituting $Y(s)$ into Eq. (3-27) we get

$$\lim_{t \to \infty} y(t) = \lim_{s \to 0} \frac{5s}{s(2s^2 + 6s + 5)} = 1$$

3-4 ROBOT ACTUATION AND FEEDBACK COMPONENTS

Control of the robot manipulator requires the application of the preceding control theory to a mechanical system. In this and the following sections we discuss some of the various types of devices commonly used as components of robot control systems. We classify these devices into four categories:

Position sensors
Velocity sensors
Actuators
Power transmission devices

Position and velocity sensors are used in robotics as feedback devices, while actuators and power transmission devices are used to accomplish the control actions indicated by the controller. Position sensors provide the necessary means for determining whether the joints have moved to correct linear or rotational locations in order to achieve the required position and orientation of the end effector. The speed with which the manipulator is moved is another performance feature which must be regulated. Many robots utilize a feedback system to ensure proper speed control. This is especially important as sophisticated control systems are being developed to fine tune the dynamic performance of the manipulator during acceleration and deceleration as it moves between points in the workspace.

Actuators and power transmission devices provide the muscle to move the robot arm. Actuators include hydraulic, electric, and pneumatic devices corresponding to the three basic robot drive systems described in Chapter Two. The power developed by these actuators must be transmitted from the actuator to the robot joint via a power transmission device, except in the case where the actuator is directly coupled to the robot joint. Transmission mechanisms, such as pulley systems, gears, and screws, are used for this purpose.

The following four sections discuss the four types of actuation and feedback components used in robotics.

3-5 POSITION SENSORS

In most cases in robotics a primary interest is to control the position of the arm. There is a large variety of devices available for sensing position. We will discuss the following devices: potentiometers, resolvers, and encoders.

Potentiometers

Potentiometers are analog devices whose output voltage is proportional to the position of a wiper. Figure 3-11 illustrates a typical pot. A voltage is applied across the resistive element. The voltage between the wiper and ground is proportional to the ratio of the resistance on one side of the wiper to the total resistance of the resistive element. Essentially the pot acts as a voltage divider network. That is, the voltage across the resistive element is divided into two parts by a wiper. Measuring this voltage gives the position of the wiper. The function of the potentiometer can be represented by the following function

$$V_o(t) = K_p \theta(t)$$
(3-28)

where $V_o(t)$ is the output voltage, K_p is the voltage constant of the pot in volts per radian (or volts per inch in the case of a linear pot) and $\theta(t)$ is the position of the pot in radians (or inches). Since a pot requires an excitation voltage, in order to calculate V_o, we can use

$$V_o = V_{ex} \frac{\theta_{act}}{\theta_{tot}}$$

$$\therefore K_p = \frac{V_{ex}}{\theta_{tot}}$$
(3-29)

where V_{ex} is the excitation voltage, θ_{tot} is the total travel available of the wiper, and θ_{act} is the actual position of the wiper.

> **Example 3-5** Find the output voltage of a potentiometer with the following characteristics. Also determine the K_p. The excitation voltage = 12 V; total wiper travel = 320°; wiper position = 64°.

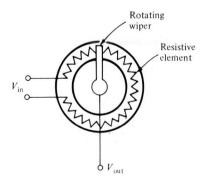

Figure 3-11 Potentiometer.

SOLUTION The $K_p = V_{ex}/\theta_{tot}$ which is $12\,V/320° = 0.0375\,V/deg$. The output voltage is

$$(64°)(0.0375\,V/deg) = 2.4\,V$$

Resolvers

A resolver is another type of analog device whose output is proportional to the angle of a rotating element with respect to a fixed element. In its simplest form a resolver has a single winding on its rotor and a pair of windings on its stator. The stator windings are 90° apart as shown in Fig. 3-12. If the rotor is excited with a signal of the type $A\sin(\omega t)$ the voltage across the two pairs of stator terminals will be

$$V_{s1}(t) = A\sin(\omega t)\sin\theta \qquad (3\text{-}30)$$

and

$$V_{s2}(t) = A\sin(\omega t)\cos\theta \qquad (3\text{-}31)$$

where θ is the angle of the rotor with respect to the stator. This signal may be used directly, or it may be converted into a digital representation using a device known as a "resolver-to-digital" converter. Since a resolver is essentially a rotating transformer it is important to remember that an ac signal must be used for excitation. If a dc signal were used there would be no output signal.

Example 3-6 At time t the excitation voltage to a resolver is 24 V. The shaft angle is 90°. What is the output signal from the resolver?

Figure 3-12 Resolver (Courtesy of Litton Systems, Incorporated, Clifton Precision Division).

SOLUTION

$$V_{s1} = (24 \text{ V})(\sin 90°) = 24 \text{ V}$$

$$V_{s2} = (24 \text{ V})(\cos 90°) = 0 \text{ V}$$

Example 3-7 At time t the excitation voltage to a resolver is 24 V and $V_{s1} = 17$ V and $V_{s2} = -17$ V. What is the angle?

SOLUTION

$$\arcsin\left(\frac{17}{24}\right) = 45° \quad \text{or} \quad 135°$$

$$\arccos\left(-\frac{17}{24}\right) = 135° \quad \text{or} \quad 225°$$

The shaft angle must be 135°.

Encoders

As more systems become controlled by computers and related devices the use of digital position encoders is increasing. Encoders are available as two basic types: incremental and absolute. This refers to the type of data available from the encoder. There are various categories of encoding devices, but we will limit our discussion to those that are most commonly used in robots. These are optical encoders.

A simple incremental encoder is illustrated in Fig. 3-13. It consists of a

Figure 13-13 Incremental optical encoder (Courtesy of Litton Systems, Incorporated, Encoder Division).

Figure 3-14 Absolute optical encoder (Courtesy of Litton Systems, Incorporated, Encoder Division).

glass disk marked with alternating transparent and opaque stripes aligned radially. A phototransmitter (a light source) is located on one side of the disk and a photoreceiver is on the other. As the disk rotates, the light beam is alternately completed and broken. The output from the photoreceiver is a pulse train whose frequency is proportional to the speed of rotation of the disk. In a typical encoder, there are two sets of phototransmitters and receivers aligned 90° out of phase. This phasing provides direction information; that is, if signal A leads signal B by 90° the encoder disk is rotating in one direction, if B leads A then it is going in the other direction. By counting the pulses and by adding or subtracting based on the sign, it is possible to use the encoder to provide position information with respect to a known starting location.

In some cases it is desirable to know the position of an object in absolute terms, that is, not with respect to a starting position. For this an absolute encoder could be used. Absolute encoders employ the same basic construction as incremental encoders except that there are more tracks of stripes and a corresponding number of receivers and transmitters. Usually the stripes are arranged to provide a binary number proportional to the shaft angle. The first track might have 2 stripes, the second 4, the third 8 and so on. In this way the angle can be read directly from the encoder without any counting being necessary. Figure 3-14 illustrates an absolute encoder. The resolution of an absolute encoder is dependent on the number of tracks and is given by

$$\text{resolution} = 2^n \tag{3-32}$$

where n is the number of tracks on the disk.

Example 3-8 What is the resolution, in degrees, of an encoder with 10 tracks?

The number of increments per revolution is 2^{10}

$$= 1024 \text{ increments/rev}$$

The angular width of each control increment is therefore

$$= \frac{360°}{2^{10}} = \frac{360°}{1024} = 0.3515°$$

The output of an absolute encoder or of an incremental encoder and counter combination is represented by

$$\text{out}(t) = K_e \theta(t) \tag{3-33}$$

where out is a number, K_e is the number of pulses per radian and θ is the shaft angle, expressed in radians.

Example 3-9 What is the output value of an absolute encoder if the shaft angle is 1 rad and the encoder has 8 tracks?

The resolution is 256 parts/rev. There are 2π rad/rev. Therefore, the output is

$$\frac{256}{2\pi} = 41$$

3-6 VELOCITY SENSORS

One of the most commonly used devices for the feedback of velocity information is the dc tachometer. A tachometer is essentially a dc generator providing an output voltage proportional to the angular velocity of the armature. Figure 3-15 illustrates a typical dc tachometer. A tachometer can be

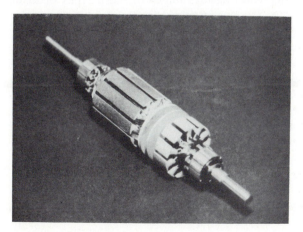

Figure 3-15 Tachometer mounted on a motor armature (Courtesy of Litton Systems, Incorporated, Clifton Precision Division).

described by the relation

$$V_o(t) = K_t(t)\omega \qquad (3\text{-}34)$$

where $V_o(t)$ is the output voltage of the tachometer in volts, $K_t(t)$ is the tachometer constant, usually in V/rad/s and ω is the angular velocity in radians per second.

Tachometers are generally used to provide velocity information to the controller. This can be used for performing velocity control of a device or, in many cases, to increase the value of K_d in a system, thereby improving the stability of the system and its response to disturbances.

Direct current tachometers provide a voltage output proportional to the armature rotational velocity, hence they are analog devices. There is a digital equivalent of the dc tachometer which provides a pulse train output of a frequency proportional to the angular velocity. They are, in effect, encoders which were described in the previous section.

3-7 ACTUATORS

Actuators are the devices which provide the actual motive force for the robot joints. They commonly get their power from one of three sources: compressed air, pressurized fluid, or electricity. They are called, respectively, pneumatic, hydraulic, or electric actuators. We will discuss all three types in this section.

Pneumatic and Hydraulic Actuators

Pneumatic and hydraulic actuators are both powered by moving fluids. In the first case the fluid is compressed air and in the second case the fluid is usually pressurized oil. The operation of these actuators is generally similar except in their ability to contain the pressure of the fluid. Pneumatic systems typically operate at about 100 lb/in.^{2} and hydraulic systems at 1000 to 3000 lb/in.^{2} We discussed the relative advantages and disadvantages of these types of drive systems in Chapter Two.

The simplest fluid power device is the cylinder as illustrated in Fig. 3-16, which could be used to actuate a linear joint by means of a moving piston. This example is called a single-ended cylinder as the piston rod only comes out of the cylinder at one end. Other types of cylinders include double-ended cylinders and rodless cylinders.

There are two relationships of particular interest when discussing actuators: the velocity of the actuator with respect to the input power and the force of the actuator with respect to the input power. For the cylinder type actuator these relationships are given by

$$V(t) = \frac{f(t)}{A} \qquad (3\text{-}35)$$

$$F(t) = P(t)A \qquad (3\text{-}36)$$

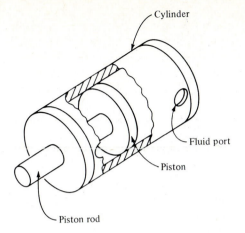

Figure 3-16 Cylinder and piston.

where $V(t)$ is the velocity of the piston, $f(t)$ is the fluid flow rate (volumetric), $F(t)$ is the force, $P(t)$ is the pressure of the fluid, and A is the area of the piston. Since the requirements of a robot are to carry a payload at a given speed we can use the relations described for choosing the appropriate actuator.

> **Example 3-10** What is the velocity of the piston and the force generated by the piston if the fluid pressure is $1500\ \text{lb/in.}^2$ inside the cylinder, the piston is 2.0 in. in diameter, and the flow rate is $10\ \text{in.}^3/\text{min}$?
> The piston area is $3.14\ \text{in.}^2$.
>
> $$F = (1500\ \text{lb/in.}^2)(3.14\ \text{in.}^2) = 4712\ \text{lb}$$
>
> $$V = \frac{10\ \text{in.}^3/\text{min}}{3.14\ \text{in.}^2} = 3.18\ \text{in./min}$$

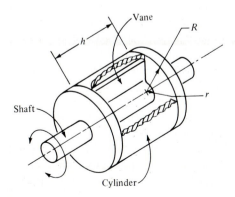

Figure 3-17 Vane actuator.

Another type of fluid power actuator is the rotary vane actuator, shown in Fig. 3-17. In a rotary actuator we are interested in the angular velocity, ω, and the torque, T. The relations describing the output of a rotary actuator are

$$\omega(t) = \frac{2f(t)}{(R^2 - r^2)h} \tag{3-37}$$

$$T(t) = \tfrac{1}{2}P(t)h(R - r)(R + r) \tag{3-38}$$

where R is the outer radius of the vane, r is the inner radius of the vane, h is the thickness of the vane, ω is the angular velocity in radians per second, and T is the torque.

Electric Motors

As their capabilities improve, electric motors are becoming more and more the actuator of choice in the design of robots. They provide excellent controllability with a minimum of maintenance required. There are a variety of types of motors in use in robots; the most common are dc servomotors, stepper motors, and ac servomotors.

Figure 3-18 shows typical dc servomotors. The main components of the dc servomotor are the rotor and the stator. Usually, the rotor includes the armature and the commutator assembly and the stator includes the permanent magnet and brush assemblies. When current flows through the windings of the armature it sets up a magnetic field opposing the field set up by the magnets. This produces a torque on the rotor. As the rotor rotates, the brush and commutator assemblies switch the current to the armature so that the field remains opposed to the one set up by the magnets. In this way the torque produced by the rotor is constant throughout the rotation. Since the field strength of the rotor is a function of the current through it, it can be shown that for a dc servomotor

$$T_m(t) = K_m I_a(t) \tag{3-39}$$

where T_m is the torque of the motor, I_a is the current flowing through the armature, and K_m is the motor's torque constant.

Another effect associated with a dc servomotor is the back-emf. A dc motor is similar to a dc generator or tachometer. Spinning the armature in the presence of a magnetic field produces a voltage across the armature terminals. This voltage is proportional to the angular velocity of the rotor:

$$e_b(t) = K_b \omega(t) \tag{3-40}$$

where e_b is the back-emf (voltage), K_b is called the voltage constant of the motor, and ω is the angular velocity. The effect of the back-emf is to act as viscous damping for the motor: as the velocity increases the damping increases proportionately. If we were to supply a voltage across the motor terminals of V_{in}, and the resistance of the armature were R_a, then the current through the

Figure 3-18 DC servomotors (Courtesy of Litton Systems, Incorporated, Clifton Precision Division).

armature would be V_{in}/R_a. This current produces a torque on the rotor and causes the motor to spin. As the armature spins it generates a back-emf equal to $K_b\omega(t)$, or $e_b(t)$. This voltage must be subtracted from V_{in} in order to calculate the armature current. The actual armature current is therefore

$$I_a(t) = \frac{V_{\text{in}}(t) - e_b(t)}{R_a} \qquad (3\text{-}41)$$

As the motor velocity increases, and the back-emf voltage increases accordingly, the current available to the armature decreases. The decreasing current reduces the torque generated by the rotor. As the torque decreases the acceleration of the rotor decreases as well. At the point at which $e_b = V_{\text{in}}$ the motor maintains a steady-state velocity (assuming there are no external disturbances on the motor). The block diagram in Fig. 3-19 illustrates the effects of the torque constant and back-emf on the model of a motor. Note that this simplified model discounts such effects as friction or inductance of the armature windings.

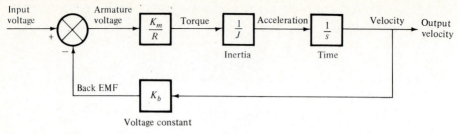

R = armature resistance

Figure 3-19 DC motor block diagram.

Example 3-11 A motor has a torque constant, $K_m = 10$ oz-in./A and a voltage constant of 12 V/Kr/min (1 Kr/min = 1000 r/min). The armature resistance is $2\,\Omega$. If 24 V were applied to the terminals what would be: (*a*) the torque at stall (0 r/min), (*b*) the speed at 0 load (torque = 0), and (*c*) the torque at 1000 r/min? Plot the results on a speed versus torque graph.

SOLUTIONS

(*a*) At 0 r/min the value of $e_b = 0$. Therefore the armature current is

$$= 24\text{ V}/2\,\Omega = 12\text{ A}$$

and the torque is

$$= (12\text{ A})(10\text{ oz-in./A}) = 120\text{ oz-in.}$$

(*b*) At no load the output voltage is equal to the input voltage so that

$$24\text{ V} = (12\text{ V/Kr/min})w(t)$$

$$w(t) = 2\text{ Kr/min} = 2000\text{ r/min}$$

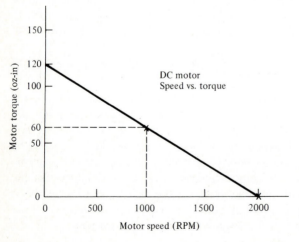

Figure 3-20 Plot for Example 3-11.

(*c*) At 1000 r/min the output voltage is 12 V. Therefore the current through the armature is

$$\frac{24 \text{ V} - 12 \text{ V}}{2 \ \Omega} = 6 \text{ A}$$

and the torque is

$$(6 \text{ A})(10 \text{ oz-in.}/\text{A}) = 60 \text{ oz-in.}$$

We can see that the relationship between speed and torque in Fig. 3-20 is a straight line. This feature is one of the desirable features of dc servomotors.

Stepper Motors

Stepper motors (also called stepping motors) are a unique type of actuator and have been used mostly in computer peripherals. A stepper motor provides output in the form of discrete angular motion increments. It is actuated by a series of discrete electrical pulses. For every electrical impulse there is a single-step rotation of the motor shaft. In robotics, stepper motors are used for relatively light duty applications. Also, stepper motors are typically used in open-loop systems rather than the closed-loop systems on which we have been concentrating in this chapter.

Figure 3-21 provides a schematic representation of one type of stepper motor. The stator is made up of four electromagnetic poles and the rotor is a two-pole permanent magnet. If the electromagnetic stator poles are activated in such a way that pole 3 is N (magnetic North) and pole 1 is S then the rotor is aligned as illustrated. If the stator is excited so that pole 4 is N and pole 2 is S, the rotor makes a 90° turn in the clockwise direction. By rapidly switching the current to the stator electronically, it is possible to make the motion of the rotor appear continuous.

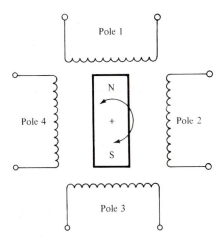

Figure 3-21 Stepper motor schematic.

Figure 3-22 Toothed stepper motor (Courtesy of Litton Systems, Incorporated, Clifton Precision Division).

The resolution (number of steps per revolution) of a stepper is determined by the number of poles in the stator and rotor. Figure 3-22 shows a commercially available stepper motor which has a notched stator and rotor. This effectively increases the number of poles and hence the resolution of the device. The relation between a stepper motor's resolution and its step angle is given by

$$n = A/360° \tag{3-42}$$

where n is the resolution and A is the step angle.

Unlike the dc servomotor, the relation between a stepper motor's speed and torque is not necessarily a straight line. Because of the discrete nature of the stepper motor's construction the torque is also a function of the angle between the stator and rotor poles. The torque is greatest when the poles are aligned. This maximum torque is known as the holding torque of the motor. It is possible to increase the resolution of a stepper by using a technique known as half-stepping or microstepping. By applying current to more than one set of field windings it is possible to make the rotor seek out an "average" position. Of course, when using this technique the holding torque is reduced.

The control of a stepper motor is dependent on the ability of the switching electronics to switch the windings at precisely the right moment. If the windings are switched too quickly, for example, it is possible that the motor will not be able to "keep up" with the command signals and will perform erratically, in some cases oscillating. With some steppers the speed-torque relation degrades badly at certain frequencies of operation, and operation of the motors at these frequencies must be avoided.

AC Servomotors and Other Types

There are numerous other aspects of electric motors which may be investigated. Recent advances in control electronics are producing ac servomotors. These motors have the advantages of being cheaper to manufacture than dc motors, they have no brushes, and they possess a high power output. With the proper electronics package, however, their performance can be made to look very much like the performance of a dc motor.

Another type of electric motor is the brushless dc motor. It is constructed like an "inside-out" dc motor. It has a permanent magnet rotor and an electromagnetic stator. Instead of using brushes, however, commutation is performed electronically using an encoder to inform the electronics of the relative positions of the stator and rotor. Also available are linear electric motors. Their construction is similar to a dc servomotor that has been cut open and flattened out.

In almost all cases of electric motors the limiting factor on power output is heat dissipation. Some of the current used in the motor must be dissipated as heat. Two ways to increase the performance of a motor is to remove heat more quickly or to reduce the current requirements. The latter may be done by increasing the magnetic flux of the permanent magnets. Recent advances in magnetic materials are allowing for performance improvements of almost 10 times with the same power requirements.

3-8 POWER TRANSMISSIONS SYSTEMS

In many cases it is not possible to find an actuator with the exact speed-force or speed-torque characteristics to perform the desired tasks. In other cases it is necessary to locate the actuator away from the intended joint of the manipulator. For these reasons it becomes necessary to use some type of power transmission. Power transmissions perform two functions: transmit power at a distance and act as a power transformer. There are a number of ways to perform mechanical power transmission. These include belts and pulleys, chains and sprockets, gears, transmission shafts, and screws. In this section we discuss a number of power transmission devices that are used for industrial robots.

Gears

The use of gears for power transmission in robots is very common. Gears are used to transmit rotary motion from one shaft to another. This transfer may be between parallel shafts, intersecting shafts, or skewed shafts. The simplest types of gears are for transmission between parallel shafts and are known as "spur gears."

Figure 3-23 illustrates a simple two-gear spur gear train. The driving gear, in this case the smaller one, is known as the pinion and the other gear is the driven gear. For example, if the pinion is one-fourth the size of the gear, for every revolution made by the pinion the driven gear turns only one-fourth of a revolution. This gear train is referred to as a speed reducer. The torque

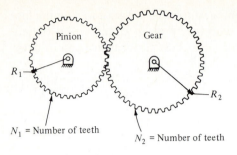

$$n = N_1/N_2$$

Figure 3.23 Spur gear train.

applied by the pinion however is multiplied by four times at the gear shaft. Since the speed is quartered and the torque is quadrupled the power out of the gear train equals the power into it.

The number of teeth in a gear is proportional to its diameter. If we let the number of teeth in a pinion be N_1 and the teeth in the gear equal N_2 then the gear ratio is given by

$$n = \frac{N_1}{N_2} \tag{3-43}$$

and the speed of the output with respect to the input is

$$\omega_o = n\omega_{in} \tag{3-44}$$

where ω_o is the output speed and ω_{in} is the input speed. The output torque is

$$T_o = \frac{T_{in}}{n}$$

Power Screws

In robotics and many other applications, power screws are often used to convert rotary motion to linear motion. The parameter of a screw which is analogous to a gear ratio is the screw pitch, p, often called the lead. The pitch defines the distance that the screw travels in a single rotation. The conversion for the screw's angular rotation to linear motion is given by

$$v(t) = p\omega(t) \tag{3-45}$$

where $v(t)$ is the linear velocity in inches per minute, $\omega(t)$ is the angular velocity in rotations per minute, and p is the screw pitch expressed in inches per rotation. In most cases the screw is rotating and a nut is moving along the length of the screw. The conversion from torque, T, applied to the screw to force, F, on the nut is obtained by the following Eq. (3-46)

$$F = \frac{2T}{d_m} \frac{\pi d_m - \mu p \sec \beta}{p + \mu \pi d_m \sec \beta} \tag{3-46}$$

where μ = the coefficient of friction between the screw threads
$\qquad \beta$ = the thread angle on an Acme or Unified thread
$\qquad d_m$ = the mean diameter of the screw

This equation applies for Acme and Unified threads, in which there is a thread angle, β. For square threads, the value of β is 0, and the secant terms = 1.0 in Eq. (3-46).

Example 3-12 A power screw mechanism is used to actuate a linear (type L) joint in a new robot design. Determine the maximum force that can be transmitted to the nut moving along the power screw if the torque available to turn the screw is 2.0 in.-lb. The screw has square threads ($\beta = 0$) whose pitch is 0.1 in., the diameter of the screw is 0.50 in., and the coefficient of friction between threads is 0.25.

SOLUTION
$$F = \frac{2(2.0)}{0.5} \frac{\pi(0.5) - 0.25(0.1)}{0.1 + 0.25\pi(0.5)} = 25.1 \text{ lb}$$

Because of the relatively high friction in a typical screw thread, ball bearing screws are often used to actuate the linear joints of a robot manipulator. In a ball bearing screw, the nut rides on ball bearings as the screw rotates, rather than directly on the screw itself. This significantly reduces the friction of the device. The conversion from screw torque T to force F resulting at the nut is given by

$$F = \frac{2\pi TE}{P} \tag{3-47}$$

where E = efficiency factor (typically around 90%) resulting from friction losses.

Example 3-13 Let us compare the force resulting from a ball bearing screw with the force in the conventional screw mechanism of Example 3-12. The same values apply: Torque = 2.0 in-lb, pitch = 0.1 in., and we will assume an efficiency factor of 0.90.

SOLUTION:
$$F = \frac{2\pi(2.0)(0.9)}{0.1} = 113.1 \text{ lb}$$

It is clear that the ball bearing screw has a significant mechanical advantage because of the lower friction.

Other Transmission Systems

Other power transmission devices include pulley systems, chain drives, and harmonic drives. Pulley systems are usually used to transmit power from actuators located in the robot's base. In some cases the rope or cord may be made out of steel fibers or from synthetic materials such as nylon. The rotational joints may be connected to a pulley which is driven by a rope attached to a rotary actuator (e.g., electric motor). Similarly, ropes may be used to activate linear joints. In either application the rope undergoes continuous flexing during operation. If the rope is incorrectly sized for the application then this may result in stretching or even failure. If the rope stretches this results in degrading the accuracy of the robot. To maintain desired performance ropes must be maintained according to the manufacturer's instructions.

Chain drives operate with a constant ratio. Due to the positive interaction between the chain and sprockets there is no slipping. The pitch of a chain is the distance between adjacent roller centers. The driving sprocket and the driven sprocket each have a number of teeth designed to match the size and pitch of the chain. The transmission of rotational speed and power between the sprockets follows relationships similar to those developed for gears which we discussed earlier. Lubrication is an important factor in chain drive maintenance. A properly lubricated chain can last 100 times longer than an identical, improperly lubricated chain.

Harmonic drives are proprietary products of USM, Inc. They can be used as speed reducers or increasers. The input and output shafts lie along the same axis, so that a harmonic drive could be mounted to the face of a motor with the output shaft coming out the same end. Harmonic drives can provide any reduction ratio from 1:1 to infinity:1, although they are typically used in the range of 100:1. Harmonic drives require little maintenance and can operate with no noticeable wear over their lifetime. They are, however, less efficient than well-designed gear trains.

Some General Comments

With any power transmission one is liable to introduce two unwanted effects to the performance of the control system. These are compliance and backlash. We used the term compliance in Chapter Two to mean the deflection of the robot arm in response to a force or torque applied against it. In our present context, compliance is defined as the deflection under load of the power transmission device, essentially the spring rate of the transmission. For example, in a gear train this might be due to the bending of the individual teeth, or due to the deflections of the bearing supporting the gears. The compliance of a robot is largely a result of the transmission system compliance.

Backlash represents the hysteresis in the transmission. In gears this is normally due to spaces between the gear teeth mesh. The result of backlash is that the output of the system does not correspond directly to the input of the system. This clearly causes difficulty when a robot manipulator consisting of five or six joints must be moved to a desired location with a precision of only a few thousandths of an inch.

Compliance, backlash, slippage of pulley systems, deflection of the manipulator links, and other imperfections inherent in the mechanical system are the principal contributors to the mechanical inaccuracies that degrade a robot's ability to position its end effector.

3-9 ROBOT JOINT CONTROL DESIGN

In this section we will look at the design of a typical robot joint and its control system using the techniques and components described in the earlier sections. The device described will be an electrically powered shoulder joint as illustrated in Fig. 3-24. The main components of the joint include a dc servomotor, a gear train, a tachometer, a position encoder, and a controller.

The block diagram of the system is shown in Fig. 3-25. The load due to gravity acting on the arm is represented as T_g and is opposing the torque generated by the motor. The term J_t represents the total inertia of the system, including the inertia of the gears, the armature, the arm, and the load. P-I-D control is used for the joint, although the controller itself is only a P-I controller. That is because damping can be controlled by adjusting the gain of the tachometer. In designing this system the parameters for K_p (proportional gain), K_i (integral gain), and K_t (tachometer gain) are adjusted to achieve the desired performance.

One implementation of the controller might be using a microprocessor. The processor would read the value for velocity and position from the feedback devices. It would then multiply them and compare them to the desired position signal. It would take the error signal and multiply it by the integrator and proportional gains to generate a command signal for the power amplifier which would power the motor. One advantage of using a microprocessor for performing this type of control is that it could adapt for varying conditions. For example, we saw earlier that in a second-order system the

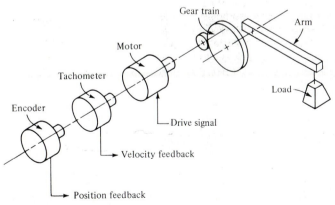

Figure 3-24 Joint schematic diagram.

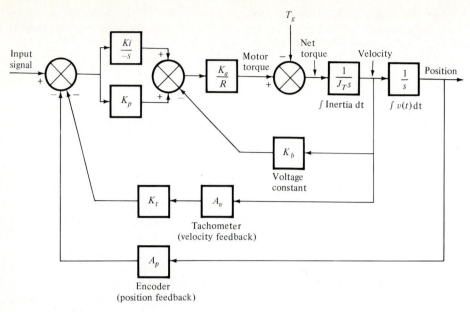

Figure 3-25 Block diagram of a robot joint.

damping is a function of the mass of the system. In a robot the mass of the load is likely to change during the work cycle. A microprocessor could be used to perform computations in support of the control function to determine the appropriate value of the damping constant for the change in mass.

REFERENCES

1. G. S. Boyes, *Synchro and Resolver Conversion*, Memory Devices Ltd., Surrey, United Kingdom, 1980.
2. Electro-Craft Corp., *DC Motors, Speed Controls, Servo Systems*, Hopkins, MN, 1975.
3. M. P. Groover, *Automation, Production Systems and Computer-Aided Manufacturing*, Prentice-Hall, Englewood Cliffs, NJ, 1980.
4. E. Kafrissen and M. Stephans, *Industrial Robots and Robotics*, Reston, Reston, VA, 1984.
5. K. Ogata, *Modern Control Engineering*, Prentice-Hall, Englewood Cliffs, NJ, 1970
6. J. E. Shigley, *Mechanical Engineering Design*, 3rd ed., McGraw-Hill, New York, 1977.
7. Stock Drive Products, *Design and Application of Small Standardized Components*, New Hyde Park, New York, 1983.

PROBLEMS

3-1 Using block diagram reduction techniques, simplify the block diagram of Fig. 3-19 in order to determine the transfer function for the dc motor.

3-2 A certain rotational joint design (including feedback controller) to be used on a new robot model has been studied to determine its response characteristics. It is known that the joint behaves very much like a second-order system. In one part of the study, measurements were taken on the position of the joint output link in response to a defined position input command. The table below presents the response data.

Table P3-2 Joint response data

Time, ms	Input position command	Output position
0	45°	0°
50	45°	24.9°
100	45°	39.2°
150	45°	45.4°
200	45°	46.9°
250	45°	46.6°
300	45°	45.9°
400	45°	45.1°
500	45°	44.9°
1000	45°	45.0°

(a) Plot the output data on a piece of graph paper.

(b) Make your best estimate as to which of the following types of system the response data come from: (1) undamped, (2) underdamped, (3) critically damped, or (4) overdamped.

(c) What additional data would be needed to determine the second order differential equation of the form of Eq. (3-2)?

3-3 A mechanical joint design for a certain robot manipulator has the following differential equation which describes the position of the output link as a function of time:

$$\frac{3.26\, d^2 y}{dt^2} + \frac{17.5\, dy}{dt} + 44.2y = X$$

where X equals the forcing function and y represents the position response of the joint.

(a) Write the characteristic equation for the differential equation above.

(b) Determine the roots of the characteristic equation.

(c) Based on the roots of the characteristic equation, will the response be (1) undamped, (2) underdamped, (3) critically damped, or (4) overdamped?

3-4 Write the transfer function of the differential equation from Prob. 3-3 above.

3-5 For the following set of equations, rewrite each equation using the s-operator notation. Then, construct the block diagram that relates the equations, using x as the input and y as the output.

$$\frac{dz}{dt} + 3.2z = w$$

$$\frac{dy}{dt} + 5.0y = 2.6z$$

$$w = x - 1.5y$$

3-6 Using the block diagram reduction techniques of Fig. 3-5 in the text, reduce the block diagram developed in Prob. 3-5 to a single block thus yielding the transfer function for the system.

3-7 For the differential equation of Prob. 3-3, calculate the natural frequency and the damping ratio of the system.

3-8 For a step input $X = 5.0$, solve the differential equation of Prob. 3-3. Plot your solution on a piece of graph paper, and determine the following transient response parameters, as we have defined them in the text of Sec. 3.3.

(a) Delay time.

(b) Rise time using the definition that rise time is the time required to go from 10 to 90 percent of the final value.

(c) Peak time.

(d) Maximum overshoot.

(e) Settling time.

3-9 Using the response data for the rotational joint design of Prob. 3-2, determine the following transient response parameters, as we have defined them in the text of Sec. 3-3.

(a) Delay time.

(b) Rise time using the definition that rise time is the time required to go from 10 to 90 percent of the final value.

(c) Peak time.

(d) Maximum overshoot.

(e) Settling time.

3-10 Using the final value theorem and your answer to Prob. 3-4 for the mechanical joint design of Prob. 3-3, determine the steady-state response of the system to a step input of $X = 5.0$. (Hint: Recall that a step input of value $X = 5.0$ would have a Laplace transform $= 5/s$.)

3-11 A certain potentiometer is to be used as the feedback device to indicate position of the output link of a rotational robot joint. The excitation voltage of the potentiometer equals 5.0 V, and the total wiper travel of the potentiometer is 300°. the wiper arm is directly connected to the rotational joint so that a given rotation of the joint corresponds to an equal rotation of the wiper arm.

(a) Determine the voltage constant of the potentiometer, K_p.

(b) The robot joint is actuated to a certain angle, causing the wiper position to be 38°. Determine the resulting output voltage of the potentiometer.

(c) In another actuation of the joint, the resulting output voltage of the potentiometer is 3.75 V. Determine the corresponding angular position of the wiper and the output link.

3-12 A resolver is used to indicate angular position of a rotational wrist joint. The excitation voltage to the resolver is 24 V. The resolver is connected to the wrist joint so that a given rotation of the output link corresponds to an equal rotation of the resolver. At a certain moment in time, the movement of the wrist joint results in voltages across the two pairs of stator terminals to be $V_{s1} = 10.0$ V and $V_{s2} = -21.82$ V. Determine the angle of the rotational joint.

3-13 What is the resolution of an absolute optical encoder that has six tracks? Nine tracks? Twelve tracks?

3-14 For an absolute optical encoder with 10 tracks, determine the value of K_e, the encoder constant. If the shaft angle of the encoder were 0.73 rad, determine its output value.

3-15 A dc tachometer is to be used as the velocity feedback device on a certain twisting joint. The joint actuator is capable of driving the joint at a maximum velocity of 0.75 rad/s, and the tachometer constant is 8.0 V/rad/s. What is the maximum output voltage that can be generated by the device, if the tachometer is geared with the joint so that it rotates with twice the angular velocity of the joint? If the joint rotates at a speed of 25°/s, determine the output voltage of the dc tachometer.

3-16 A hydraulic single-ended piston cylinder is to be used to actuate the linear arm joint for a polar configuration robot. The size of the cylinder is 10.0 in.² on the forward stroke (piston extension), and 9.0 in.² on the reverse stroke (piston retraction). The hydraulic power source can generate up to 1000 lb/in.² of pressure for delivery to the cylinder at a rate of 100 in.³/min.

(a) Determine the force that can be applied by the piston on the forward stroke and the reverse stroke.

(b) Determine the maximum velocity at which the piston can operate in the forward and reverse strokes.

3-17 A hydraulic rotary vane actuator is to be used for a twist joint with the same hydraulic power source used for Prob. 3-16. The outer and inner radii (R and r) of the vane are 2.5 in. and 0.75 in., respectively. The thickness of each vane (h) is 0.20 in. Determine the angular velocity and the torque that can be generated by the actuator.

3-18 A dc servomotor is used to actuate a robot joint. It has a torque constant of 10 in.-lb/A, and a voltage constant of 12 V/Kr/min (1 Kr/min = 1000 r/min). The armature resistance = 2.5 Ω. At a particular moment during the robot cycle, the joint is not moving and a voltage of 25 V is applied to the motor.

(*a*) Determine the torque of the motor immediately after the voltage is applied.

(*b*) As the motor accelerates, the effect of the back-emf is to reduce the torque. Determine the back-emf and the corresponding torque of the motor at 250 and 500 r/min.

3-19 A certain dc servomotor used to actuate a robot joint has a torque constant of 25 in.-lb/A, and a voltage constant of 15 V/Kr/min. The armature resistance = 3.0 Ω. At a particular moment during the robot cycle, the joint is not moving and a voltage of 30 V is applied to the motor.

(*a*) Determine the torque of the motor immediately after the voltage is applied.

(*b*) Determine the back-emf and the corresponding torque of the motor at 500 and 1000 r/min.

(*c*) If there were no resisting torques and no inductance of the armature windings operating to reduce the speed of the motor, determine the maximum theoretical speed of the motor when the input voltage is 30 V.

(*d*) If the resisting torques due to friction and the payload being carried by the robot total 72 in.-lb, determine the maximum theoretical speed of the motor when the input voltage is 30 V. Assume no effect of inductance from the armature windings.

3-20 A stepping motor is to be used to actuate one joint of a robot arm in a light duty pick-and-place application. The step angle of the motor is 10°. For each pulse received from the pulse train source, and motor rotates through a distance of one step angle.

(*a*) What is the resolution of the stepping motor?

(*b*) Relate this value to the definitions of control resolution, spatial resolution, and accuracy, as these terms were defined in Chap. Two.

3-21 For the stepping motor described in Prob. 3-20, a pulse train is to be generated by the robot controller.

(*a*) How many pulses are required to rotate the motor through a total of three complete revolutions?

(*b*) If it is desired to rotate the motor at a speed of 25 r/min, what pulse rate must be generated by the robot controller?

3-22 A power screw mechanism is used to convert rotational motion into linear motion for a robot joint. The screw has 12 threads/in. and the thread angle is 10°. The diameter of the screw is 0.375 in., and the coefficient of friction between threads on the screw and the moving nut is 0.30. If the torque applied to the screw is 10 in.-lb, determine the force that will be transmitted to the nut moving along the screw.

3-23 Solve Problem 3-22 except that a ball bearing screw will be used instead of a conventional screw thread. The applied torque is 10 in-lb, there are 12 threads/in., and the efficiency factor is assumed to be 90%.

3-24 A stepping motor is to be used to drive each of the three linear axes of a cartesian coordinate robot. The motor output shaft will be connected to a screw thread with a screw pitch of 0.125 in. It is desired that the control resolution of each of the axes be 0.025 in.

(*a*) To achieve this control resolution, how many step angles are required on the stepping motor?

(*b*) What is the corresponding step angle?

(*c*) Determine the pulse rate that will be required to drive a given joint at a velocity of 3.0 in./sec.

CHAPTER
FOUR

ROBOT MOTION ANALYSIS AND CONTROL

In the preceding chapter we discussed the control of a single robot joint. In order for a robot to perform useful work it is necessary for the arm to consist of a number of joints. The typical commercial industrial robot has five or six joints. It is also necessary to control the path which the end of the arm follows, as opposed to merely controlling the final positions of the joints. In this chapter we will discuss the problems of robot motion control and the mathematical techniques used to analyze manipulator positions and motions. Later in the chapter we explore the dynamics of robot manipulators. Highly motivated readers may wish to pursue the subject of this chapter in more detail, and we recommend Paul's book [9] which has become the standard reference in this area.

4-1 INTRODUCTION TO MANIPULATOR KINEMATICS

In order to develop a scheme for controlling the motion of a manipulator it is necessary to develop techniques for representing the position of the arm at points in time. We will define the robot manipulator using the two basic elements, joints and links. Each joint represents 1 degree of freedom. As discussed in Chap. Two, the joints may involve either linear motion (joint-type L) or rotational motion (joint-types R, T, and V) between the adjacent links. According to our definitions in Chap. Two, the links are assumed to be the rigid structures that connect the joints.

Joints are labeled J_n where n begins with 1 at the base of the manipulator, and links are labeled L_n, again with 1 being the link closest to the base. Figure 4-1 illustrates the labeling system for two different robot arms, each possessing 2 degrees of freedom. By the joint notation scheme described in Chap. Two,

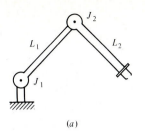
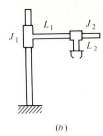

Figure 4-1 Two different 2-jointed manipulators. (a) two rotational joints (RR), (b) two linear joints (LL).

the manipulator in Fig. 4-1(*a*) has an *RR* notation and the manipulator in Fig. 4-1(*b*) has an *LL* notation.

We will also use the symbol L_n to indicate the length of the link in some of our equation derivations early in the chapter. Later in the chapter, we define a "standard" notation system used for computing joint-link transformations. This notation system uses the symbol a_n to denote the length of a manipulator link.

Position Representation

The kinematics of the *RR* robot are more difficult to analyze than the *LL* robot, and we will make frequent use of this configuration (and extensions of it) throughout the chapter. Figure 4-2 illustrates the geometric form of the *RR* manipulator. For the present discussion, our analysis will be limited to the two-dimensional case. The position of the end of the arm may be represented in a number of ways. One way is to utilize the two joint angles θ_1 and θ_2. This is known as the representation in "joint" space and we may define it as

$$P_j = (\theta_1, \theta_2)$$

Another way to define the arm position is in "world" space. This involves the use of a cartesian coordinate system that is external to the robot. The origin of

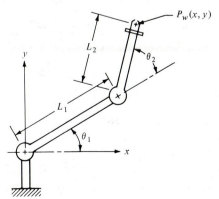

Figure 4-2 A two-dimensional 2 degree-of-freedom manipulator (type RR).

the cartesian axis system is often located in the robot's base. The end-of-arm position would be defined in world space as

$$P_w = (x, y)$$

This concept of a point definition in world space can readily be extended to three dimensions, that is, $P_w = (x, y, z)$. Representing an arm's position in world space is useful when the robot must communicate with other machines. These other machines may not have a detailed understanding of the robot's kinematics and so a "neutral" representation such as the world space must be used. In order to use both representations we must be able to transform from one to the other. Going from joint space to world space is called the forward transformation and going from world space to joint space is called the reverse transformation.

Forward Transformation of a 2-Degree of Freedom Arm

We can determine the position of the end of the arm in world space by defining a vector for link 1 and another for link 2.

$$\mathbf{r}_1 = [L_1 \cos \theta_1, L_1 \sin \theta_1] \tag{4-1}$$

$$\mathbf{r}_2 = [L_2 \cos(\theta_1 + \theta_2), L_2 \sin(\theta_1 + \theta_2)] \tag{4-2}$$

Vector addition of (4-1) and (4-2) yields the coordinates x and y of the end of the arm (point P_w) in world space

$$x = L_1 \cos \theta_1 + L_2 \cos(\theta_1 + \theta_2) \tag{4-3}$$

$$y = L_1 \sin \theta_1 + L_2 \sin(\theta_1 + \theta_2) \tag{4-4}$$

Reverse Transformation of the 2-Degree of Freedom Arm

In many cases it is more important to be able to derive the joint angles given the end-of-arm position in world space. The typical situation is where the robot's controller must compute the joint angles required to move its end-of-arm to a point in space defined by the point's coordinates. For the two-link manipulator we have developed, there are two possible configurations for reaching the point (x, y), as shown in Fig. 4-3. Some strategy must be developed to select the appropriate configuration. One approach is that employed in the control system of the Unimate PUMA robot. In the PUMA's control language, VAL, there is a set of commands called ABOVE and BELOW that determines whether the elbow is to make an angle θ_2 that is greater than or less than zero, as illustrated in Fig. 4-3. For our example, let us assume the θ_2 is positive as shown in Fig. 4-2. Using the trigonometric identities,

$$\cos(A + B) = \cos A \cos B - \sin A \sin B$$

$$\sin(A + B) = \sin A \cos B + \sin B \cos A$$

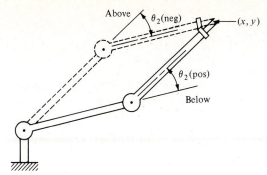

Figure 4-3 The arm at point $P(x, y)$, indicating two possible configurations to achieve the position.

we can rewrite Eqs. (4-3) and (4-4) as

$$x = L_1 \cos \theta_1 + L_2 \cos \theta_1 \cos \theta_2 - L_2 \sin \theta_1 \sin \theta_2$$

$$y = L_1 \sin \theta_1 + L_2 \sin \theta_1 \cos \theta_2 + L_2 \cos \theta_1 \sin \theta_2$$

Squaring both sides and adding the two equations yields

$$\cos \theta_2 = \frac{x^2 + y^2 - L_1^2 - L_2^2}{2 L_1 L_2} \qquad (4\text{-}5)$$

Defining α and β as in Fig. 4-4 we get

$$\tan \alpha = \frac{L_2 \sin \theta_2}{L_2 \cos \theta_2 + L_1} \qquad \alpha = \beta - \theta_1$$

$$\tan \beta = \frac{y}{x} \qquad (4\text{-}6)$$

Using the trigonometric identity

$$\tan(A - B) = \frac{\tan A - \tan B}{1 + \tan A \tan B}$$

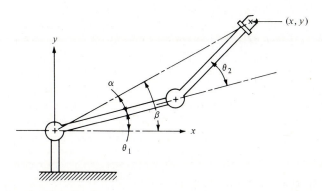

Figure 4-4 Solving for the joint angles.

we get

$$\tan \theta_1 = \frac{[y(L_1 + L_2 \cos \theta_2) - xL_2 \sin \theta_2]}{[x(L_1 + L_2 \cos \theta_2) + yL_2 \sin \theta_2]} \quad (4\text{-}7)$$

Knowing the link lengths L_1 and L_2 we are now able to calculate the required joint angles to place the arm at a position (x, y) in world space.

Adding Orientation: A 3-Degree of Freedom Arm in Two Dimensions

The arm we have been modeling is very simple; a two-jointed robot arm has little practical value except for very simple tasks. Let us add to the manipulator a modest capability for orienting as well as positioning a part or tool. Accordingly, we will incorporate a third degree of freedom into the previous configuration to develop the $RR{:}R$ manipulator shown in Fig. 4-5. This third degree of freedom will represent a wrist joint. The world space coordinates for the wrist end would be

$$\left. \begin{aligned} x &= L_1 \cos \theta_1 + L_2 \cos(\theta_1 + \theta_2) + L_3 \cos(\theta_1 + \theta_2 + \theta_3) \\ y &= L_1 \sin \theta_1 + L_2 \sin(\theta_1 + \theta_2) + L_3 \sin(\theta_1 + \theta_2 + \theta_3) \\ \psi &= (\theta_1 + \theta_2 + \theta_3) \end{aligned} \right\} \quad (4\text{-}8)$$

We can use the results that we have already obtained for the 2-degree of freedom manipulator to do the reverse transformation for the 3-degree of freedom arm. When defining the position of the end of the arm we will use x, y, and ψ. The angle ψ is the orientation angle for the wrist. Given these three values, we can solve for the joint angles (θ_1, θ_2, and θ_3) using

$$x_3 = x - L_3 \cos \psi$$

$$y_3 = y - L_3 \sin \psi$$

Having determined the position of joint 3, the problem of determining θ_1 and

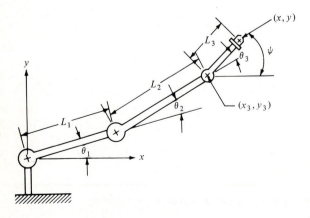

Figure 4-5 The two-dimensional 3 degree-of-freedom manipulator with orientation (type RR:R).

θ_2 reduces to the case of the 2-degree of freedom manipulator previously analyzed.

A 4-Degree of Freedom Manipulator in Three Dimensions

Figure 4-6 illustrates the configuration of a manipulator in three dimensions. The manipulator has 4 degrees of freedom: joint 1 (type T joint) allows rotation about the z axis; joint 2 (type R) allows rotation about an axis that is perpendicular to the z axis; joint 3 is a linear joint which is capable of sliding over a certain range; and joint 4 is a type R joint which allows rotation about an axis that is parallel to the joint 2 axis. Thus, we have a $TRL:R$ manipulator.

Let us define the angle of rotation of joint 1 to be the base rotation θ; the angle of rotation of joint 2 will be called the elevation angle ϕ; the length of linear joint 3 will be called the extension L (L represents a combination of links 2 and 3); and the angle that joint 4 makes with the $x - y$ plane will be called the pitch angle ψ. These features are shown in Fig. 4-6.

The position of the end of the wrist, P, defined in the world coordinate system for the robot, is given by

$$x = \cos \theta(L \cos \phi + L_4 \cos \psi) \tag{4-9}$$

$$y = \sin \theta(L \cos \phi + L_4 \cos \psi) \tag{4-10}$$

$$z = L_1 + L \sin \phi + L_4 \sin \psi \tag{4-11}$$

Given the specification of point P (x, y, z) and pitch angle ψ, we can find any

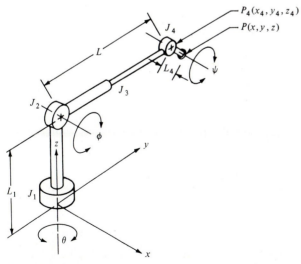

Figure 4-6 A three-dimensional 4 degree-of-freedom manipulator (type TRL:R).

of the joint positions relative to the world coordinate system. Using P_4 (x_4, y_4, z_4), which is the position of joint 4, as an example,

$$x_4 = x - \cos \theta (L_4 \cos \psi) \qquad (4\text{-}12)$$

$$y_4 = y - \sin \theta (L_4 \cos \psi) \qquad (4\text{-}13)$$

$$z_4 = z - L_4 \sin \psi \qquad (4\text{-}14)$$

The values of L, ϕ, and θ can next be computed:

$$L = [x_4^2 + y_4^2 + (z_4 - L_1)^2]^{-1} \qquad (4\text{-}15)$$

$$\sin \phi = \frac{z_4 - L_1}{L} \qquad (4\text{-}16)$$

$$\cos \theta = \frac{y_4}{L} \qquad (4\text{-}17)$$

The example we have just done is simple but not unrealistic. In order for a robot controller to be able to perform the calculations necessary quickly enough to maintain good performance they must be kept as simple as possible. The manipulator kinematics described in this example are very similar to those of the MAKER robot, by U.S. Robots. The only real difference is that the MAKER's wrist mechanism has more than a single joint.

One facet of our approach in the preceding analysis which should be noted by the reader is that we separated the orientation problem from the positioning problem. This approach of separating the two problems greatly simplifies the task of arriving at a solution.

4-2 HOMOGENEOUS TRANSFORMATIONS AND ROBOT KINEMATICS

The approach used in the previous section becomes quite cumbersome when a manipulator with many joints must be analyzed. Another, more general method for solving the kinematic equations of a robot arm makes use of homogeneous transformations. We describe this technique in this section, assuming the reader has at least some familiarity with the mathematics of vectors and matrices. Let us begin by defining the notation to be used.

A point vector, $\mathbf{v} = a\mathbf{i} + b\mathbf{j} + c\mathbf{k}$ can be represented in three-dimensional space by the column matrix

$$\begin{bmatrix} x \\ y \\ z \\ w \end{bmatrix}$$

where $a = x/w$, $b = y/w$, $c = z/w$, and w is a scaling factor. For example, any of the following matrices can be used to represent the vector $\mathbf{v} = 25\mathbf{i} + 10\mathbf{j} + 20\mathbf{k}$.

$$\begin{bmatrix} 25 \\ 10 \\ 20 \\ 1 \end{bmatrix} \quad \text{or} \quad \begin{bmatrix} 50 \\ 20 \\ 40 \\ 2 \end{bmatrix} \quad \text{or} \quad \begin{bmatrix} 12.5 \\ 5.0 \\ 10.0 \\ 0.5 \end{bmatrix}$$

Vectors of the above form can be used to define the end-of-arm position for a robot manipulator. (If $w = 0$, then the vector represents direction only.)

A vector can be translated or rotated in space by means of a transformation. The transformation is accomplished by a 4×4 matrix **H**. For instance the vector **v** is transformed into the vector **u** by the following matrix operation:

$$\mathbf{u} = \mathbf{Hv} \tag{4-18}$$

The transformation to accomplish a translation of a vector in space by a distance a in the x direction, b in the y direction, and c in the z direction is given by

$$\mathbf{H} = \mathbf{Trans}(a, b, c) = \begin{bmatrix} 1 & 0 & 0 & a \\ 0 & 1 & 0 & b \\ 0 & 0 & 1 & c \\ 0 & 0 & 0 & 1 \end{bmatrix} \tag{4-19}$$

Example 4-1 For the vector $\mathbf{v} = 25\mathbf{i} + 10\mathbf{j} + 20\mathbf{k}$, perform a translation by a distance of 8 in the x direction, 5 in the y direction, and 0 in the z direction. The translation transformation would be

$$\mathbf{H} = \mathbf{Trans}(a, b, c) = \begin{bmatrix} 1 & 0 & 0 & 8 \\ 0 & 1 & 0 & 5 \\ 0 & 0 & 1 & 0 \\ 0 & 0 & 0 & 1 \end{bmatrix}$$

The translated vector would be

$$\mathbf{Hv} = \begin{bmatrix} 1 & 0 & 0 & 8 \\ 0 & 1 & 0 & 5 \\ 0 & 0 & 1 & 0 \\ 0 & 0 & 0 & 1 \end{bmatrix} \begin{bmatrix} 25 \\ 10 \\ 20 \\ 1 \end{bmatrix} = \begin{bmatrix} 33 \\ 15 \\ 20 \\ 1 \end{bmatrix}$$

Rotations of a vector about each of the three axes by an angle θ can be accomplished by rotation transformations. About the x axis, the rotation transformation is

$$\mathbf{Rot}(x, \theta) = \begin{bmatrix} 1 & 0 & 0 & 0 \\ 0 & \cos\theta & -\sin\theta & 0 \\ 0 & \sin\theta & \cos\theta & 0 \\ 0 & 0 & 0 & 1 \end{bmatrix} \tag{4-20}$$

About the y axis,

$$\mathbf{Rot}(y, \theta) = \begin{bmatrix} \cos\theta & 0 & \sin\theta & 0 \\ 0 & 1 & 0 & 0 \\ -\sin\theta & 0 & \cos\theta & 0 \\ 0 & 0 & 0 & 1 \end{bmatrix} \qquad (4\text{-}21)$$

About the z axis,

$$\mathbf{Rot}(z, \theta) = \begin{bmatrix} \cos\theta & -\sin\theta & 0 & 0 \\ \sin\theta & \cos\theta & 0 & 0 \\ 0 & 0 & 1 & 0 \\ 0 & 0 & 0 & 1 \end{bmatrix} \qquad (4\text{-}22)$$

In concept it is possible to develop the rotation transformation about any vector $\mathbf{K}$, where $\mathbf{K}$ is not one of the major axes x, y, or z of the coordinate system. This rotation transformation is defined as $\mathbf{Rot}(\mathbf{K}, \theta)$. We present the concept here but leave the derivation of the transformation itself to books such as R. P. Paul's [9].

Example 4-2 Rotate the vector $\mathbf{v} = 5\mathbf{i} + 3\mathbf{j} + 8\mathbf{k}$ by an angle of 90° about the x axis. The rotation transformation is given by

$$\mathbf{H} = \mathbf{Rot}(x, 90) = \begin{bmatrix} 1 & 0 & 0 & 0 \\ 0 & \cos 90 & -\sin 90 & 0 \\ 0 & \sin 90 & \cos 90 & 0 \\ 0 & 0 & 0 & 1 \end{bmatrix}$$

$$\mathbf{Hv} = \begin{bmatrix} 1 & 0 & 0 & 0 \\ 0 & 0 & -1 & 0 \\ 0 & 1 & 0 & 0 \\ 0 & 0 & 0 & 1 \end{bmatrix} \begin{bmatrix} 5 \\ 3 \\ 8 \\ 1 \end{bmatrix} = \begin{bmatrix} 5 \\ -8 \\ 3 \\ 1 \end{bmatrix}$$

In both of the examples, matrix multiplication was carried out to determine the results of the transformations. In subsequent discussions, we will be performing a series of matrix multiplications in order to accomplish more complex transformations. It is important to note that performing two or more transformations in a row will only yield the same result if the transformations are carried out in the same sequence. In general,

$$\mathbf{AB} \text{ does not equal } \mathbf{BA}$$

One final concept important in our discussion of robotics is the concept of the inverse transformation. Given a transformation of the form,

$$\mathbf{T} = \begin{bmatrix} n_x & o_x & a_x & p_x \\ n_y & o_y & a_y & p_y \\ n_z & o_z & a_z & p_z \\ 0 & 0 & 0 & 1 \end{bmatrix} \qquad (4\text{-}23)$$

The inverse transformation of **T**, denoted by $\mathbf{T}^{-1}$, is defined as follows:

$$\mathbf{T}^{-1} = \begin{bmatrix} n_x & n_y & n_z & -p \cdot n \\ o_x & o_y & o_z & -p \cdot o \\ a_x & a_y & a_z & -p \cdot a \\ 0 & 0 & 0 & 1 \end{bmatrix} \tag{4-24}$$

where $\mathbf{p} \cdot \mathbf{n}$, $\mathbf{p} \cdot \mathbf{o}$ and $\mathbf{p} \cdot \mathbf{a}$ represent the dot products of the column vectors **n**, **o**, **a**, and **p**. The dot product $\mathbf{p} \cdot \mathbf{n}$ is the scalar $p_x n_x + p_y n_y + p_z n_z$. Similar interpretations apply to $\mathbf{p} \cdot \mathbf{o}$ and $\mathbf{p} \cdot \mathbf{a}$. As the reader might anticipate, the effect of an inverse transformation $\mathbf{T}^{-1}$ is to undo the operation accomplished by the transformation itself.

Kinematic Equations Using Homogeneous Transformations

The transformation **T** in the previous subsection is of the form

$$\mathbf{T} = \begin{bmatrix} n_x & o_x & a_x & p_x \\ n_y & o_y & a_y & p_y \\ n_z & o_z & a_z & p_z \\ 0 & 0 & 0 & 1 \end{bmatrix}$$

We can consider **T** to consist of four column vectors. These four vectors can be used to define the position and orientation of the end of a robot manipulator. This definition is illustrated in Fig. 4-7 for a hand or end effector attached to the robot arm. The vector **p** represents the position of the end of the arm with respect to the base frame. The three vectors, **n**, **o**, and **a**, specify the orientation of the end effector. The **a** vector is called the approach vector, and it points in the direction of the wrist. The **o** vector is called the orientation vector. It specifies the orientation of the hand from one finger to the other. The third vector, **n**, is the normal vector to **o** and **a**. Together the three vectors form a coordinate frame relative to the base frame. Respectively, the **n**, **o**, and **a** vectors constitute the x, y, and z axis of the end effector frame.

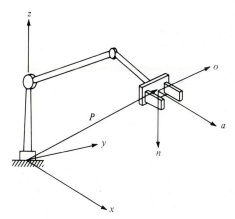

Figure 4-7 The vectors **o**, **a**, **n**, and **p** for a robot manipulator.

The position and orientation of the end of a manipulator can be described as the product of n homogeneous transformations, one for each of the n joints of the manipulator. Similarly, the position of each and any joint in the robot arm can be considered as the product of transformations describing the rotations and translations of the preceding joint–link combinations.

Figure 4-8 illustrates the four variables affected by the joint–link combination. The length of the link, a_n, is defined as the distance along the line that is mutually perpendicular to the axes of two adjacent joints. The twist between the axes of the joints in a plane perpendicular to a_n is the angle t_n. The distance d_n is the distance between the normal a_n and a_{n-1} of the two links. This is sometimes referred to as the joint offset. In the case of a linear joint, this is the joint variable. The angle θ_n is the angle between the links measured as the angle between the normals a_n and a_{n-1} in the plane normal to the axis of the joint. The position and orientation of joint n is completely defined with respect to joint $n-1$ by the four parameters a_n, t_n, θ_n, and d_n.

We can assign a coordinate frame to each link and establish the relationships between succeeding links using transformations for each of the four variables. The origin of the coordinate frame for link n is imbedded at the intersection of the axis, z_n, for the joint $n-1$ and the common normal, a_n from joint n to joint $n+1$. That is, the frame is imbedded at the end of the link in the successive joint. This is accomplished using:

A rotation of angle θ_n about the z_{n-1} axis
A translation of distance d_n along z_{n-1}
A translation along x_n by a length a_n
A rotation about x_n by a twist angle t_n

Using the transformations defined previously for rotation and translation, we can develop the transformation which relates the coordinate frame of link n

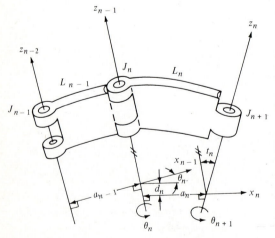

Figure 4-8 The variables in a link using the notation of Paul [9]. The rules used to define the notation are: (1) Axis z_{n-1} defines the position of the axis of rotation for joint J_n, z_n for joint J_{n+1}, and so forth. (2) Axis x_{n-1} is selected to be an extension for the common perpendicular line of length a_{n-1} between consecutive joints z_{n-2} and z_{n-1}. (3) The axis y_{n-1} is selected to provide a right-hand coordinate system with the other axes. (4) Axis x_n is an extension of the common perpendicular line of length a_n.

with link $n-1$. For example, the homogeneous transformation $\mathbf{A}_2$ describing the position and orientation of the coordinate frame of the second link with respect to the coordinate frame of the first link is written

$$\mathbf{A}_n = \mathbf{Rot}(z_1, \theta_2)t\, \mathbf{Trans}(0, 0, d_2)\, \mathbf{Trans}(a_2, 0, 0)\ \mathbf{Rot}(x_2, t_2)$$

In general, this transformation is referred to as the $\mathbf{A}$ matrix. In practice, the subscripts in the preceding equation are usually suppressed and the equation for the $\mathbf{A}$ matrix is written

$$\mathbf{A}_2 = \mathbf{Rot}(z_1, \theta_2)\, \mathbf{Trans}(0, 0, d_2)\, \mathbf{Trans}(a_2, 0, 0)\, \mathbf{Rot}(x_2, t_2)$$

For a rotational joint, this transformation reduces to the following:

$$\mathbf{A}_n = \begin{bmatrix} \cos\theta & -\sin\theta\cos t & \sin\theta\sin t & a\cos\theta \\ \sin\theta & \cos\theta\cos t & -\cos\theta\sin t & a\sin\theta \\ 0 & \sin t & \cos t & d \\ 0 & 0 & 0 & 1 \end{bmatrix} \tag{4-25}$$

For a linear joint, the value of $a = 0$ in Eq. (4-25).

The description of the end of a link with respect to the robot reference frame (world space) can be described as the link coordinate frame $\mathbf{T}_n$ where

$$\mathbf{T}_n = \mathbf{A}_1 \mathbf{A}_2 \cdots \mathbf{A}_n \tag{4-26}$$

In order to calculate the $\mathbf{T}$ matrix we must first find the n $\mathbf{A}$ matrices. This is accomplished by substituting for the values of a, θ, d, and t in Eq. (4-25). Let us do this for the three-jointed arm shown in Fig. 4-9. First, the parameters for each link must be defined. These values are presented in Table 4-1. The $\mathbf{A}$ matrices are

$$\mathbf{A}_1 = \begin{bmatrix} \cos\theta_1 & 0 & -\sin\theta_1 & 0 \\ \sin\theta_1 & 0 & \cos\theta_1 & 0 \\ 0 & -1 & 0 & 0 \\ 0 & 0 & 0 & 1 \end{bmatrix} \tag{4-27}$$

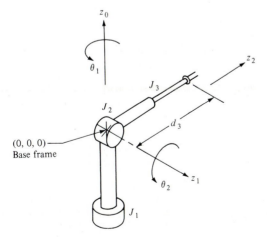

Figure 4-9 A three-dimensional 3 degree-of-freedom manipulator (type TRL).

Table 4-1 Link parameters

Link	Variable	t	a	d
1	θ_1	$-90°$	0	0
2	θ_2	$90°$	0	0
3	d_3	0	0	d_3

$$\mathbf{A}_2 = \begin{bmatrix} \cos\theta_2 & 0 & \sin\theta_2 & 0 \\ \sin\theta_2 & 0 & -\cos\theta_2 & 0 \\ 0 & 1 & 0 & 0 \\ 0 & 0 & 0 & 1 \end{bmatrix} \tag{4-28}$$

$$\mathbf{A}_3 = \begin{bmatrix} 1 & 0 & 0 & 0 \\ 0 & 1 & 0 & 0 \\ 0 & 0 & 1 & d_3 \\ 0 & 0 & 0 & 1 \end{bmatrix} \tag{4-29}$$

In a similar way, we can define the frame of any link n with respect to any other preceding link m.

$$^m\mathbf{T}_n = \mathbf{A}_{m+1} \cdots \mathbf{A}_{n-1}\mathbf{A}_n \tag{4-30}$$

To illustrate, we can write the following for the manipulator in our example:

$$^2\mathbf{T}_3 = \begin{bmatrix} 1 & 0 & 0 & 0 \\ 0 & 1 & 0 & 0 \\ 0 & 0 & 1 & d_3 \\ 0 & 0 & 0 & 1 \end{bmatrix} \tag{4-31}$$

Since

$$^1\mathbf{T}_3 = \mathbf{A}_2{}^2\mathbf{T}_3 \tag{4-32}$$

we get

$$^1\mathbf{T}_3 = \begin{bmatrix} \cos\theta_2 & 0 & \sin\theta_2 & d_3\sin\theta_2 \\ \sin\theta_2 & 0 & -\cos\theta_2 & -d_3\cos\theta_2 \\ 0 & 1 & 0 & 0 \\ 0 & 0 & 0 & 1 \end{bmatrix} \tag{4-33}$$

and

$$^0\mathbf{T}_3 = \begin{bmatrix} \cos\theta_1\cos\theta_2 & -\sin\theta_1 & \cos\theta_1\sin\theta_2 & d_3\cos\theta_1\sin\theta_2 \\ \sin\theta_1\cos\theta_2 & \cos\theta_1 & \sin\theta_1\sin\theta_2 & d_3\sin\theta_1\sin\theta_2 \\ -\sin\theta_2 & 0 & \cos\theta_2 & d_3\cos\theta_2 \\ 0 & 0 & 0 & 1 \end{bmatrix} \tag{4-34}$$

In the preceding example the position and orientation of the end-of-arm were determined with respect to the base frame. It is possible to determine the

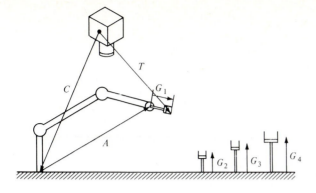

Figure 4-10 A simple robot workcell.

location of the end of the arm with respect to other coordinate frames by "premultiplying" Eq. (4-34) by the transformation relating the arm's base frame to the new base frame. Additionally, we can develop a more complicated arm by "postmultiplying" Eq. (4-34) with the transformations describing the additional links that might be contained in the wrist. In some cases, we may wish to define the specific tool at the end of the arm. This can also be described by a transformation. Figure 4-10 illustrates a robot workcell including a manipulator, a camera, and a set of tools. The camera is positioned in one coordinate frame and the arm is positioned in another. The arm's base frame is related to the camera's by the transformation **C**. The arm is described by the transformation **A** and the tools by $\mathbf{G}_n$. The position of the end of a tool with respect to the camera's base frame would be given by

$$\mathbf{T} = \mathbf{CAG}_n \qquad (4\text{-}35)$$

Solving the Kinematic Equations

It is desirable to be able to determine the joint positions required by the manipulator to reach a desired endpoint. Before explaining the procedure for doing this, let us introduce an analysis technique known as the transform graph.

Figure 4-11 illustrates the transform graph for the manipulator shown in Fig. 4-9 and described in the previous subsection. The transform graph shows the two ways of getting from the base frame **0** to the end-of-arm frame. One way is along the path represented by $\mathbf{T}_3$ and the other is along the path $\mathbf{A}_1\mathbf{A}_2\mathbf{A}_3$. We always travel along the direction of the arrows in a transform graph. If we wish to travel against the arrows, we must use the inverse of the transform. That is, in our example, we could travel along the path $\mathbf{A}_1^{-1}\mathbf{T}_3$ to get to the end of the arm.

If we begin at one point on the graph and end at a second point there are always two paths that we can follow. The transform products represented by these two paths are equivalent. That is

$$\mathbf{A}_1^{-1}\mathbf{T}_3 = {}^1\mathbf{T}_3 \qquad (4\text{-}36)$$

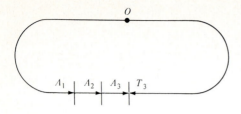

Figure 4-11 Transform graph for manipulator of Figure 4-9.

or

$$\mathbf{A}_2^{-1}\mathbf{A}_1^{-1}\mathbf{T}_3 = {}^2\mathbf{T}_3 \tag{4-37}$$

and so forth.

We can make use of these equivalents to solve for the joint positions given the end-of-arm position in the form of $\mathbf{T}$ in Eq. (4-23). Substituting Eq. (4-27) for $\mathbf{A}_1$ the inverse is

$$\mathbf{A}_1^{-1} = \begin{bmatrix} \cos\theta_1 & \sin\theta_1 & 0 & 0 \\ 0 & 0 & -1 & 0 \\ -\sin\theta_1 & \cos\theta_1 & 0 & 0 \\ 0 & 0 & 0 & 1 \end{bmatrix} \tag{4-38}$$

Substituting Eqs. (4-38), (4-23), and (4-33) into (4-36) we get

$$\begin{bmatrix} q1 & q2 & q3 & q4 \\ -n_z & -o_z & -a_z & -p_z \\ q5 & q6 & q7 & q8 \\ 0 & 0 & 0 & 1 \end{bmatrix} = \begin{bmatrix} \cos\theta_2 & 0 & \sin\theta_2 & d_3\sin\theta_2 \\ \sin\theta_2 & 0 & -\cos\theta_2 & -d_3\cos\theta_2 \\ 0 & 1 & 0 & 0 \\ 0 & 0 & 0 & 1 \end{bmatrix} \tag{4-39}$$

where $q1 = n_x \cos\theta_1 + n_y \sin\theta_1$
$q2 = o_x \cos\theta_1 + o_y \sin\theta_1$
$q3 = a_x \cos\theta_1 + a_y \sin\theta_1$
$q4 = p_x \cos\theta_1 + p_y \sin\theta_1$
$q5 = -n_x \sin\theta_1 + n_y \cos\theta_1$
$q6 = -o_x \sin\theta_1 + o_y \cos\theta_1$
$q7 = -a_x \sin\theta_1 + a_y \cos\theta_1$
$q8 = -p_x \sin\theta_1 + p_y \cos\theta_1$

We can use this to solve for θ_1, θ_2, and d_3. From Eq. (4-39) we see that

$$q2 = o_x \cos\theta_1 + o_y \sin\theta_1 = 0 \tag{4-40}$$

so that

$$\tan\theta_1 = -\frac{o_x}{o_y} \tag{4-41}$$

We can also see that

$$\sin\theta_2 = -n_z \tag{4-42}$$

and

$$-\cos \theta_2 = -a_z \tag{4-43}$$

so that

$$\tan \theta_2 = -\frac{n_z}{a_z} \tag{4-44}$$

and finally,

$$d_3 = \frac{p_z}{\cos \theta_2} \tag{4-45}$$

The solution was obtained in a relatively straightforward manner because of the simple configuration of our 3-degree of freedom arm. In cases where the manipulator is more complex kinematically it may be necessary to use more of the equivalents obtained using the transform graph.

A Discussion on Orientation

The use of the **o**, **a**, and **n** terms as direction vectors to describe the orientation of the wrist is sometimes difficult. Another technique for describing the orientation of the robot end effector involves the use of roll, pitch, and yaw. These are the same terms defined in Chap. Two as the three possible degrees of freedom associated with the wrist motion. They were illustrated in Fig. 2-11. In our present discussion we can define these terms more precisely as the angles that are associated with these degrees of freedom. Specifically, roll, pitch, and yaw are, respectively, the angles of rotation of the wrist assembly about the z, y, and x axes of the end-of-arm connection to the wrist. The definitions are illustrated in Fig. 4-12. The end-of-arm connection constitutes a coordinate reference frame.

Given **a**, **o**, and **n**, we can find the roll, pitch, and yaw for the wrist mechanism using the following equations:

$$\text{Roll} = \arctan \frac{n_y}{n_x} \tag{4-46}$$

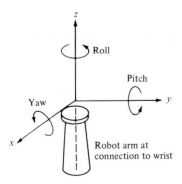

Figure 4-12 Roll, pitch, and yaw for a manipulator wrist mechanism (Refer to Figure 2-11 in chapter two).

$$\text{Pitch} = \arctan \frac{-n_z}{n_x \cos(\text{roll}) + n_y \sin(\text{roll})} \quad\quad (4\text{-}47)$$

$$\text{Yaw} = \arctan \frac{a_x \sin(\text{roll}) - a_y \cos(\text{roll})}{o_y \cos(\text{roll}) - o_x \sin(\text{roll})} \quad\quad (4\text{-}48)$$

A derivation and explanation of these results is presented in Paul [9].

4-3 MANIPULATOR PATH CONTROL

In controlling the manipulator we are not only interested in the endpoints reached by the robot joints, but also in the path followed by the arm in traveling from one point to another in the workspace.

Motion Types

There are three common types of motion that a robot manipulator can make in traveling from point to point. These are slew motion, joint-interpolated motion, and straight line motion.

Slew motions represent the simplest type of motion. As the robot is commanded to travel from point *A* to point *B*, each axis of the manipulator travels as quickly as possible from its respective initial position to its required final position. Therefore, all axes begin moving at the same time, but each axis ends its motion in an elapsed time that is proportional to the product of its distance moved and its top speed (allowing for acceleration and deceleration). Slew motion generally results in unnecessary wear on the joints and often leads to unanticipated results in terms of the path taken by the manipulator.

> **Example 4-3** Determine the time required for each joint of a three-axis *RRR* manipulator to travel the following distances using slew motion: joint 1, 30°; joint 2, 60°; and joint 3, 90°. All joints travel at a rotational velocity of 30°/s, neglecting effects of acceleration and deceleration.
> Joint 1 will complete its move in 30°/(30°/s) = 1 s, joint 2 will take 60°/(30°/s) = 2 s, and joint 3 will take 90°/(30°/s) = 3 s.

Joint-interpolated motion requires the robot controller to calculate the amount of time it will take each joint to reach its destination at the commanded speed. It then selects the maximum time among these, and uses this value as the time for all the axes. This means that a separate velocity is calculated for each axis. The advantage of joint-interpolated motion over slew motion is that the joints are generally driven at less than their respective maximum velocities, thus reducing maintenance problems for the robot. Also, the path that is followed by the manipulator is repeatable and predictable regardless of the total time and commanded velocity. An example computation will illustrate the operation of joint-interpolated motion.

Example 4-4 Determine the time required to complete the move and the velocity of each joint for the three-axis *RRR* manipulator of Example 4-3 to travel the following distances under joint-interpolated motion: joint 1, 30°; joint 2, 60°; and joint 3, 90°. The maximum velocity of any joint is 30°/s; however, no joint may travel at greater than 90 percent of maximum velocity. Neglect any effects of acceleration and deceleration.

Joint 3 has the longest distance to travel and will therefore result in the maximum time to complete its move. The time for joint 3 to travel the 90° is

$$90°/(0.9 \times 30°/s) = 3.33 \text{ s}$$

The velocity for joint 2 is

$$60°/3.33 \text{ s} = 18.0°/s$$

and for joint 1 the velocity is

$$30°/3.33 \text{ s} = 9.0°/s$$

Straight line interpolation motion requires the end of the manipulator (or end effector) to travel along a straight path defined in cartesian coordinates. This is the most demanding type of motion for the controller to execute, except for a cartesian coordinate (*LLL*) robot. For manipulators with rotational joints, most straight line motions are unnatural and the controller must compute the sequence of incremental joint rotations required for the end-of-arm to move in a linear fashion. The transformation developed in the last section are useful in determining the joint angles required to follow the straight line path. To command the arm to travel from point *A* to point *B* along a linear path, the series of intermediate transformations along the path must be computed. These transformations would be a certain distance apart, the distance determined to allow enough time for the robot controller to calculate the transformations and to solve the arm kinematics. The time between calculations has several implications regarding the dynamics of the arm. Since each calculation results in a new command position for the end-of-the-arm, these calculations must be carried out quickly enough so that the new commands do not set up impulses that could result in oscillation of the arm. Generally, this means that the calculations should be performed at a rate of about 10 times the natural frequency of the arm (typically, at about 50 Hz). In addition, since the arm will be joint-interpolating the moves between calculations, the accuracy with which the end-of-arm will follow the desired straight line trajectory is a function of how frequently the points along the straight line path are computed relative to the arm velocity.

Straight line interpolation is very useful in applications such as arc welding, laying adhesives along a straight path, and inserting a peg into a hole in an assembly operation. We shall return to the issue of motion interpolation in Chap. Eight on robot programming.

4-4 ROBOT DYNAMICS

Accurate control of the manipulator requires precise control of each joint. The control of the joint depends on knowledge of the forces that will be acting on the joint and the inertias reflected at the joint (the masses of the joints and links of the manipulator). While these forces and masses are relatively easy to determine for a single joint, it becomes more difficult to determine them as the complexity of the manipulator increases. We will explore these issues using the two-axis manipulator developed earlier. Our purpose is to introduce this problem area to the reader rather than to analyze its complexities in detail.

Static Analysis

Let us begin by considering the torques required by the joints to produce a force **F** at the tip of the robot arm as shown in Fig. 4-13. By balancing the forces on each link we get

$$\mathbf{F}_1 - \mathbf{F}_2 = 0 \tag{4-49}$$

and

$$\mathbf{F}_2 - \mathbf{F} = 0 \tag{4-50}$$

That is,

$$\mathbf{F}_1 = \mathbf{F}_2 = \mathbf{F} \tag{4-51}$$

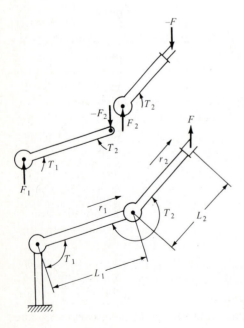

Figure 4-13 Two-link arm forces and torques.

The torques are the vector cross-products of the forces and the link vectors **r** as developed earlier in Eqs. (4-1) and (4-2) so that

$$\mathbf{T}_1 = \mathbf{T}_2 + \mathbf{r}_1 \times \mathbf{F} \tag{4-52}$$

and

$$\mathbf{T}_2 = \mathbf{r}_2 \times \mathbf{F} \tag{4-53}$$

Therefore we can state

$$\mathbf{T}_1 = (\mathbf{r}_1 + \mathbf{r}_2) \times \mathbf{F} \tag{4-54}$$

If $\mathbf{F} = (F_x, F_y)$, then Eq. (4-57) can be written as

$$\mathbf{T}_1 = [L_1 \cos \theta_1 + L_2 \cos(\theta_1 + \theta_2)], [L_1 \sin \theta_1 + L_2 \sin(\theta_1 + \theta_2)] \times (F_x, F_y) \tag{4-55}$$

Since the vector cross-product $(a, b) \times (c, d)$ is $ad - bc$ we get

$$\mathbf{T}_1 = [L_1 \cos \theta_1 + L_2 \cos(\theta_1 + \theta_2)] \times F_x - [L_1 \sin \theta_1 + L_2 \sin(\theta_1 + \theta_2)] \times F_y \tag{4-56}$$

and

$$\mathbf{T}_2 = L_2 \cos(\theta_1 + \theta_2) \times F_x - L_2 \sin(\theta_1 + \theta_2) \times F_y \tag{4-57}$$

Therefore, given a force vector $\mathbf{F} = (F_x, F_y)$, the torques required by each joint to generate that force at the tip can be computed using the preceding analysis.

In order to calculate the force generated by the end of the arm given the joint torques we must solve the equations for F_x and F_y. The solution is obtained by first subtracting Eq. (4-57) from Eq. (4-56). We can then substitute for F_y using the value in Eq. (4-57). This results in an expression for F_x, T_1, T_2, θ_1, and θ_2. After simplification using trigonometric identities as outlined earlier we get

$$F_x = \frac{T_1 L_2 \cos(\theta_1 + \theta_2) - T_2(L_1 \cos \theta_1 + L_2 \cos(\theta_1 + \theta_2))}{L_1 L_2 \sin \theta_2} \tag{4-58}$$

Similarly, F_y is determined as

$$F_y = \frac{T_1 L_2 \sin(\theta_1 + \theta_2) - T_2(L_1 \sin \theta_1 + L_2 \sin(\theta_1 + \theta_2))}{L_1 L_2 \sin \theta_2} \tag{4-59}$$

The technique previously described for kinematic analysis of torque can be extended to more joints. The ability to determine the joint torques and the components of forces applied at the wrist or tool are useful for controlling the arm when it must contact the environment. In Chap. Six we consider various sensors that can be used to measure forces and torques in robotic applications.

Compensating for Gravity

Gravity was ignored in the previous analysis. It can be included by adding the forces exerted by gravity on each link to the force balance equations in the previous section. Figure 4-14 shows the manipulator with the force vectors $\mathbf{F}_{g1}$ and $\mathbf{F}_{g2}$ acting at the center of each link, where the links have masses m_1 and

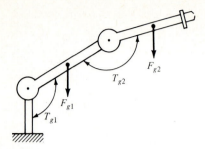

Figure 4-14 Gravity forces and torques.

m_2. Balancing the forces due to gravity only we get

$$\mathbf{F}_2 = m_2\mathbf{g} \tag{4-60}$$

$$\mathbf{F}_1 = \mathbf{F}_2 + m_1\mathbf{g} \tag{4-61}$$

As we did before, the cross-products of the link vectors $\mathbf{r}$ and the forces give us the torque due to gravity on each joint so that

$$\mathbf{T}_{g2} = \frac{-m_2 r_2 \times \mathbf{g}}{2}$$

$$= \frac{g[m_2 L_2 \cos(\theta_1 + \theta_2)]}{2} \tag{4-62}$$

and

$$\mathbf{T}_{g1} = g\left[\left(\frac{m_1}{2} + m_2\right) L_1 \cos\theta_1 + \frac{m_2 L_2 \cos(\theta_1 + \theta_2)}{2}\right] \tag{4-63}$$

If the manipulator is acting under gravity loads then these torques must be added to the joint torques to provide a desired output force. The obvious point of this is that if the arm can be constructed with as low a mass as possible more of the motor torque will be used to provide forces at the end of the arm for doing useful work, rather than just supporting the mass of the links.

Robot Arm Dynamics

The topic of robot dynamics is concerned with the analysis of the torques and forces due to acceleration and deceleration. Torques experienced by the joints due to acceleration of the links, as well as forces experienced by the links due to torques applied by the joints, are included within the scope of dynamic analysis. Solving for the accelerations of the links is difficult due to a number of factors. For one, the acceleration is dependent on the inertia of the arm. However, the inertia is dependent on the configuration of the arm, and this is continually changing as the joints are moved. An additional factor that influences the inertia is the mass of the payload and its position with respect to the joints. This also changes as the joints are moved. Figure 4-15 shows the two-link arm in the maximum and minimum inertia configurations.

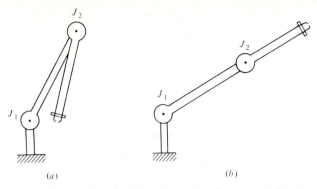

Figure 4-15 Arm inertias. (a) Minimum inertia about $J1$. (b) Maximum inertia about J_1.

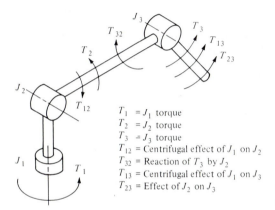

T_1 = J_1 torque
T_2 = J_2 torque
T_3 = J_3 torque
T_{12} = Centrifugal effect of J_1 on J_2
T_{32} = Reaction of T_3 by J_2
T_{13} = Centrifugal effect of J_1 on J_3
T_{23} = Effect of J_2 on J_3

Figure 4-16 Dynamic forces and torques for a TRR robot.

The torques required to drive the robot arm are not only determined by the static and dynamic forces described above; each joint must also react to the torques of other joints in the manipulator, and the effects of these reactions must be included in the analysis. Also, if the arm moves at a relatively high speed, the centrifugal effects may be significant enough to consider. The various torques applied to the two-jointed manipulator are illustrated in Figure 4-16. The picture becomes substantially more complicated as the number of joints is increased.

A detailed analysis of manipulator dynamics is beyond the scope and purpose of this chapter, and we leave it to the interested reader to pursue the subject in some of the references listed at the end of the chapter.

4-5 CONFIGURATION OF A ROBOT CONTROLLER

If the control requirements outlined in Chap. Three and the present chapter are combined, we can develop a general configuration for a robot controller. The elements needed in the controller include: joint servocontrollers, joint power amplifiers, mathematical processor, executive processor, program memory, and input device. The number of joint servocontrollers and joint power amplifiers would correspond to the number of joints in the manipulator. These elements might be organized in the robot controller as shown in Fig. 4-17.

Motion commands are executed by the controller from two possible sources: operator input or program memory. Either an operator inputs commands to the system using an input device such as a teach pendant or a CRT terminal, or the commands are downloaded to the system from program memory under control of the executive processor. In the second case, the set of commands have been previously programmed into memory using the operator input device(s). For each motion command, the executive processor informs the mathematical processor of the coordinate transformation calculations that must be made. When the transformation computations are completed, the executive processor downloads the results to the joint controllers as position commands. Each joint controller then drives its corresponding joint actuator by means of the power amplifier.

Microprocessors are typically utilized in several of the components of a modern robot controller. These components include the mathematical proces-

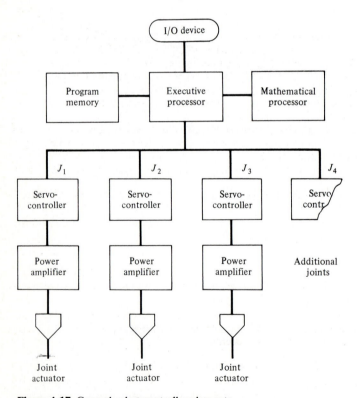

Figure 4-17 General robot controller elements.

sor, the executive processor, the servocontrollers, and the input device. Each of the control boards makes use of a common data buss and address buss. The microprocessors communicate with each other by sending messages into common areas in the system memory. This architecture provides several advantages in the design of the controller. These advantages include commonality of components, expansion of the system to more joints, and information flow between joint control elements. For example, the control configuration would permit the sharing of feedback information among the various joints, thus providing the opportunity to develop algorithms for improving the individual joint dynamics.

REFERENCES

1. M. Brady, J. M. Hollerbach, T. C. Johnson, T. Lozano Perez, and M. T. Mason, *Robot Motion: Planning and Control*, MIT Press, Cambridge, MA, 1982.
2. T. A. W. Dwyer and G. K. F. Lee, "Design of Microprocessor-Based Real-Time Controllers for Manipulator Systems," *Conference Papers*, Robotics Research—The Next Five Years and Beyond, Soc. Mfg. Engrs, Lehigh University, August 1984.
3. B. K. P. Horn, "Kinematics, Statics, and Dynamics of Two-Dimensional Manipulators," *Artificial Intelligence: An MIT Perspective*, vol. 2, MIT Press, Cambridge, MA, 1979.
4. B. Huang and V. Milenkovic, "Kinematics of Minor Robot Linkage," *Conference Papers*, Robotics Research—The Next Five Years and Beyond, Soc. Mfg. Engrs, Lehigh University, August 1984.
5. Y. Koren, *Robotics for Engineers*, McGraw-Hill, New York, 1985, chaps. 4 and 5.
6. C. S. G. Lee and M. Ziegler, "A Geometric Approach in Solving the Inverse Kinematics of PUMA Robots," *Conference Proceedings*, 13th International Symposium on Industrial Robots and Robots 7, Soc. Mfg. Engrs, Chicago, IL, April 1983, pp. 16-1–16-18.
7. M. C. Leu and R. Mahajan, "Computer Graphic Simulation of Robot Kinematics and Dynamics," *Conference Papers*, Robotics Research—The Next Five Years and Beyond, Soc. Mfg. Engrs, Lehigh University, August 1984.
8. V. Milenkovic and B. Huang, "Kinematics of Major Robot Linkage," *Conference Proceedings*, 13th International Symposium on Industrial Robots and Robots 7, Soc. Mfg. Engrs, Chicago, IL, April 1983, pp. 16-31–16-47.
9. R. P. Paul, *Robot Manipulators: Mathematics, Programming, and Control*, MIT Press, Cambridge, MA, 1981.
10. W. E. Snyder, *Industrial Robots*: Computer Interfacing and Control, Prentice-Hall, Englewood Cliffs, NJ, 1985.

PROBLEMS

Problems 4-1 through 4-12 deal with the issues discussed in Sec. 4-1 related to determining the positions in space of a manipulator with multiple joints.

4-1 Consider the forward transformation of the two-joint manipulator shown in Fig. 4-2 in the text. Given that the length of joint 1, $L_1 = 12$ in., the length of joint 2, $L_2 = 10$ in., the angle $\theta_1 = 30°$, and the angle $\theta_2 = 45°$, compute the coordinate position (x and y coordinates) for the end-of-the-arm P_w.

4-2 It is desired to determine the values to which the angles θ_1 and θ_2 must be set in order to achieve a certain point in space for the manipulator shown in Fig. 4-2. The length of joint 1, $L_1 = 12$ in., the length of joint 2, $L_2 = 10$ in. The point P_w which the robot must achieve is defined by the coordinates $x = 15.7$ and $y = 12.6$. Using the reverse transformation methods described in Sec. 4-1, determine the angles θ_1 and θ_2 required to achieve the point in the ABOVE configuration as pictured in Fig. 4-3.

4-3 Solve Prob. 4-2 except the manipulator is to be in the BELOW configuration as pictured in Fig. 4-3.

4-4 It is desired to determine the values to which the angles θ_1 and θ_2 must be set in order to achieve a certain point in space for the manipulator shown in Fig. 4-2. The length of joint 1, $L_1 = 12$ in., the length of joint 2, $L_2 = 10$ in. The point P_w which the robot must achieve is defined by the coordinates $x = 6.0$ and $y = 12.0$. Using the reverse transformation methods described in Sec. 4-1, determine the angles θ_1 and θ_2 required to achieve the point in the ABOVE configuration as pictured in Fig. 4-3.

4-5 Solve Prob. 4-4 except the manipulator is to be in the BELOW configuration as pictured in Fig. 4-3.

4-6 Consider the manipulator shown in Fig. 4-6. Suppose the links and joints of the manipulator had the following settings:

$$\text{Length of link } L_1 = 10.0 \text{ in.}$$
$$\text{Length of the extension link } L = 15.0 \text{ in.}$$
$$\text{Length of link 4, } L_4 = 3.0 \text{ in.}$$
$$\text{Base angle } \theta = 0°$$
$$\text{Elevation angle } \phi = 20°$$
$$\text{Pitch angle } \psi = 34°$$

Determine the coordinates of the resulting point P at which the end-of-arm would be located.

4-7 In Prob. 4-6, determine the x, y, and z coordinates of joint 4.

4-8 In the manipulator shown in Fig. 4-6, the end-of-the-arm is positioned at the point defined by $x = 18.0$ in., $y = 0$, and $z = 21.0$ in. If the pitch angle $\psi = 45°$, and link 4 is 2.0 in. long, the length of link 1 is 12.0 in., determine the position of joint 4, defined as point P_4. Also, determine the values of the other manipulator parameters, length L, and angles θ and ϕ.

4-9 For the manipulator shown in Fig. 4-6, the end-of-the-arm is positioned at a point defined by $x = 16.0$ in., $y = 5.0$, and $z = 21.0$ in. If the pitch angle $\psi = 45°$, and link 4 is 2.0 in. long, the length of link 1 is 12.0 in., determine the position of joint 4, defined as point P_4. Also, determine the values of the other manipulator parameters, length L, and angles θ and ϕ.

4-10 Write a computer program that will calculate and print out the x, y, and z coordinates for the manipulator illustrated in Fig. 4-6, given the specification of the following parameters: the length of link L_1, the length of the extension link L, the length of link 4, L_4, the base angle θ, the elevation angle ϕ, and the pitch angle ψ.

4-11 Write a computer program that will calculate and print out the values of the manipulator parameters L, θ, and ϕ, given the specification of the end-of-arm position (x, y, and z coordinates) and the pitch angle ψ. The length of the link 1, $L_1 = 12.0$ in., and the length of link 4, $L_4 = 4.0$ in.

4-12 Do Prob. 4-11 except that it is known that the range of values which the length L (links 2 and 3) can assume is from 12.0 to 20 in. Include in the program a calculation routine that will determine whether the input values of x, y, and z constitute a point that is outside the work volume of the manipulator. If the input values are outside of the work volume, the program should print out an appropriate error message.

Problems 4-13 through 4-24 are concerned with the homogeneous transformations discussed in Sec. 4-2.

4-13 For the point $3i + 7j + 5k$ perform the following operations:
  (a) Rotate 30° about the X axis.
  (b) Rotate 45° about the Y axis.
  (c) Rotate 90° about the Z axis.
  (d) Translate 8 units along the Y axis.
  (e) Rotate 30° about X, then translate 6 along Y.
  (f) Translate 6 along Y, then rotate 30° about X.

4-14 Perform the same operations as in Prob. 4-13 for the point $-3i+7j-5k$.

4-15 Find the inverse of the matrices given below:

(a) $\begin{bmatrix} 0 & 10 & 5 & 3 \\ 1 & 5 & 3 & 2 \\ 0 & 1 & 5 & 1 \\ 0 & 0 & 0 & 1 \end{bmatrix}$

(b) $\begin{bmatrix} 2 & 3 & 4 & 5 \\ 1 & 2 & 3 & 4 \\ 1 & 3 & 5 & 7 \\ 0 & 0 & 0 & 1 \end{bmatrix}$

4-16 Multiply the matrices in Prob. 4-15 with the solutions. The result should be the identity matrix.

4-17 Develop the **A** matrices for Prob. 4-13(e) and 4-13(f).

4-18 Find the inverse matrices for the solutions to Prob. 4-17.

4-19 Find the **A** matrices for the manipulator in Fig. P4-19 using the link parameters given in Table P4-19.

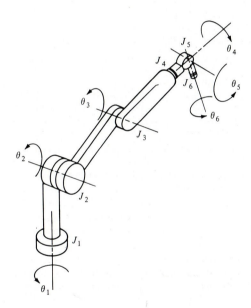

Figure P4-19

Table P4-19 Link parameters for manipulator in Fig. P4-19

Link	Variable	t	a	d
1	θ_1	$-90°$	0	0
2	θ_2	0	0	d_2
3	θ_3	0	a_3	d_3
4	θ_4	90°	a_4	0
5	θ_5	$-90°$	0	0
6	θ_6	90°	0	0

4-20 Find the inverses of the **A** matrices in Prob. 4-19.

4-21 Find the following for the manipulator in Fig. P4-19:

(a) $^5\mathbf{T}_6$ (b) $^4\mathbf{T}_6$ (c) $^3\mathbf{T}_6$

(d) $^2\mathbf{T}_6$ (e) $^1\mathbf{T}_6$ (f) $^0\mathbf{T}_6$

4-22 Consider the coordinate frames **O**, **A**, **B**, and **C** shown in Fig. P4-22, where **O** is the reference frame. Determine, either by inspection or by means of Eq. (4-25) in the text, the homogeneous transforms corresponding to frames **A**, **B**, and **C** with respect to **O**.

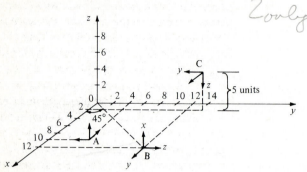

Figure P4-22

4-23 Consider a jointed-arm robot manipulator with its x, y, and z axes aligned with a reference cartesian coordinate frame but located at $(x, y) = (10\,\text{ft}, -5\,\text{ft})$. The end-of-arm of the robot is currently at $(x, y, z) = (12\,\text{ft}, 2\,\text{ft}, 2.5\,\text{ft})$ relative to the reference coordinate frame. An end effector of 10 in. in length is attached to the end-of-arm and is pointing vertically down. Relative to the tip of the end effector is a cube, with 6 in. on a side, and with its nearest corner positioned $+1$ ft in the x direction, $+2$ ft in the y direction, and 0 ft in the z direction from the tip of the end effector.

(a) Make a sketch of the workcell.

(b) Identify all transforms numerically.

(c) Show by means of the transform graph how you would solve for the transform for the cube relative to the end effector. That is, determine all transforms needed to find the transform of the cube relative to the end effector.

4-24 Consider the prism shown in Fig. P4-24. The positions of the prism vertices have been indicated relative to the reference axis system. Positions are given in meters. From its current position, the prism is rotated 90° about the z axis, followed by a 90° rotation about the y axis, followed by a translation of -2 m in the x direction.

(a) Define the transformation which describes the change in position of the prism. That is, determine the 4×4 homogeneous transform for the move.

(b) What are the new coordinates of the vertices of the prism after the move?

(c) What is the inverse transform and how should it be interpreted?

The following problems are concerned with motion of multiple joints in a robot manipulator (Sec. 4-3).

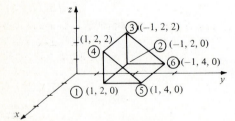

Figure P4-24

4-25 A jointed-arm robot of configuration *VVR* is to move all three axes so that the first joint is rotated through 50°, the second joint is rotated through 90°, and the third joint is rotated through 25°. Maximum speed of any of these rotational joints is 10°/s. Ignore effects of acceleration and deceleration.

(*a*) Determine the time required to move each joint if slew motion is used.

(*b*) Determine the time required to move the arm to the desired position and the rotational velocity of each joint, if joint-interpolated motion is used.

4-26 Solve Prob. 4-25 under the condition that the three joints move at different rotational velocities. The first joint moves at 10°/s, the second joint moves at 25°/s, and the third joint moves at 30°/s.

4-27 A cartesian coordinate robot of configuration *LLL* is to move its three axes from position $(x, y, z) = (0, 5, 5)$ to position $(x, y, z) = (20, 35, 15)$. All distance measures are given in inches. The maximum velocities for the three joints are, respectively, 20 in./sec, 15 in./sec, and 10 in./sec.

(*a*) Determine the time required to move each joint if slew motion is used.

(*b*) Determine the time required to move the arm to the new position and the velocity of each joint, if joint-interpolated motion is used.

4-28 Do Prob. 4-27 except that acceleration and deceleration must be considered. Assume that the rate of acceleration and deceleration is known to be 30 in./sec/sec for each of the the three linear joints.

#4-23 Assume that all coordinate frames are aligned (I.E. the corresponding axes are parallel.

FIVE

ROBOT END EFFECTORS

An end effector is a device that attaches to the wrist of the robot arm and enables the general-purpose robot to perform a specific task. It is sometimes referred to as the robot's "hand." Most production machines require special-purpose fixtures and tools designed for a particular operation, and a robot is no exception. The end effector is part of that special-purpose tooling for a robot. Usually, end effectors must be custom engineered for the particular task which is to be performed. This can be accomplished either by designing and fabricating the device from scratch, or by purchasing a commercially available device and adapting it to the application. The company installing the robot can either do the engineering work itself or it can contract for the services of a firm that does this kind of work. Most robot manufacturers have special engineering groups whose function is to design end effectors and to provide consultation services to their customers. Also, there are a growing number of robot systems firms which perform some or all of the engineering work to install robot systems. Their services would typically include end effector design.

5-1 TYPES OF END EFFECTORS

There are a wide assortment of end effectors required to perform the variety of different work functions. The various types can be divided into two major categories:

1. Grippers
2. Tools

Grippers are end effectors used to grasp and hold objects. The objects are generally workparts that are to be moved by the robot. These part-handling applications include machine loading and unloading, picking parts from a conveyor, and arranging parts onto a pallet. In addition to workparts, other objects handled by robot grippers include cartons, bottles, raw materials, and tools. We tend to think of grippers as mechanical grasping devices, but there are alternative ways of holding objects involving the use of magnets, suction cups, or other means. In this chapter we will divide grippers according to whether they are mechanical grasping devices or some other physical principle is used to retain the object. Section 5-2 examines the different types of mechanical gripper used in robotics, and Sec. 5-3 explores the various grippers that use a means of retention other than mechanical.

Grippers can be classified as single grippers or double grippers although this classification applies best to mechanical grippers. The single gripper is distinguished by the fact that only one grasping device is mounted on the robot's wrist. A double gripper has two gripping devices attached to the wrist and is used to handle two separate objects. The two gripping devices can be actuated independently. The double gripper is especially useful in machine loading and unloading applications. To illustrate, suppose that a particular job calls for a raw workpart to be loaded from a conveyor onto a machine and the finished part to be unloaded onto another conveyor. With a single gripper, the robot would have to unload the finished part before picking up the raw part. This would consume valuable time in the production cycle because the machine would have to remain open during these handling motions. With a double gripper, the robot can pick the part from the incoming conveyor with one of the gripping devices and have it ready to exchange for the finished part. When the machine cycle is completed, the robot can reach in for the finished part with the available grasping device, and insert the raw part into the machine with the other grasping device. The amount of time that the machine is open is minimized.

The term multiple gripper is applied in the case where two or more grasping mechanisms are fastened to the wrist. Double grippers are a subset of multiple grippers. The occasions when more than two grippers would be required are somewhat rare. There is also a cost and reliability penalty which accompanies an increasing number of gripper devices on one robot arm.

Another way of classifying grippers depends on whether the part is grasped on its exterior surface or its internal surface, for example, a ring-shaped part. The first type is called an external gripper and the second type is referred to as an internal gripper.

In the context of this chapter, tools are end effectors designed to perform work on the part rather than to merely grasp it. By definition, the tool-type end effector is attached to the robot's wrist. One of the most common applications of industrial robots is spot welding, in which the welding electrodes constitute the end effector of the robot. Other examples of robot applications in which tools are used as end effectors include spray painting and arc welding. Section 5-4 discusses the various end effectors in this category.

It was mentioned above that grippers are sometimes used to hold tools rather than workparts. The reason for using a gripper instead of attaching the tool directly to the robot's wrist is typically because the job requires several tools to be manipulated by the robot during the work cycle. An example of this kind of application would be a deburring operation in which several different sizes and geometries of deburring tool must be held in order to reach all surfaces of the workpart. The gripper serves as a quick change device to provide the capability for a rapid changeover from one tool to the next.

In addition to end effectors, other types of fixturing and tooling are required in many industrial robot applications. These include holding fixtures, welding fixtures, and other forms of holding devices to position the workpart or tooling during the work cycle.

5-2 MECHANICAL GRIPPERS

Basic Definitions and Operation

A mechanical gripper is an end effector that uses mechanical fingers actuated by a mechanism to grasp an object. The fingers, sometimes called jaws, are the appendages of the gripper that actually make contact with the object. The fingers are either attached to the mechanism or are an integral part of the mechanism. If the fingers are of the attachable type, then they can be detached and replaced. The use of replaceable fingers allows for wear and interchangeability. Different sets of fingers for use with the same gripper mechanism can be designed to accommodate different part models. An example of this interchangeability feature is illustrated in Fig. 5-1, in which the gripper is designed to accommodate fingers of varying sizes. In most applications, two fingers are sufficient to hold the workpart or other object. Grippers with three or more fingers are less common.

The function of the gripper mechanism is to translate some form of power input into the grasping action of the fingers against the part. The power input is supplied from the robot and can be pneumatic, electric, mechanical, or hydraulic. We will discuss the alternatives in Sec. 5-5. The mechanism must be able to open and close the fingers and to exert sufficient force against the part when closed to hold it securely.

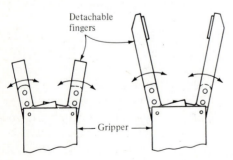

Detachable fingers

Gripper

Figure 5-1 Interchangeable fingers can be used with the same gripper mechanism.

There are two ways of constraining the part in the gripper. The first is by physical constriction of the part within the fingers. In this approach, the gripper fingers enclose the part to some extent, thereby constraining the motion of the part. This is usually accomplished by designing the contacting surfaces of the fingers to be in the approximate shape of the part geometry. This method of constraining the part is illustrated in Fig. 5-2. The second way of holding the part is by friction between the fingers and the workpart. With this approach, the fingers must apply a force that is sufficient for friction to retain the part against gravity, acceleration, and any other force that might arise during the holding portion of the work cycle. The fingers, or the pads attached to the fingers which make contact with the part, are generally fabricated out of a material that is relatively soft. This tends to increase the coefficient of friction between the part and the contacting finger surface. It also serves to protect the part surface from scratching or other damage.

The friction method of holding the part results in a less complicated and therefore less expensive gripper design, and it tends to be readily adaptable to a greater variety of workparts. However, there is a problem with the friction method that is avoided with the physical constriction method. If a force of sufficient magnitude is applied against the part in a direction parallel to the friction surfaces of the fingers as shown in Fig. 5-3(a), the part might slip out of the gripper. To resist this slippage, the gripper must be designed to exert a

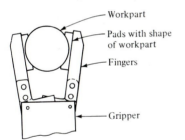

Figure 5-2 Physical constriction method of finger design.

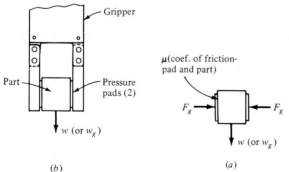

Figure 5-3 Force against part parallel to finger surfaces tending to pull part out of gripper.

force that depends on the weight of the part, the coefficient of friction between the part surface and the finger surface, the acceleration (or deceleration) of the part, and the orientation between the direction of motion during acceleration and the direction of the fingers. Referring to Fig. 5-3(b), the following force equations, Eqs. (5-1) and (5-2), can be used to determine the required magnitude of the gripper force as a function of these factors. Equation (5-1) covers the simpler case in which weight alone is the force tending to cause the part to slip out of the gripper.

$$\mu n_f F_g = w \tag{5-1}$$

where μ = coefficient of friction of the finger contact surface against the part surface
n_f = number of contacting fingers
F_g = gripper force
w = weight of the part or object being gripped

This equation would apply when the force of gravity is directed parallel to the contacting surfaces. If the force tending to pull the part out of the fingers is greater than the weight of the object, then Eq. (5-1) would have to be altered. For example, the force of acceleration would be a significant factor in fast part-handling cycles. Engelberger [3] suggests that in a high-speed handling operation the acceleration (or deceleration) of the part could exert a force that is twice the weight of the part. He reduces the problem to the use of a g factor in a revised version of Eq. (5-1) as follows:

$$\mu n_f F_g = wg \tag{5-2}$$

where g = the g factor. The g factor is supposed to take account of the combined effect of gravity and acceleration. If the acceleration force is applied in the same direction as the gravity force, then the g value = 3.0. If the acceleration is applied in the opposite direction, then the g value = 1.0 (2 × the weight of the part due to acceleration minus 1 × the weight of the part due to gravity). If the acceleration is applied in a horizontal direction, then use $g = 2.0$. The following example will illustrate the use of the equations.

Example 5-1 Suppose a stiff cardboard carton weighing 10 lb is held in a gripper using friction against two opposing fingers. The coefficient of friction between the finger contacting surfaces and the carton surface is 0.25. The orientation of the carton is such that the weight of the carton is directed parallel to finger surfaces. A fast work cycle is anticipated so that a g factor of 3.0 should be applied to calculate the required gripper force. Determine the required gripper force for the conditions given. We can make use of Eq. (5-2) to solve for F_g.

$$(0.25)(2)F_g = (10)(3.0)$$

$$F_g = \frac{30}{0.5} = 60 \text{ lb}$$

The gripper must cause a force of 60 lb to be exerted by the fingers against the carton surface. There is an assumption implicit in the preceding calculations that should be acknowledged. In particular, it is assumed that the robot grasps the carton at its center of mass, so that there are no moments that would tend to rotate the carton in the gripper.

In using Eq. (5-2) as a design formula, a factor of safety would typically be applied to provide a higher computed value of gripper force. For example, a safety factor (SF) of 1.5 in Example 5-1 would result in a calculated value of $1.5 \times 60 = 90$ lb as the required gripper force. This safety factor would help to compensate for the potential problem of the carton being grasped at a position other than its center of mass.

It is possible to determine the actual value of the g factor without resorting to the rough rule of thumb suggested above. However, data must be available concerning the accelerations and decelerations of the part and the direction of the acceleration (or deceleration) force.

Example 5-2 To illustrate, assume that the conditions of Example 5-1 apply except it is known that the carton will experience a maximum acceleration of 40 ft/sec/sec in a vertical direction when it is lifted by the gripper fingers. We assume that the force of acceleration will therefore be applied in the same direction as the weight, and that this direction is parallel to the contacting surfaces between the carton and the fingers. Determine the value of the g factor for this situation.

The value of g would be $1.0 +$ the ratio of the actual acceleration divided by gravity acceleration of 32.2 ft/sec/sec. This ratio is $40/32.2 = 1.24$. The g value would be $1.0 + 1.24 = 2.24$.

Types of Gripper Mechanisms

There are various ways of classifying mechanical grippers and their actuating mechanisms. One method is according to the type of finger movement used by the gripper. In this classification, the grippers can actuate the opening and closing of the fingers by one of the following motions:

1. Pivoting movement
2. Linear or translational movement

These gripper actions are shown in the photographs of Figs. 5-4 and 5-5. In the pivoting movement, the fingers rotate about fixed pivot points on the gripper to open and close. The motion is usually accomplished by some kind of linkage mechanism. In the linear movement, the fingers open and close by moving in parallel to each other. This is accomplished by means of guide rails so that each finger base slides along a guide rail during actuation. The translational finger movement might also be accomplished by means of a

Figure 5-4 Mechanical gripper finger with pivoting movement (Photo courtesy of Phd, Inc.)

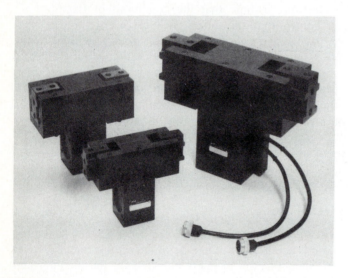

Figure 5-5 Mechanical gripper finger with linear movement using guide rails (Photo courtesy of Phd, Inc.)

linkage which would maintain the fingers in a parallel orientation to each other during actuation.

Mechanical grippers can also be classed according to the type of kinematic device used to actuate the finger movement [2]. In this classification we have the following types:

1. Linkage actuation
2. Gear-and-rack actuation
3. Cam actuation

4. Screw actuation
5. Rope-and-pulley actuation
6. Miscellaneous

The linkage category covers a wide range of design possibilities to actuate the opening and closing of the gripper. A few examples are illustrated in Fig. 5-6. The design of the linkage determines how the input force F_a to the gripper is converted into the gripping force F_g applied by the fingers. The linkage configuration also determines other operational features such as how wide the gripper fingers will open and how quickly the gripper will actuate.

Figure 5-7 illustrates one method of actuating the gripper fingers using a gear-and-rack configuration. The rack gear would be attached to a piston or

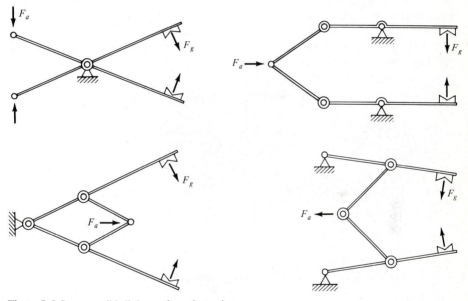

Figure 5-6 Some possible linkages for robot grippers.

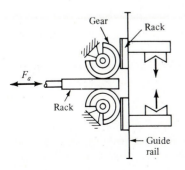

Figure 5-7 Gear-and-rack method of actuating the gripper.

some other mechanism that would provide a linear motion. Movement of the rack would drive two partial pinion gears, and these would in turn open and close the fingers.

The cam actuated gripper includes a variety of possible designs, one of which is shown in Fig. 5-8. A cam-and-follower arrangement, often using a spring-loaded follower, can provide the opening and closing action of the gripper. For example, movement of the cam in one direction would force the gripper to open, while movement of the cam in the opposite direction would cause the spring to force the gripper to close. The advantage of this arrangement is that the spring action would accommodate different sized parts. This might be desirable, for example, in a machining operation where a single gripper is used to handle the raw workpart and the finished part. The finished part might be significantly smaller after machining.

An example of the screw-type actuation method is shown in Fig. 5-9. The screw is turned by a motor, usually accompanied by a speed reduction mechanism. When the screw is rotated in one direction, this causes a threaded block to be translated in one direction. When the screw is rotated in the opposite direction, the threaded block moves in the opposite direction. The threaded block is, in turn, connected to the gripper fingers to cause the corresponding opening and closing action.

Rope-and-pulley mechanisms can be designed to open and close a

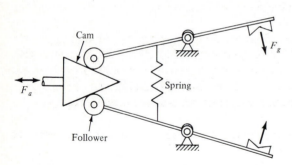

Figure 5-8 Cam-actuated gripper.

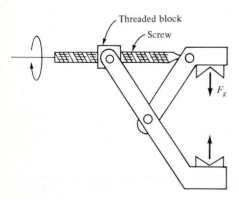

Figure 5-9 Screw-type gripper actuation.

mechanical gripper. Because of the nature of these mechanisms, some form of tension device must be used to oppose the motion of the rope or cord in the pulley system. For example, the pulley system might operate in one direction to open the gripper, and the tension device would take up the slack in the rope and close the gripper when the pulley system operates in the opposite direction.

The miscellaneous category is included in our list to allow for gripper-actuating mechanisms that do not logically fall into one of the above categories. An example might be an expandable bladder or diaphragm that would be inflated and deflated to actuate the gripper fingers.

Gripper Force Analysis

As indicated previously, the purpose of the gripper mechanism is to convert input power into the required motion and force to grasp and hold an object. Let us illustrate the analysis that might be used to determine the magnitude of the required input power in order to obtain a given gripping force. We will assume that a friction-type grasping action is being used to hold the part, and we will therefore use the gripper force calculated in Example 5-1 as our starting point in Example 5-3 below. We will demonstrate the analysis by several examples given below. A detailed study of mechanism analysis is beyond the scope of this text, and the reader might refer to other books such as Beer and Johnson [1] and Shigley and Mitchell [6].

Example 5-3 Suppose the gripper is a simple pivot-type device used for holding the cardboard carton, as pictured in Fig. 5-10. The gripper force, calculated in our previous Example 5-1, is 60 lb. The gripper is to be actuated by a piston device to apply an actuating force F_a. The corresponding lever arms for the two forces are shown in the diagram of the figure.

The analysis would require that the moments about the pivot arms be summed and made equal to zero.

$$F_g L_g - F_a L_a = 0$$

$$(60 \text{ lb})(12 \text{ in.}) - (F_a)(3 \text{ in.}) = 0$$

$$F_a = \frac{720}{3} = 240 \text{ lb}$$

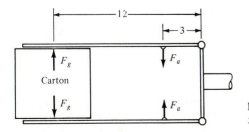

Figure 5-10 Pivot type gripper used in Example 5-3.

The piston device would have to provide an actuating force of 240 lb to close the gripper with a force against the carton of 60 lb.

Example 5-4 The diagram in Fig. 5-11 shows the linkage mechanism and dimensions of a gripper used to handle a workpart for a machining operation. Suppose it has been determined that the gripper force is to be 25 lb. What is required is to compute the actuating force to deliver this force of 25 lb.

Figure 5-12(a) shows how the symmetry of the gripper can be used to

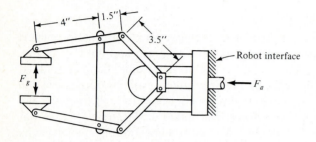

Figure 5-11 Gripper considered in Example 5-4.

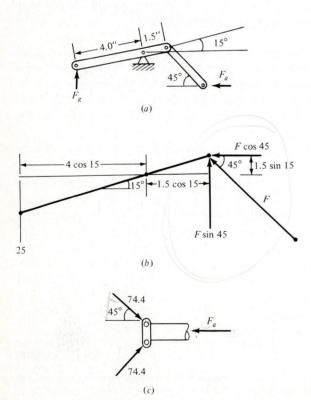

Figure 5-12 Linkage analysis of Example 5-4.

advantage so that only one-half of the mechanism needs to be considered. Part (*b*) of the figure shows how the moments might be summed about the pivot point for the finger link against which the 25-lb gripper force is applied. We find

$$25(4 \cos 15°) = F \sin 45°(1.5 \cos 15°) + F \cos 45°(1.5 \sin 15°)$$

$$96.6 = F(1.0246 + 0.2745) = 1.2991F$$

$$F = 74.4 \text{ lb}$$

The actuating force applied to the plunger to deliver this force of 74.4 lb to each finger is pictured in Fig. 5-12(*c*) and can be calculated as

$$F_a = 2 \times 74.4 \times \cos 45°$$

$$F_a = 105.2 \text{ lb}$$

Some power input mechanism would be required to deliver this actuating force of 105.2 lb to the gripper.

5-3 OTHER TYPES OF GRIPPERS

In addition to mechanical grippers there are a variety of other devices that can be designed to lift and hold objects. Included among these other types of grippers are the following:

1. Vacuum cups
2. Magnetic grippers
3. Adhesive grippers
4. Hooks, scoops, and other miscellaneous devices

Vacuum Cups

Vacuum cups, also called suction cups, can be used as gripper devices for handling certain types of objects. The usual requirements on the objects to be handled are that they be flat, smooth, and clean, conditions necessary to form a satisfactory vacuum between the object and the suction cup. An example of a vacuum cup used to lift flat glass is pictured in Fig. 5-13.

The suction cups used in this type of robot gripper are typically made of elastic material such as rubber or soft plastic. An exception would be when the object to be handled is composed of a soft material. In this case, the suction cup would be made of a hard substance. The shape of the vacuum cup, as shown in the figure, is usually round. Some means of removing the air between the cup and the part surface to create the vacuum is required. The vacuum pump and the venturi are two common devices used for this purpose. The vacuum pump is a piston-operated or vane-driven device powered by an electric motor. It is capable of creating a relatively high vacuum. The venturi is a simpler device as pictured in Fig. 5-14 and can be driven by means of

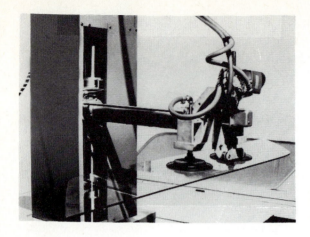

Figure 5-13 Vacuum cup gripper lifting glass plates (Photo courtesy of Prab Conveyors, Inc.)

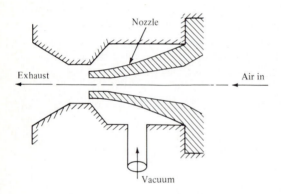

Figure 5-14 Venturi device used to operate a suction cup.

"shop air pressure." Its initial cost is less than that of a vacuum pump and it is relatively reliable because of its simplicity. However, the overall reliability of the vacuum system is dependent on the source of air pressure.

The lift capacity of the suction cup depends on the effective area of the cup and the negative air pressure between the cup and the object. The relationship can be summarized in the following equation

$$F = PA \qquad (5\text{-}3)$$

where F = the force or lift capacity, lb
P = the negative pressure, lb/in.^{2}
A = the total effective area of the suction cup(s) used to create the vacuum, in.^{2}

The effective area of the cup during operation is approximately equal to the undeformed area determined by the diameter of the suction cup. The squashing action of the cup as it presses against the object would tend to make the effective area slightly larger than the undeformed area. On the other hand, if the center portion of the cup makes contact against the object during deformation, this would reduce the effective area over which the vacuum is applied. These two conditions tend to cancel each other out. The negative air pressure is the pressure differential between the inside and the outside of the vacuum cup. The following example will illustrate the operation of the vacuum cup as a robotic gripper device.

Example 5-5 A vacuum cup gripper will be used to lift flat plates of 18-8 stainless steel. Each piece of steel is $\frac{1}{4}$ in. thick and measures 2.0 by 3.0 ft. The gripper will utilize two suction cups separated by about 1.5 ft for stability. Each suction cup is round and has a diameter of 5.0 in. Two cups are considered a requirement to overcome the problem that the plates may be off center with respect to the gripper. Because of variations in the positioning of the end effector or in the positions of the steel plates before pickup, the suction cups will not always operate on the center of mass of the plates. Consequently, static moments and inertia will result which must be considered in the design of the end effector. We are attempting to compensate for these moments by providing two pressure points on the part, separated by a substantial distance.

The negative pressure (compared to atmospheric pressure of 14.7 lb/in.^{2}) required to lift the stainless steel plates is to be determined. A safety factor of 1.6 is to be used to allow for acceleration of the plate and for possible contact of the suction cup against the plate which would reduce the effective area of the cup.

We must begin by calculating the weight of the stainless steel plate. Stainless steel has a density of 0.28 lb/in.^{3}. The weight of the plate would therefore be

$$w = 0.28 \times \tfrac{1}{4} \times 24 \times 36 = 60.48 \text{ lb}$$

This would be equal to the force F which must be applied by the two suction cups, ignoring for the moment any effects of gravity (the g factor used before). The area of each suction cup would be

$$A = 3.142(\tfrac{5}{2})^2 = 19.63 \text{ in.}^2$$

The area of the two cups would be $2 \times 19.63 = 39.26$ in.^{2} From Eq. (5-3), the negative pressure required to lift the weight can be determined by dividing the weight by the combined area of the two suction cups.

$$P = \frac{w}{A} = 60.48 \text{ lb}/39.26 \text{ in.}^2 = 1.54 \text{ lb/in.}^2$$

Applying the safety factor of 1.6, we have

$$P = 1.6 \times 1.54 \text{ lb/in.}^2 = 2.46 \text{ lb/in.}^2 \text{ negative pressure}$$

Some of the features and advantages that characterize the operation of suction cup grippers used in robotics applications are:

Requires only one surface of the part for grasping
Applies a uniform pressure distribution on the surface of the part
Relatively light-weight gripper
Applicable to a variety of different materials

Magnetic Grippers

Magnetic grippers can be a very feasible means of handling ferrous materials. The stainless steel plate in Example 5-3 would not be an appropriate application for a magnetic gripper because 18-8 stainless steel is not attracted by a magnet. Other steels, however, including certain types of stainless steel, would be suitable candidates for this means of handling, especially when the materials are handled in sheet or plate form.

In general, magnetic grippers offer the following advantages in robotic-handling applications:

Pickup times are very fast.
Variations in part size can be tolerated. The gripper does not have to be designed for one particular workpart.
They have the ability to handle metal parts with holes (not possible with vacuum grippers).
They require only one surface for gripping.

Disadvantages with magnetic grippers include the residual magnetism remaining in the workpiece which may cause a problem in subsequent handling, and the possible side slippage and other errors which limit the precision of this means of handling. Another potential disadvantage of a magnetic gripper is the problem of picking up only one sheet from a stack. The magnetic attraction tends to penetrate beyond the top sheet in the stack, resulting in the possibility that more than a single sheet will be lifted by the magnet. This problem can be confronted in several ways. First, magnetic grippers can be designed to limit the effective penetration to the desired depth, which would correspond to the thickness of the top sheet. Second, the stacking device used to hold the sheets can be designed to separate the sheets for pickup by the robot. One such type of stacking device is called a "fanner," and it makes use of a magnetic field to induce a charge in the ferrous sheets in the stack. Each sheet toward the top of the stack is given a magnetic charge, causing them to possess the same polarity and repel each other. The sheet most affected is the one at the top of the stack. It tends to rise above the remainder of the stack, thus facilitating pickup by the robot gripper.

Magnetic grippers can be divided into two categories, those using electromagnets, and those using permanent magnets. Electromagnetic grippers are easier to control, but require a source of dc power and an appropriate controller unit. As with any other robotic-gripping device, the part must be

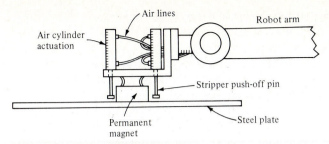

Figure 5-15 Stripper device operated by air cylinders used with a permanent magnet gripper.

released at the end of the handling cycle. This is easier to accomplish with an electromagnet than with a permanent magnet. When the part is to be released, the controller unit reverses the polarity at a reduced power level before switching off the electromagnet. This procedure acts to cancel the residual magnetism in the workpiece and ensures a positive release of the part.

Permanent magnets have the advantage of not requiring an external power source to operate the magnet. However, there is a loss of control that accompanies this apparent advantage. For example, when the part is to be released at the end of the handling cycle, some means of separating the part from the magnet must be provided. The device which accomplishes this is called a stripper or stripping device. Its function is to mechanically detach the part from the magnet. One possible stripper design is illustrated in Fig. 5-15.

Permanent magnets are often considered for handling tasks in hazardous environments requiring explosion proof apparatus. The fact that no electrical circuit is needed to operate the magnet reduces the danger of sparks which might cause ignition in such an environment.

Adhesive Grippers

Gripper designs in which an adhesive substance performs the grasping action can be used to handle fabrics and other lightweight materials. The requirements on the items to be handled are that they must be gripped on one side only and that other forms of grasping such as a vacuum or magnet are not appropriate. One of the potential limitations of an adhesive gripper is that the adhesive substance loses its tackiness on repeated usage. Consequently, its reliability as a gripping device is diminished with each successive operation cycle. To overcome this limitation, the adhesive material is loaded in the form of a continuous ribbon into a feeding mechanism that is attached to the robot wrist. The feeding mechanism operates in a manner similar to a typewriter ribbon mechanism.

Hooks, Scoops, and Other Miscellaneous Devices

A variety of other devices can be used to grip parts or materials in robotics applications. Hooks can be used as end effectors to handle containers of parts and to load and unload parts hanging from overhead conveyors. Obviously, the items to be handled by a hook must have some sort of handle to enable the hook to hold it.

Scoops and ladles can be used to handle certain materials in liquid or powder form. Chemicals in liquid or powder form, food materials, granular substances, and molten metals are all examples of materials that can be handled by a robot using this method of holding. One of its limitations is that the amount of material being scooped by the robot is sometimes difficult to control. Spillage during the handling cycle is also a problem.

Other types of grippers include inflatable devices, in which an inflatable bladder or diaphragm is expanded to grasp the object. The inflatable bladder is fabricated out of rubber or other elastic material which makes it appropriate for gripping fragile objects. The gripper applies a uniform grasping pressure against the surface of the object rather than a concentrated force typical of a mechanical gripper. An example of the inflatable bladder type gripper is shown in Fig. 5-16. Part (a) of the figure shows the bladder fully expanded. Part (b) shows the bladder used to grasp the inside diameter of a bottle.

Research and development is being carried out with the objective of designing a universal gripper capable of grasping and handling a variety of objects with differing geometries. If such a universal device could be developed and marketed at a relatively low cost, it would save the time and expense of designing a specific end effector for each new robot application. Most of the gripper models under consideration are patterned after the human hand which turns out to possess considerable versatility. A gripping device with the number of joints and axes of controlled motion as the human hand is mechanically very complex. Accordingly, these research end effectors typically have only three fingers rather than five. This reduces the complexity of the hand without a significant loss of functionality. One possible design of the universal hand is illustrated in Fig. 5-17. In Chap. Nineteen, we explore the research and development directions in end effector design.

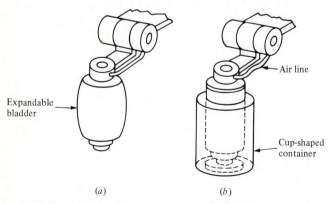

Expandable bladder

Air line

Cup-shaped container

(a)   (b)

Figure 5-16 Expansion bladder used to grasp inside of a cup-shaped container.

Figure 5-17 The Stanford/JPL three-fingered anthropomorphic hand. (Photo courtesy of Salisbury Robotics, Inc.)

5-4 TOOLS AS END EFFECTORS

In many applications, the robot is required to manipulate a tool rather than a workpart. In a limited number of these applications, the end effector is a gripper that is designed to grasp and handle the tool. The reason for using a gripper in these applications is that there may be more than one tool to be used by the robot in the work cycle. The use of a gripper permits the tools to be exchanged during the cycle, and thus facilitates this multitool handling function. References 8 and 9 examine the design problems involved in a tool exchange mechanism.

In most of the robot applications in which a tool is manipulated, the tool is attached directly to the robot wrist. In these cases the tool is the end effector. Some examples of tools used as end effectors in robot applications include:

Spot-welding tools
Arc-welding torch
Spray-painting nozzle
Rotating spindles for operations such as:
    drilling
    routing
    wire brushing
    grinding
Liquid cement applicators for assembly
Heating torches
Water jet cutting tool

In each case, the robot must control the actuation of the tool. For example, the robot must coordinate the actuation of the spot-welding operation as part of its work cycle. This is controlled much in the same manner as the opening and closing of a mechanical gripper. We will discuss the interface between the robot and its end effector in the following section. Design and application considerations of most of the robot tools listed above will be considered in Chaps. Fourteen and Fifteen of the book.

5-5 THE ROBOT/END EFFECTOR INTERFACE

An important aspect of the end effector applications engineering involves the interfacing of the end effector with the robot. This interface must accomplish at least some of the following functions:

Physical support of the end effector during the work cycle must be provided.
Power to actuate the end effector must be supplied through the interface.
Control signals to actuate the end effector must be provided. This is often accomplished by controlling the actuating power.
Feedback signals must sometimes be transmitted back through the interface to the robot controller.

In addition, certain other general-design objectives should be met. These include high reliability of the interface, protection against the environment, and overload protection in case of disturbances and unexpected events during the work cycle.

Physical Support of the End Effector

The physical support of the end effector is achieved by the mechanical connection between the end effector and the robot wrist. This mechanical connection often consists of a faceplate at the end of the wrist to which the end effector is bolted. In other cases, a more complicated wrist socket is used. Ideally, there should be three characteristics taken into consideration in the design of the mechanical connection [7]; strength, compliance, and overload protection. The strength of the mechanical connection refers to its ability to withstand the forces associated with the operation of the end effector. These forces include the weight of the end effector, the weight of the objects being held by the end effector if it is a gripper, acceleration and deceleration forces, and any applied forces during the work cycle (e.g., thrust forces during a drilling operation). The wrist socket must provide sufficient strength and rigidity to support the end effector against these various forces.

The second consideration in the design of the mechanical connection is compliance. Compliance refers to the wrist socket's ability to yield elastically when subjected to a force. In effect, it is the opposite of rigidity. In some applications, it is desirable to design the mechanical interface so that it will yield during the work cycle. A good example of this is found in robot assembly work. Certain assembly operations require the insertion of an object into a hole where there is very little clearance between the hole and the object to be inserted. If an attempt is made to insert the object off center, it is likely that the object will bind against the sides of the hole. Human assembly workers can make adjustments in the position of the object as it enters the hole using hand–eye coordination and their sense of "touch." Robots have difficulty with this kind of insertion task because of limitations on their accuracy. To overcome these limitations, remote center compliance (RCC) devices have been designed to provide high lateral compliance for centering the object relative to the hole in response to sideways forces encountered during insertion. We will discuss the RCC device in Chap. Fifteen on assembly and inspection applications.

The third factor which must be considered relative to the mechanical interface between the robot wrist and the end effector is overload protection. An overload results when some unexpected event happens to the end effector such as a part becoming stuck in a die, or a tool getting caught in a moving conveyor. Whatever the cause, the consequences involve possible damage to the end effector or maybe even the robot itself. Overload protection is intended to eliminate or reduce this potential damage. The protection can be provided either by means of a breakaway feature in the wrist socket or by using sensors to indicate that an unusual event has occurred so as to somehow take preventive action to reduce further overloading of the end effector.

A breakaway feature is a mechanical device that will either break or yield when subjected to a high force. Such a device is generally designed to accomplish its breakaway function when the force loading exceeds a certain specific level. A shear pin is an example of a device that is designed to fail if subjected to a shear force above a certain value. It is relatively inexpensive and its purpose as a component in the mechanical interface is to be sacrificed in order to save the end effector and the robot. The disadvantage of a device that breaks is that it must be replaced and this generally involves downtime and the attention of a human operator. Some mechanical devices are designed to yield or give under unexpected load rather than fail. Examples of these devices include spring-loaded detents and other mechanisms used to hold structural components in place during normal operation. When abnormal conditions are encountered, these mechanisms snap out of position to release the structural components. Although more complicated than shear pins and other similar devices that fail, their advantage is that they can be reused and in some cases reset by the robot without human assistance.

Sensors are sometimes used either as an alternative to a breakaway device or in conjunction with such a device. The purpose of the sensors in this context is to signal the robot controller that an unusual event is occurring in the operation of the end effector and that some sort of evasive action should be taken to avoid or reduce damage. Of course, the kinds of unusual events must be anticipated in advance so that the robot controller can be programmed to respond in the appropriate way. For example, if the robot is working in a cell with a moving conveyor, and the end effector becomes caught in a part that is fastened to the conveyor, the most appropriate response might be simply to stop the conveyor and call for help. In other cases the robot might be programmed to perform motions that would remove the end effector from the cause of the unusual force loading.

Power and Signal Transmission

End effectors require power to operate. They also require control signals to regulate their operation. The principal methods of transmitting power and control signals to the end effector are:

Pneumatic
Electric
Hydraulic
Mechanical

The method of providing the power to the end effector must be compatible with the capabilities of the robot system. For example, it makes sense to use a pneumatically operated gripper if the robot has incorporated into its arm design the facility to transmit air pressure to the end effector. The control signals to regulate the end effector are often provided simply by controlling the transmission of the actuating power. The operation of a pneumatic gripper is generally accomplished in this manner. Air pressure is supplied to either open the gripper or to close it. In some applications, greater control is required to operate the end effector. For example, the gripper might possess a range of open/close positions and there is the need to exercise control over these positions. In more complicated cases, feedback signals from sensors in the end effector are required to operate the device. These feedback signals might indicate how much force is being applied to the object held in the gripper, or they might show whether an arc-welding operation was following the seam properly. In the paragraphs below, we will explore the four methods of power and signal transmission to the end effector.

Pneumatic power using shop air pressure is one of the most common methods of operating mechanical grippers. Actuation of the gripper is controlled by regulating the incoming air pressure. A piston device is typically used to actuate the gripper. Two air lines feed into opposite ends of the piston, one to open the gripper and the other to close it. This arrangement can be accomplished with a single shop air line by providing a pneumatic valve to switch the air pressure from one line to the other. A schematic diagram of a piston is illustrated in Fig. 5-18. When air pressure enters the left portion of the piston chamber the piston ram is extended, and when air is forced into the opposite end of the chamber, the piston ram is retracted. The force supplied by the piston on the extension stroke is equal to the air pressure multiplied by the area of the piston diameter. Because of the diameter of the piston ram, the force supplied by the piston on the retraction stroke is less than on the

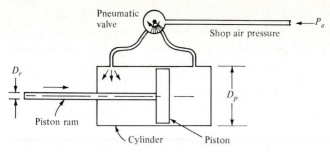

Figure 5-18 Schematic diagram of a piston.

extension stroke. These piston forces can be calculated as follows:

$$F_{exten} = P_a \frac{D_p^2}{4} \pi \qquad (5\text{-}4)$$

$$F_{retract} = \frac{P_a}{4} (D_p^2 - D_r^2) \pi \qquad (5\text{-}5)$$

where F_{exten} = the piston force on the extension stroke, lb
$F_{retract}$ = the piston force on the retraction stroke, lb
D_p = the piston diameter, in.
D_r = the ram diameter, in.
P_a = air pressure, lb/in.^{2}

We will illustrate the kind of engineering analysis required to design a pneumatic piston for a mechanical gripper. Our example will refer back to our previous problem in Example 5-3.

Example 5-6 The gripper is used for holding cartons and is pictured in Fig. 5-10. It was determined in the previous Example 5-3 that an actuating force of 240 lb is required to provide the desired gripper force. We are now concerned with the problem of designing a piston which can supply the acuating force of 240 lb. It is logical to orient the piston device in the gripper so that this force is supplied on the extension stroke of the piston. Suppose that the shop air pressure is 75 lb/in.^{2} Our problem, therefore, is to determine the required diameter of the piston in order to provide the desired actuating force of 240 lb. Using Eq. (5-4), we can solve for this diameter.

$$240 \text{ lb} = 75 \text{ lb/in.}^2 \frac{D_p^2}{4}$$

$$D_p^2 = \frac{4 \times 240}{75}$$

$$D_p = 2.02 \text{ in.}$$

Another use of pneumatic power in end effector design is for vacuum cup grippers. When a venturi device is used to provide the vacuum, the device can be actuated by shop air pressure. Otherwise, some means of developing and controlling the vacuum must be provided to operate the suction cup gripping device.

A second method of power transmission to the end effector is electrical. Pneumatic actuation of the gripper is generally limited to two positions, open and closed. The use of an electric motor can allow the designer to exercise a greater degree of control over the actuation of the gripper and of the holding force applied. Instead of merely two positions, the gripper can be controlled to any number of partially closed positions. This feature allows the gripper to be used to handle a variety of objects of different sizes, a likely requirement in assembly operations. By incorporating force sensors into the gripper fingers, a feedback control system can be built into the gripper to regulate the holding force applied by the fingers rather than their position. This would be useful, for example, if the objects being grasped are delicate or if the objects vary in size and the proper finger positions for gripping are not known.

Other uses of electric power for end effectors include electromagnet grippers, spot-welding and arc-welding tools, and powered spindle tools used as robot end effectors.

Hydraulic and mechanical power transmission are less common means of actuating the end effector in current practice. Hydraulic actuation of the gripper has the potential to provide very high holding forces, but its disadvantage is the risk of oil leaks. Mechanical power transmission would involve an arrangement in which a motor (e.g., pneumatic, electric, hydraulic, etc.) is mounted on the robot arm and connected mechanically to the gripper, perhaps by means of a flexible cable or the use of pulleys. The possible advantage of this arrangement is a reduction of the weight and mass at the robot's wrist.

5-6 CONSIDERATIONS IN GRIPPER SELECTION AND DESIGN

Most of this chapter has been concerned with grippers rather than tools as end effectors. As indicated in Sec. 5-4, tools are used for spot welding, arc welding, rotating spindle operations, and other processing applications. We will examine the tooling used with these operations when we discuss the corresponding applications in Chaps. Fourteen and Fifteen. In this section, let us summarize our discussion of grippers by enumerating some of the considerations in their selection and design.

Certainly one of the considerations deals with determining the grasping requirements for the gripper. Engelberger [3] defines many of the factors that should be considered in assessing gripping requirements. The following list is based on Engelberger's discussion of these factors:

1. The part surface to be grasped must be reachable. For example, it must not be enclosed within a chuck or other holding fixture.

Table 5-1 Checklist of factors in the selection and design of grippers

Factor	Consideration
Part to be handled	Weight and size Shape Changes in shape during processing Tolerances on the part size Surface condition, protection of delicate surfaces
Actuation method	Mechanical grasping Vacuum cup Magnet Other methods (adhesives, scoops, etc.)
Power and signal transmission	Pneumatic Electrical Hydraulic Mechanical
Gripper force (mechanical gripper)	Weight of the object Method of holding (physical constriction or friction) Coefficient of friction between fingers and object Speed and acceleration during motion cycle
Positioning problems	Length of fingers Inherent accuracy and repeatability of robot Tolerances on the part size
Service conditions	Number of actuations during lifetime of gripper Replaceability of wear components (fingers) Maintenance and serviceability
Operating environment	Heat and temperature Humidity, moisture, dirt, chemicals
Temperature protection	Heat shields Long fingers Forced cooling (compressed air, water cooling, etc.) Use of heat-resistant materials
Fabrication materials	Strength, rigidity, durability Fatigue strength Cost and ease of fabrication Friction properties for finger surfaces Compatibility with operating environment
Other considerations	Use of interchangeable fingers Design standards Mounting connections and interfacing with robot Risk of product design changes and their effect on the gripper design Lead time for design and fabrication Spare parts, maintenance, and service Tryout of the gripper in production

2. The size variation of the part must be accounted for, and how this might influence the accuracy of locating the part. For example, there might be a problem in placing a rough casting or forging into a chuck for machining operations.
3. The gripper design must accommodate the change in size that occurs between part loading and unloading. For example, the part size is reduced in machining and forging operations.
4. Consideration must be given to the potential problem of scratching and distorting the part during gripping, if the part is fragile or has delicate surfaces.
5. If there is a choice between two different dimensions on a part, the larger dimension should be selected for grasping. Holding the part by its larger surface will provide better control and stability of the part in positioning.
6. Gripper fingers can be designed to conform to the part shape by using resilient pads or self-aligning fingers. The reason for using self-aligning fingers is to ensure that each finger makes contact with the part in more than one place. This provides better part control and physical stability. Use of replaceable fingers will allow for wear and also for interchangeability for different part models.

A related issue is the problem of determining the magnitude of the grasping force that can be applied to the object by the gripper. The important factors that determine the required grasping force are:

The weight of the object.
Consideration of whether the part can be grasped consistently about its center of mass. If not, an analysis of the possible moments from off-center grasping should be considered.
The speed and acceleration with which the robot arm moves (acceleration and deceleration forces), and the orientational relationship between the direction of movement and the position of the fingers on the object (whether the movement is parallel or perpendicular to the finger surface contacting the part).
Whether physical constriction or friction is used to hold the part.
Coefficient of friction between the object and the gripper fingers.

We have discussed the methods for dealing with these factors and analyzing the gripping forces in Secs. 5-2 and 5-3. Table 5-1 provides a checklist of the many different issues and factors that must be considered in the selection and design of robot gripper.

REFERENCES

1. F. P. Beer and E. R. Johnson, Jr., *Vector Mechanics for Engineers*, 3rd ed., McGraw-Hill, New York, 1977.
2. F. Y. Chen, "Gripping Mechanisms for Industrial Robots," *Mechanism and Machine Theory* **17**(5), 299–311 (1982).
3. J. F. Engelberger, *Robotics in Practice*, AMACOM (American Management Association), New York, 1980, chap. 3.

4. M. P. Groover and E. W. Zimmers, Jr., *CAD/CAM: Computer-Aided Design and Manufacturing*, Prentice-Hall, Englewood Cliffs, NJ, 1984, chap. 10.
5. G. Lundstrom, B. Glemme, and B. W. Rocks, *Industrial Robots—Gripper Review*, International Fluidics Services Ltd., Bedford, England.
6. J. E. Shigley and L. D. Mitchell, *Mechanical Engineering Design*, McGraw-Hill, New York, 1983.
7. L. L. Toepperwein, M. T. Blackman, et al., "ICAM Robotics Application Guide," *Technical Report AFWAL-TR-80-4042*, vol. II, Materials Laboratory, Air Force Wright Aeronautical Laboratories, Ohio, April 1980.
8. J. M. Vranish, "Quick Change System for Robots," SME paper MS84-418, *Conference Papers*, Robotics Research—The Next Five Years and Beyond, Lehigh University, Bethlehem, Pennsylvania, August 1984.
9. A. J. Wright, "Light Assembly Robots—An End Effector Exchange Mechanism," *Mechanical Engineering*, July 1983, pp. 29–35.

PROBLEMS

5-1 A part weighing 8 lb is to be held by a gripper using friction against two opposing fingers. The coefficient of friction between the fingers and the part surface is estimated to be 0.3. The orientation of the gripper will be such that the weight of the part will be applied in a direction parallel to the contacting finger surfaces. A fast work cycle is anticipated so that the g factor to be used in force calculations should be 3.0. Compute the required gripper force for the specifications given.

5-2 Solve Prob. 5-1 except using a safety factor of 1.5 in the calculations.

5-3 Solve Prob. 5-1 except that instead of using a g factor of 3.0, the following information is given to make the force computation. The robot motion cycle has been analyzed and it has been determined that the largest acceleration experienced by the part held in the gripper is immediately after the pickup. The maximum acceleration is measured as 53 ft/sec/sec. Using a safety factor of 1.5, compute the required gripper force.

5-4 A part weighing 15 lb is to be grasped by a mechanical gripper using friction between two opposing fingers. The coefficient of static friction is 0.35 and the coefficient of dynamic friction is 0.20. The direction of the acceleration force is parallel to the contacting surfaces of the gripper fingers. Which value of coefficient of friction is appropriate to use in the force calculations? Why? Compute the required gripper force assuming that a g factor of 2.0 is applicable.

5-5 For the information given in the mechanical gripper design of Fig. P5-5, determine the required actuating force if the gripper force is to be 25 lb.

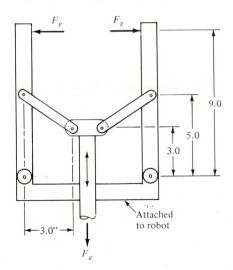

Figure P5-5

5-6 For the information given in the mechanical gripper design of Fig. P5-6, calculate the required actuating force if the gripper force is to be 20 lb.

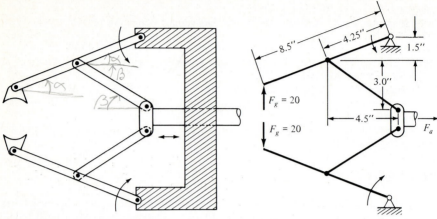

Figure P5-6

5-7 A vacuum gripper is to be designed to handle flat plate glass in an automobile windshield plant. Each plate weighs 28 lb. A single suction cup will be used and the diameter of the suction cup is 6.0 in. Determine the negative pressure required (compared to atmospheric pressure of 14.7 lb/in.^{2}) to lift each plate. Use a safety factor of 1.5 in your calculations.

5-8 A vacuum pump to be used in a robot vacuum gripper application is capable of drawing a negative pressure of 4.0 lb/in.^{2} compared to atmospheric pressure. The gripper is to be used for lifting stainless steel plates, each plate having dimensions of 15 by 35 in. and weighing 52 lb. Determine the diameter of the suctions cups to be used for the robot gripper if it has been decided that two suction cups will be used for the gripper for greater stability. A factor of safety of 1.5 should be used in the design computations.

5-9 A piston is to be designed to exert an actuation force of 120 lb on its extension stroke. The inside diameter of the piston is 2.0 in. and the ram diameter is 0.375 in. What shop air pressure will be required to provide this actuation force? Use a safety factor of 1.3 in your computations.

5-10 The mechanical gripper in Fig. P5-10 uses friction to grasp a part weighing 10 lb. The

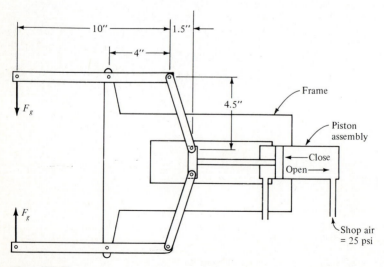

Figure P5-10

coefficient of friction between the fingers and the part is 0.25 and the anticipated g factor is 2.0. The gripper is to be operated by a piston whose diameter is 2.5 in. as shown in the drawing. The ram diameter is 0.375 in.

(*a*) Determine the required gripping force to retain the part. *40 #*

(*b*) Determine the actuation force that must be applied to achieve this gripping force for this mechanical design. *506 #*

(*c*) Determine the air pressure needed to operate the piston so as to apply the required actuation force. *161 psi*

(*d*) If a safety factor of 1.5 were to be used for this design, at what point in the computations would it be appropriate to apply it? *241.6 psi*

#5-6

$$\Sigma M_A = 20 \,(8.5\cos\alpha) - F_1\,(4.25\sin(\alpha+\beta))$$

$$F = 2\,F_1\,\cos\beta$$

SIX

SENSORS IN ROBOTICS

In Chap. Three, we discussed various sensor devices used as components of the robot control system. These devices included position sensors such as resolvers and encoders, and velocity sensors such as tachometers. Sensors can also be used as peripheral devices for the robot, just as end effectors discussed in Chap. Five are used as peripheral devices. Most industrial applications require that sensors be employed in this way in order for the robot to operate with other pieces of equipment in the workcell.

In this chapter we consider the various types of sensors that are used as peripheral devices in robotics and the reasons why they are needed. We begin with a general discussion of transducers and sensors, then examine the sensors that are most commonly employed in robot applications. The final section in the chapter describes the four major categories of uses of sensor systems in robotics.

6-1 TRANSDUCERS AND SENSORS

A transducer is a device that converts one type of physical variable (e.g., force, pressure, temperature, velocity, flow rate, etc.) into another form. A common conversion is to electrical voltage, and the reason for making the conversion is that the converted signal is more convenient to use and evaluate. A sensor is a transducer that is used to make a measurement of a physical variable of interest. Some of the common sensors and transducers include strain gauges (used to measure force and pressure), thermocouples (temperatures), speedometers (velocity), and Pitot tubes (flow rates).

Any sensor or transducer requires calibration in order to be useful as a measuring device. Calibration is the procedure by which the relationship between the measured variable and the converted output signal is established.

Transducers and sensors can be classified into two basic types depending on the form of the converted signal. The two types are:

1. Analog transducers
2. Digital transducers

Analog transducers provide a continuous analog signal such as electrical voltage or current. This signal can then be interpreted as the value of the physical variable that is being measured. Digital transducers produce a digital output signal, either in the form of a set of parallel status bits or a series of pulses that can be counted. In either form, the digital signal represents the value of the measured variable. Digital transducers are becoming more popular because of the ease with which they can be read as separate measuring instruments. In addition, they offer the advantage in automation and process control that they are generally more compatible with the digital computer than analog-based sensors.

In order to be useful as measuring devices, in robotics and in other applications, sensors must possess certain features. Some of the desirable engineering features of sensors and transducers are presented in Table 6-1. Few if any sensors have all of these desirable features, and a compromise must be made among them to select the best sensor for a given application.

Table 6-1 Desirable features of sensors

1. *Accuracy.* The accuracy of the measurement should be as high as possible. Accuracy is interpreted to mean that the true value of the variable can be sensed with no systematic positive or negative errors in the measurement. Over many measurements of the variable, the average error between the actual value and the sensed value will tend to be zero.

2. *Precision.* The precision of the measurement should be as high as possible. Precision means that there is little or no random variability in the measured variable. The dispersion in the values of a series of measurements will be minimized.

3. *Operating range.* The sensor should possess a wide operating range and should be accurate and precise over the entire range.

4. *Speed of response.* The transducer should be capable of responding to changes in the sensed variable in minimum time. Ideally, the response would be instantaneous.

5. *Calibration.* The sensor should be easy to calibrate. The time and trouble required to accomplish the calibration procedure should be minimum. Further, the sensor should not require frequent recalibration. The term "drift" is commonly applied to denote the gradual loss in accuracy of the sensor with time and use, and which would necessitate recalibration.

6. *Reliability.* The sensor should possess a high reliability. It should not be subject to frequent failures during operation.

7. *Cost and ease of operation.* The cost to purchase, install, and operate the sensor should be as low as possible. Further, the ideal circumstance would be that the installation and operation of the device would not require a specially trained, highly skilled operator.

Table 6-2 Sensor devices used in robot Workcells†

Ammeter—(miscellaneous). Electrical meter used to measure electrical current.

Eddy current detectors—(proximity sensor). Device that emits an alternating magnetic field at the tip of a probe, which induces eddy currents in any conductive object in the range of the device. Can be used to indicate presence or absence of a conductive object.

Electrical contact switch—(touch sensor). Device in which an electrical potential is established between two objects, and when the potential becomes zero, this indicates contact between the two objects. Not a commercial device. Can be used to indicate presence or absence of a conductive object.

Infrared sensor—(proximity sensor). Transducer which measures temperatures by the infrared light emitted from the surface of an object. Can be used to indicate presence or absence of a hot object.

Limit switch—(touch sensor). Electrical on–off switch actuated by depressing a mechanical lever or button on the device. Can be used to measure presence or absence of an object.

Linear variable differential transformer—(miscellaneous). Electromechanical transducer used to measure linear or angular displacement.

Microswitch—(touch sensor). Small electrical limit switch (see limit switch). Can be used to indicate presence or absence of an object.

Ohmmeter—(miscellaneous). Meter used to measure electrical resistance.

Optical pyrometer—(proximity sensor, miscellaneous). Device used to measure high temperatures by sensing the brightness of an object's surface. Can be used to indicate presence or absence of a hot object.

Photometric sensors—(proximity sensor, miscellaneous). Various transducers used to sense light. Category includes photocells, photoelectric transducers, phototubes, photodiodes, phototransistors, and photoconductors. Can be used to indicate presence or absence of an object.

Piezoelectric accelerometer—(miscellaneous). Sensor used to indicate or measure vibration.

Potentiometer—(miscellaneous). Electrical meter used to measure voltage.

Pressure transducers—(miscellaneous). Various transducers used to indicate air pressure and other fluid pressures.

Radiation pyrometer—(proximity sensor, miscellaneous). Device used to measure high temperatures by sensing the thermal radiation emitting from the surface of an object. Can be used to indicate presence or absence of a hot object.

Strain gauge—(force sensor). Common transducer used to measure force, torque, pressure, and other related variables. Can be used to indicate force applied to grasp an object.

Thermistor—(miscellaneous). Device based on electrical resistance used to measure temperatures.

Thermocouple—(miscellaneous). Commonly used device used to measure temperatures. Based on the physical principle that a junction of two dissimilar metals will emit an emf which can be related to temperature.

Vacuum switches—(proximity sensor, miscellaneous). Device used to indicate negative air pressures. Can be used with a vacuum gripper to indicate presence or absence of an object.

Vision sensors—(vision system). Advanced sensor system used in conjunction with pattern recognition and other techniques to view and interpret events occurring in the robot workplace.

Voice sensors—(voice and speech recognition). Advanced sensor system used to communicate commands or information orally to the robot.

---

† Adapted from refs. 1, 2, and 6.

146

6-2 SENSORS IN ROBOTICS

The sensors used in robotics include a wide range of devices which can be divided into the following general categories:

1. Tactile sensors
2. Proximity and range sensors
3. Miscellaneous sensors and sensor-based systems
4. Machine vision systems

We will discuss the first three of these categories in the sections that follow. The fourth category, machine vision, is examined in the next chapter. Table 6-2 provides a listing of some of the common sensors that are applicable in robotic workcells.

6-3 TACTILE SENSORS

Tactile sensors are devices which indicate contact between themselves and some other solid object. Tactile sensing devices can be divided into two classes: touch sensors and force sensors. Touch sensors provide a binary output signal which indicates whether or not contact has been made with the object. Force sensors (also sometimes called stress sensors) indicate not only that contact has been made with the object but also the magnitude of the contact force between the two objects.

Touch Sensors

Touch sensors are used to indicate that contact has been made between two objects without regard to the magnitude of the contacting force. Included within this category are simple devices such as limit switches, microswitches, and the like. The simpler devices are frequently used in the design of interlock systems in robotics. For example, they can be used to indicate the presence or absence of parts in a fixture or at the pickup point along a conveyor. Another use for a touch-sensing device would be as part of an inspection probe which is manipulated by the robot to measure dimensions on a workpart. A robot with 6 degrees of freedom would be capable of accessing surfaces on the part that would be difficult for a three-axis coordinate measuring machine, the inspection system normally considered for such an inspection task. Unfortunately, the robot's accuracy would be a limiting factor in contact inspection work.

Force Sensors

The capacity to measure forces permits the robot to perform a number of tasks. These include the capability to grasp parts of different sizes in material handling, machine loading, and assembly work, applying the appropriate level of force for the given part. In assembly applications, force sensing could be used to determine if screws have become cross-threaded or if parts are jammed.

Force sensing in robotics can be accomplished in several ways. A commonly used technique is a "force-sensing wrist." This consists of a special load-cell mounted between the gripper and the wrist. Another technique is to measure the torque being exerted by each joint. This is usually accomplished by sensing motor current for each of the joint motors. Finally, a third technique is to form an array of force-sensing elements so that the shape and other information about the contact surface can be determined. We discuss these three possibilities in the paragraphs that follow.

Force-sensing wrist The purpose of a force-sensing wrist is to provide information about the three components of force (F_x, F_y, and F_z) and the three moments (M_x, M_y, and M_z) being applied at the end-of-the-arm. One possible construction of a force-sensing wrist is illustrated in Fig. 6-1. The device consists of a metal bracket fastened to a rigid frame. The frame is mounted to the wrist of the robot and the tool is mounted to the center of the bracket. The figure shows how the sensors might react to a moment applied to the bracket due to forces and moments on the tool.

Since the forces are usually applied to the wrist in combinations it is necessary to first resolve the forces and moments into their six components. This kind of computation can be carried out by the robot controller (if it has the required computational capability) or by a specialized amplifier designed for this purpose. Based on these calculations, the robot controller can obtain the required information of the forces and moments being applied at the wrist. This information could be used for a number of applications. As an example, an insertion operation (e.g., inserting a peg into a hole in an assembly application) requires that there are no side forces being applied to the peg. Another example is where the robot's end effector is required to follow along an edge or contour of an irregular surface. This is called force accommodation. With this technique, certain forces are set equal to zero while others are set equal to specific values. Using force accommodation, one could

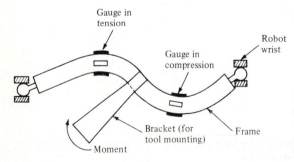

Gauge in tension

Gauge in compression

Robot wrist

Bracket (for tool mounting)

Frame

Moment

Figure 6-1 Possible configuration of sensing device used for a force sensing wrist, showing deflection (exaggerated) due to a moment about one of the axes.

command the robot to follow the edge or contour by maintaining a fixed velocity in one direction and fixed forces in other directions.

The robot equipped with a force-sensing wrist plus the proper computing capacity could be programmed to accomplish these kinds of applications. The procedure would begin by deciding on the desired force to be applied in each axis direction. The controller would perform the following sequence of operations, with the resulting offset force calculated as illustrated in Fig. 6-2:

1. Measure the forces at the wrist in each axis direction.
2. Calculate the force offsets required. The force offset in each direction is determined by subtracting the desired force from the measured force.
3. Calculate the torques to be applied by each axis to generate the desired force offsets at the wrist. These are moment calculations which take into account the combined effects of the various joints and links of the robot.
4. Then the robot must provide the torques calculated in step 3 so that the desired forces are applied in each direction.

Force-sensing wrists are usually very rigid devices so that they will not deflect undesirably while under load. When designing a force-sensing wrist there are several problems that may be encountered. The end-of-the-arm is often in a relatively hostile environment. This means that the device must be sufficiently rugged to withstand the environment. For example, it must be capable of tolerating an occasional crash of the robot arm. At the same time

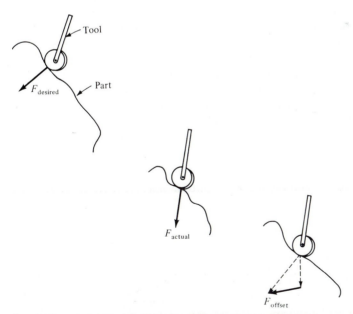

Figure 6-2 Force accommodation, showing how the required offset force would compensate for the difference between actual force and desired force.

the device must be sensitive enough to detect small forces. This design problem is usually solved by using overtravel limits. An overtravel limit is a physical stop designed to prevent the force sensor from deflecting so far that it would be damaged.

The calculations required to utilize a force-sensing wrist are complex and require considerable computation time. Also, for an arm traveling at moderate-to-high speeds, the level of control over the arm just as it makes contact with an object is limited by the dynamic performance of the arm. The momentum of the arm makes it difficult to stop its forward motion quickly enough to prevent a crash.

Joint sensing If the robot uses dc servomotors then the torque being exerted by the motors is proportional to the current flowing through the armature. A simple way to measure this current is to measure the voltage drop across a small precision resistor in series with the motor and power amplifier. This simplicity makes this technique attractive; however, measuring the joint torque has several disadvantages. First, measurements are made in joint space, while the forces of interest are applied by the tool and would be more useful if made in tool space. The measurements therefore not only reflect the forces being applied at the tool, but also the forces and torques required to accelerate the links of the arm and to overcome the friction and transmission losses of the joints. In fact, if the joint friction is relatively high (and it usually is), it will mask out the small forces being applied at the tool tip. One area where joint torque sensing shows promise of working well is with direct-drive robots. Direct-drive robots are a relatively new innovation in which the drive motors are located at the joints of the manipulator. In torque sensing, this configuration reduces the friction and transmission losses, and the problems of torque measurement which accompany these losses are thereby reduced. We will discuss direct-drive robots in more detail in Chap. Nineteen.

Tactile array sensors A tactile array sensor is a special type of force sensor composed of a matrix of force-sensing elements. The force data provided by this type of device may be combined with pattern recognition techniques to describe a number of characteristics about the impression contacting the array sensor surface. Among these characteristics are (1) the presence of an object, (2) the object's contact area, shape, location, and orientation, (3) the pressure and pressure distribution, and (4) force magnitude and location. Tactile array sensors can be mounted in the fingers of the robot gripper or attached to a work table as a flat touch surface. Figures 6-3 and 6-4 illustrate these two possible mountings for the sensor device.

The device is typically composed of an array of conductive elastomer pads. As each pad is squeezed its electrical resistance changes in response to the amount of deflection in the pad, which is proportional to the applied force. By measuring the resistance of each pad, information about the shape of the object against the array of sensing elements can be determined. The operation of a tactile array sensor (with an 8×8 matrix of pressure-sensitive pads) is

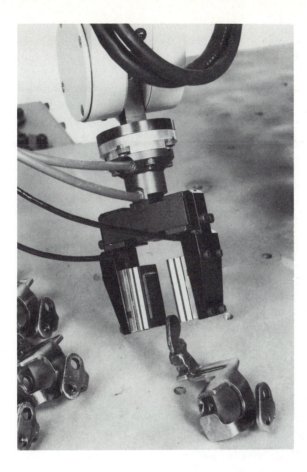

Figure 6-3 Tactile array sensor device mounted in a mechanical gripper. (Photo courtesy of Lord Corporation.)

pictured in Fig. 6-5. In the background is the CRT monitor display of the tactile impression made by the object placed on the surface of the sensor device. As the number of pads in the array is increased the resolution of the displayed information improves.

   Example 6-1 A possible use of a tactile array sensor is to measure the force and moments being applied to an object by the robot. A typical application of this capability is the case of a robot inserting a peg into a hole. Suppose the peg misses the hole slightly creating a binding action between the peg and the hole. Figure 6-6 illustrates the situation showing the likely pattern of forces on the tactile array sensor surface.

   The problems in designing and using a tactile array sensor include wear of the contacting surface and the possible need for multiplexing to minimize the number of interconnections if the number of sensors in the array is very large. Figure 6-4 shows the Lord Corporation's LTS 300 Series product which contains a large touch field surface (approximately 6 in. square) composed of

Figure 6-4 Tactile array sensor mounted on a flat work surface. (Photo courtesy of Lord Corporation.)

Figure 6-5 Tactile array sensor using 8×8 sensor array with display of tactile impression. (Photo courtesy of Lord Corporation.)

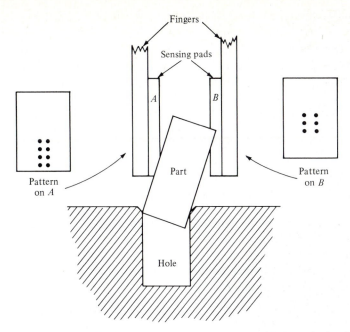

Figure 6-6 Possible binding action in an insertion task of Example 6-1 showing pattern of forces of the tactile array sensor surfaces.

an 80 by 80 pattern of sensor elements, making a total of 6400 sensitive sites. Each sensor has gray scale capability to allow force magnitude to be measured at each sensor location on the surface. The principal application foreseen for this product is for location and orientation of components in robotic assembly operations.

Research into potential materials for tactile sensors has lead to the development of a force-sensing skin of Polyvinylidene di-Fluoride. This is a piezoelectric material which means that it generates an output voltage when it is squeezed. Advantages of Polyvinylidene di-Fluoride include its availability in sheets only a few ten-thousandths of an inch and the fact that it is impervious to most chemicals. Another interesting feature of this material is that it is pyroelectric, meaning that it generates a voltage when heated. This feature suggests the possibility for designing a tactile sensor that provides simultaneous force and temperature sensing.

6-4 PROXIMITY AND RANGE SENSORS

Proximity sensors are devices that indicate when one object is close to another object. How close the object must be in order to activate the sensor is dependent on the particular device. The distances can be anywhere between several millimeters and several feet. Some of these sensors can also be used to measure the distance between the object and the sensor, and these devices are called range sensors. Proximity and range sensors would typically be located on the wrist or end effector since these are the moving parts of the robot. One practical use of a proximity sensor in robotics would be to detect the presence or absence of a workpart or other object. Another important application is for sensing human beings in the robot workcell. Range sensors would be useful for determining the location of an object (e.g., the workpart) in relation to the robot.

A variety of technologies are available for designing proximity and range sensors. These technologies include optical devices, acoustics, electrical field techniques (e.g., eddy currents and magnetic fields), and others. We will survey only a few of the possibilities in the following paragraphs.

Optical proximity sensors can be designed using either visible or invisible (infrared) light sources. Infrared sensors may be active or passive. The active sensors send out an infrared beam and respond to the reflection of the beam against a target. The infrared-reflectance sensor using an incandescent light source is a common device that is commercially available. The active infrared sensor can be used to indicate not only whether or not a part is present, but also the position of the part. By timing the interval from when the signal is sent and the echo is received, a measurement of the distance between the object and the sensor can be made. This feature is especially useful for locomotion and guidance systems. Passive infrared sensors are simply devices which detect the presence of infrared radiation in the environment. They are often utilized in security systems to detect the presence of bodies giving off heat within the range of the sensor. These sensor systems are effective at covering large areas in building interiors.

Another optical approach for proximity sensing involves the use of a collimated light beam and a linear array of light sensors. By reflecting the light beam off the surface of the object, the location of the object can be determined from the position of its reflected beam on the sensor array. This scheme is illustrated in Fig. 6-7. The formula for the distance between the object and

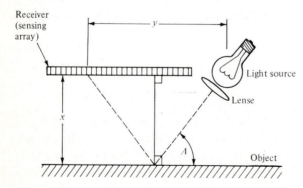

Figure 6-7 Scheme for a proximity sensor using reflected light against a sensor array.

the sensor is given as follows:

$$x = 0.5 y \tan(A)$$

where x = the distance of the object from the sensor

   y = the lateral distance between the light source and the reflected light beam against the linear array. This distance corresponds to the number of elements contained within the reflected beam in the sensor array

   A = the angle between the object and the sensor array as illustrated in Fig. 6-7

Use of this device in the configuration shown relies on the fact that the surface of the object must be parallel to the sensing array.

Acoustical devices can be used as proximity sensors. Ultrasonic frequencies (above 20,000 Hz) are often used in these devices because the sound is beyond the range of human hearing. One type of acoustical proximity sensor uses a cylindrical open-ended chamber with an acoustic emitter at the closed end of the chamber. The emitter sets up a pattern of standing waves in the cavity which is altered by the presence of an object near the open end. A microphone located in the wall of the chamber is used to sense the change in the sound pattern. This kind of device can also be used as a range sensor.

Proximity and range sensors based on the use of electrical fields are commercially available. Two of the types in this category and eddy-current sensors and magnetic field sensors. Eddy-current devices create a primary alternating magnetic field in the small region near the probe. This field induces eddy currents in an object placed in the region so long as the object is made of a conductive material. These eddy currents produce their own magnetic field which interacts with the primary field to change its flux density. The probe detects the change in the flux density and this indicates the presence of the object.

Magnetic field proximity sensors are relatively simple and can be made using a reed switch and a permanent magnet. The magnet can be made a part of the object being detected or it can be part of the sensor device. In either case, the device can be designated so that the presence of the object in the region of the sensor completes the magnetic circuit and activates the reed switch. This type of proximity sensor design is attractive because of its relative simplicity and because no external power supply is required for its operation.

6-5 MISCELLANEOUS SENSORS AND SENSOR-BASED SYSTEMS

The miscellaneous category covers the remaining types of sensors and transducers that might be used for interlocks and other purposes in robotic workcells. This category includes devices with the capability to sense variables such as temperature, pressure, fluid flow, and electrical properties. Many of the common transducers and sensors used for these variables are listed in Table 6-2.

An area of robotics research that might be included in this chapter is voice sensing or voice programming. Voice-programming systems can be used in robotics for oral communication of instructions to the robot. Voice sensing relies on the techniques of speech recognition to analyze spoken words uttered by a human and compare those words with a set of stored word patterns. When the spoken word matches the stored word pattern, this indicates that the robot should perform some particular actions which correspond to the word or series of words. We will explore this technology in more detail in Chap. Nineteen.

6-6 USES OF SENSORS IN ROBOTICS

The major uses of sensors in industrial robotics and other automated manufacturing systems can be divided into four basic categories:

1. Safety monitoring
2. Interlocks in workcell control
3. Part inspection for quality control
4. Determining positions and related information about objects in the robot cell

One of the important applications of sensor technology in automated manufacturing operations is safety or hazard monitoring which concerns the protection of human workers who work in the vicinity of the robot or other equipment. This subject of safety monitoring and other methods of ensuring worker safety in robotics is explored in Chap. Seventeen.

The second major use of sensor technology in robotics is to implement interlocks in workcell control. As mentioned in Chap. Two, interlocks are used to coordinate the sequence of activities of the different pieces of equipment in the workcell. In the execution of the robot program, there are certain elements of the work cycle whose completion must be verified before proceeding with the next element in the cycle. Sensors, often very simple devices, are utilized to provide this kind of verification. We will discuss interlocks and their uses in Chap. Eleven on workcell control.

The third category is quality control. Sensors can be used to determine a variety of part quality characteristics. Traditionally, quality control has been performed using manual inspection techniques on a statistical sampling basis. The use of sensors permits the inspection operation to be performed automatically on a 100 percent basis, in which every part is inspected. The limitation on the use of automatic inspection is that the sensor system can only inspect for a limited range of part characteristics and defects. For example, a sensor probe designed to measure part length cannot detect flaws in the part surface. Many applications of automated inspection are accomplished without the use of robotics. The reason for including this category in our discussion of robotic sensors is that robots are, in fact, often used to implement automatic inspection systems by means of sensors. We will discuss robot inspection applications in Chap. Fifteen.

The fourth major use of sensors in robotics is to determine the positions and other information about various objects in the workcell (e.g., workparts, fixtures, people, equipment, etc.). In addition to positional data about a particular object, other information required to properly execute the work cycle might include the object's orientation, color, size, and other characteristics. Reasons why this kind of data would need to be determined during the program execution include:

Workpart identification.
Random position and orientation of parts in the workcell.
Accuracy requirements in a given application exceed the inherent capabilities
    of the robot. Feedback information is required to improve the accuracy of
    the robot's positioning.

An example of workpart identification would be in a workcell in which the robot processes several types of workparts, each requiring a different sequence of actions by the robot. Each part presented to the robot would have to be properly identified so that the correct subroutine could be called for execution. This type of identification problem arises in automobile body spot-welding lines where the line is designed to weld several different body styles (e.g., coupes, sedans, wagons). Each welding robot along the line must execute the welding cycle for the particular body style at that station. Simple optical sensors are typically used to indicate the presence or absence of specific body style features in order to make the proper identification.

An example of the part position and orientation problem is where a robot would be required to pick up parts moving along a conveyor in random orientation and position, and place them into a fixture. To accomplish the task, the exact location of each part would have to be sensed as it came down the line. In addition, for the robot to use a mechanical gripper to grasp the particular workpart, the orientation of the part on the conveyor would have to be determined. All of this information would have to be processed by the workcell controller (or other computer) in real time in order to guide the robot in the execution of its programmed work cycle. Vision systems represent an important category of sensor system that might be employed to determine such characteristics as part location and orientation. We consider machine vision to be a topic in robotics which is significant enough to merit its own chapter. That chapter follows this one.

In some applications, the accuracy requirements in the application are more stringent than the inherent accuracy and repeatability of the robot. Certain assembly operations represent examples of this case. The robot is required to assemble two parts whose alignment must be very close, closer than the accuracy of the robot. One possible solution that might be used is a remote center compliance (RCC) device. The use of the RCC device in assembly will be described in Chap. Fifteen.

All four categories of sensor applications (safety monitoring, interlocks, inspection, and positional data) are instances where the sensor constitutes a component of a control system used in the robot work cell to accomplish some specific control function. That control system, in turn, is a component of a larger control system which we are calling the workcell control system. All of the control functions which takes place in the workcell are coordinated and regulated by this larger system. Our discussion of robot workcell control will resume in Chap. Eleven, after we have examined robot programming, an obvious prerequisite for workcell control.

REFERENCES

1. J. F. Engelberger, *Robotics in Practice*, AMACOM (American Management Association), New York, 1980, chap. 4.
2. M. P. Groover, *Automation, Production Systems, and Computer-Aided Manufacturing*, Prentice-Hall, Englewood Cliffs, NJ, 1980, chap. 11.
3. E. Kafrissen and M. Stephans, *Industrial Robots and Robotics*, Reston, Reston, VA, 1984, chap. 5.
4. Lord Corporation, *Lord Tactile Sensors*, Marketing/Technical Brochure, Cary, NC.
5. R. N. Stauffer, "Progress in Tactile Sensor Development," *Robotics Today*, June 1983, pp. 43–49.
6. L. L. Toepperwein, M. T. Blackman, et al., "ICAM Robotics Application Guide," *Technical Report AFWAL-TR-80-4042*, vol. II, Materials Laboratory, Air Force Wright Aeronautical Laboratories, Ohio, April 1980.

PROBLEMS

6-1 It is desired to design a safety monitoring system for a robot cell in which a robot loads an automatic production machine with parts arriving on a conveyor. The robot is large and it is considered dangerous for workers to wander into the work volume of the robot. Aside from the other safety precautions that might be taken to ensure the safety of the workers, the safety monitoring system will use one or more sensors to detect the presence of humans in the cell.

(*a*) Write the detailed "functional specifications" for the sensor system. That is, make up a list of the things the sensor system must do, and the ways in which it will have to operate.

(*b*) From the list of sensors in Table 6-2, select several alternative sensors that will satisfy the functional specifications, and compare its features against the specifications. For each sensor, explain how it will be configured in the safety monitoring system. Use sketches if necessary to illustrate the configuration.

(*c*) Select the best sensor alternative, and justify your selection.

6-2 Using the list of sensors in Table 6-2 in the text, describe several methods for determining the presence or absence of a metallic part in a fixture. Assume the dimensions of the part are: length = 127.0 mm (5 in.), width = 36.0 mm (1.4 in.), and thickness = 20.0 mm (0.8 in.). The fixture is a mechanical vise with two jaws for holding the part. For each alternative, make a sketch of how the sensor would be positioned relative to the part.

6-3 Using the list of sensors in Table 6-2 in the text, describe several methods for determining the presence or absence of a nonmetallic part in a fixture. Assume the dimensions of the part are: length = 127.0 mm (5 in.), width = 36.0 mm (1.4 in.), and thickness = 20.0 mm (0.8 in.). The fixture is a mechanical vise with two jaws for holding the part. For each alternative, make a sketch of how the sensor would be positioned relative to the part.

6-4 Using the list of sensors in Table 6-2 in the text, describe several sensors that might be used for determining the dimensions of a workpart of a flat worktable in a robot cell. Assume that the robot could be used to implement the sensor system, or that the sensor operation could be independent of the robot. Use sketches as necessary to illustrate your proposal.

6-5 Two differently sized parts share the same conveyor. Both parts are made of steel and have a square base which measures 46 mm (1.8 in.) on a side. One part is 33 mm (1.3 in.) in height while the other part is 25 mm (1.0 in.) high. Both parts travel with their bases down against the conveyor in the same orientation and position. (They are not randomly oriented on the conveyor.) It is required that the parts be picked up by a robot and placed in two separate bins depending on their size. Select a sensor system from Table 6-2 that could be used to distinguish between the two part sizes, and describe how the workplace would be set up to implement the operation (make sketches as needed). Before selecting the sensor, consider several alternative sensor candidates and compare their relative attributes for the application.

6-6 With reference to Prob. 6-5 above, discuss how the problem of distinguishing between the two part sizes would become more complicated if the parts were to travel down the conveyor in random orientations. How would you deal with the problem? Describe more than one possible approach.

6-7 A linear array of light sensors is to be used to determine the distance x in the setup illustrated in Fig. 6-7 in the text. If the angle A representing the orientation of the collimated light source is 60 degrees and the reflection of the object reaches a position $y = 42$ mm (1.65 in.), determine the value of x. What will be the effect on the accuracy of the measurement, if the surface of the object is not parallel to the linear sensor array as illustrated in Fig. 6-7?

CHAPTER
SEVEN *Lambertian Scattering*

MACHINE VISION

Machine vision (other names include computer vision and artificial vision) is an important sensor technology with potential applications in many industrial operations. Many of the current applications of machine vision are in inspection; however, it is anticipated that vision technology will play an increasingly significant role in the future of robotics.

Vision systems designed to be utilized with robot or manufacturing systems must meet two important criteria which currently limit the influx of vision systems to the manufacturing community. The first of these criteria is the need for a relatively low-cost vision system, typically under $30,000. The second criterion is the need for relatively rapid response time needed for robot or manufacturing applications, typically a fraction of a second.

Nevertheless, there has been a significant influx of vision systems into the manufacturing world. The systems are used to perform tasks which include selecting parts that are randomly oriented from a bin or conveyer, parts identification, and limited inspection. These capabilities are selectively used in traditional applications to reduce the cost of part and tool fixturing, and to allow the robot program to test for and adapt to limited variations in the environment.

Advances in vision technology for robotics are expected to broaden the capabilities of robotic vision systems to allow for vision-based guidance of the robot arm, complex inspection for close dimensional tolerances, and improved recognition and part location capabilities. These will result from the constantly reducing cost of computational capability, increased speed, and new and better algorithms currently being developed.

Predictions are that the field of computer vision will be one of the fastest growing commercial areas in the remainder of the twentieth century. Computer vision is a complex and multidisciplinary field and is still in its early stages of development. Advances in vision technology and related disciplines

160

are expected within the next decade which will permit applications not only in manufacturing, but also in photointerpretation, warehousing, robotic operations in hazardous environments, autonomous navigation, cartography, and medical image analysis.

The purpose of this chapter is to explore the field of machine vision as a fundamental sensor technology in robotics.

7-1 INTRODUCTION TO MACHINE VISION

Machine vision is concerned with the sensing of vision data and its interpretation by a computer. The typical vision system consists of the camera and digitizing hardware, a digital computer, and hardware and software necessary to interface them. This interface hardware and software is often referred to as a preprocessor. The operation of the vision system consists of three functions:

1. Sensing and digitizing image data
2. Image processing and analysis
3. Application

The relationships between the three functions are illustrated in the diagram of Fig. 7-1.

The sensing and digitizing functions involve the input of vision data by means of a camera focused on the scene of interest. Special lighting techniques are frequently used to obtain an image of sufficient contrast for later processing. The image viewed by the camera is typically digitized and stored in computer memory. The digital image is called a frame of vision data, and is frequently captured by a hardware device called a frame grabber. These devices are capable of digitizing images at the rate of 30 frames per second. The frames consist of a matrix of data representing projections of the scene sensed by the camera. The elements of the matrix are called picture elements, or pixels. The number of pixels are determined by a sampling process performed on each image frame. A single pixel is the projection of a small portion of the scene which reduces that portion to a single value. The value is a measure of the light intensity for that element of the scene. Each pixel intensity is converted into a digital value. (We are ignoring the additional complexities involved in the operation of a color video camera.) We will examine the details of machine vision sensing in Sec. 7-2.

The digitized image matrix for each frame is stored and then subjected to image processing and analysis functions for data reduction and interpretation of the image. These steps are required in order to permit the real-time application of vision analysis required in robotic applications. Typically an image frame will be thresholded to produce a binary image, and then various feature measurements will further reduce the data representation of the image. This data reduction can change the representation of a frame from several

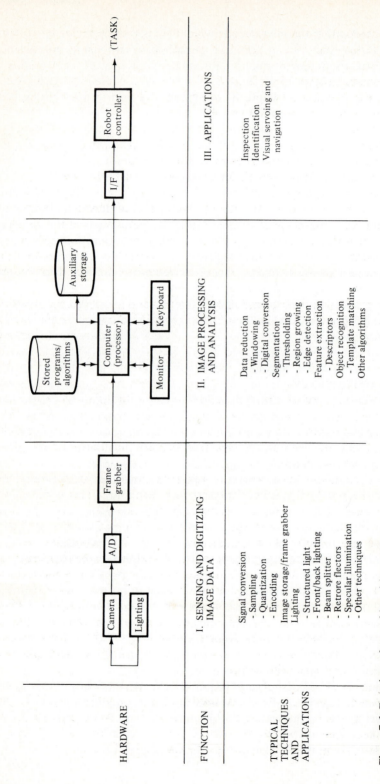

Figure 7-1 Functions of a machine vision system.

HARDWARE

FUNCTION

TYPICAL
TECHNIQUES
AND
APPLICATIONS

I. SENSING AND DIGITIZING
   IMAGE DATA

II. IMAGE PROCESSING
    AND ANALYSIS

III. APPLICATIONS

Camera

Lighting

A/D

Frame
grabber

Stored
programs/
algorithms

Computer
(processor)

Auxiliary
storage

Monitor

Keyboard

I/F

Robot
controller

(TASK)

Signal conversion
 - Sampling
 - Quantization
 - Encoding
Image storage/frame grabber
Lighting
 - Structured light
 - Front/back lighting
 - Beam splitter
 - Retrore flectors
 - Specular illumination
 - Other techniques

Data reduction
 - Windowing
 - Digital conversion
Segmentation
 - Thresholding
 - Region growing
 - Edge detection
Feature extraction
 - Descriptors
Object recognition
 - Template matching
Other algorithms

Inspection
Identification
Visual servoing and
  navigation

hundred thousand bytes of raw image data to several hundred bytes of feature value data. The resultant feature data can be analyzed in the available time for action by the robot system.

Various techniques to compute the feature values can be programmed into the computer to obtain feature descriptors of the image which are matched against previously computed values stored in the computer. These descriptors include shape and size characteristics that can be readily calculated from the thresholded image matrix. Some of the important techniques used in vision system image processing and analysis will be discussed in Sec. 7-3.

To accomplish image processing and analysis, the vision system frequently must be trained. In training, information is obtained on prototype objects and stored as computer models. The information gathered during training consists of features such as the area of the object, its perimeter length, major and minor diameters, and similar features. During subsequent operation of the system, feature values computed on unknown objects viewed by the camera are compared with the computer models to determine if a match has occurred. Section 7-4 will discuss training of a vision system.

The third function of a machine vision system is the applications function. The current applications of machine vision in robotics include inspection, part identification, location, and orientation. Research is ongoing in advanced applications of machine vision for use in complex inspection, guidance, and navigation. In Sec. 7-5 we discuss these applications.

Vision systems can be classified in a number of ways. One obvious classification is whether the system deals with a two-dimensional or three-dimensional model of the scene. Some vision applications require only a two-dimensional analysis. Examples of two-dimensional vision problems include checking the dimensions of a part or verifying the presence of components on a subassembly.

Many two-dimensional vision systems can operate on a binary image which is the result of a simple thresholding technique. This is based on an assumed high contrast between the object (s) and the background. The desired contrast can often be accomplished by using a controlled lighting system.

Three-dimensional vision systems may require special lighting techniques and more sophisticated image processing algorithms to analyze the image. Some systems require two cameras in order to achieve a stereoscopic view of the scene, while other three-dimensional systems rely on the use of structured light and optical triangulation techniques with a single camera. An example of a structured light system is one that projects a controlled band of light across the object. The light band is distorted according to the three-dimensional shape of the object. The vision system sees the distorted band and utilizes triangulation to deduce the shape.

Another way of classifying vision systems is according to the number of gray levels (light intensity levels) used to characterize the image. In a binary image the gray level values are divided into either of two categories, black or white. Other systems permit the classification of each pixel's gray level into various levels, the range of which is called a gray scale.

7-2 THE SENSING AND DIGITIZING FUNCTION IN MACHINE VISION

Our description of the typical machine vision system in the preceding section identified three functions: sensing and digitizing, image processing and analysis, and application. This and the following sections in the chapter will elaborate on these functions. The present section is concerned with the sensing and digitizing aspects of machine vision.

Image sensing requires some type of image formation device such as a camera and a digitizer which stores a video frame in the computer memory. We divide the sensing and digitizing functions into several steps. The initial step involves capturing the image of the scene with the vision camera. The image consists of relative light intensities corresponding to the various portions of the scene. These light intensities are continuous analog values which must be sampled and converted into digital form.

The second step, digitizing, is achieved by an analog-to-digital (A/D) converter. The A/D converter is either a part of a digital video camera or the front end of a frame grabber. The choice is dependent on the type of hardware in the system. The frame grabber, representing the third step, is an image storage and computation device which stores a given pixel array. The frame grabber can vary in capability from one which simply stores an image to significant computation capability. In the more powerful frame grabbers, thresholding, windowing, and histogram modification calculations can be carried out under computer control. The stored image is then subsequently processed and analyzed by the combination of the frame grabber and the vision controller.

Imaging Devices

There are a variety of commercial imaging devices available. Camera technologies available include the older black-and-white vidicon camera, and the newer, second-generation, solid state cameras. Solid state cameras used for robot vision include charge-coupled devices (CCD), charge injection devices (CID), and silicon bipolar sensor cameras. For our purposes, we review two such devices in this subsection, the vidicon camera and the charge-coupled device (CCD).

Figure 7-2 illustrates the vidicon camera. In the operation of this system, the lens forms an image on the glass faceplate of the camera. The faceplate has an inner surface which is coated with two layers of material. The first layer consists of a transparent signal electrode film deposited on the faceplate of the inner surface. The second layer is a thin photosensitive material deposited over the conducting film. The photosensitive layer consists of a high density of small areas. These areas are similar to the pixels mentioned previously. Each area generates a decreasing electrical resistance in response to increasing illumination. A charge is created in each small area upon illumination. An electrical charge pattern is thus generated corresponding to the image formed

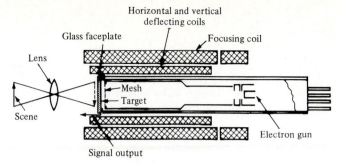

Horizontal and vertical
deflecting coils

Glass faceplate

Focusing coil

Lens

Mesh

Target

Scene

Electron gun

Signal output

Figure 7-2 Cross section of a videcon tube and its associated deflection and focusing tube. (Reprinted with permission of McGraw-Hill, Inc. [10].)

on the faceplate. The charge accumulated for an area is a function of the intensity of impinging light over a specified time.

Once a light sensitive charge is built up, this charge is read out to produce a video signal. This is accomplished by scanning the photosensitive layer by an electron beam. The scanning is controlled by a deflection coil mounted along the length of the tube. For an accumulated positive charge the electron beam deposits enough electrons to neutralize the charge. An equal number of electrons flow to cause a current at the video signal electrode. The magnitude of the signal is proportional to the light intensity and the amount of time with which an area is scanned. The current is then directed through a load resistor which develops a signal voltage which is further amplified and analyzed. Raster scanning eliminates the need to consider the time at each area by making the scan time the same for all areas. Only the intensity of the impinging light is considered. In the United States, the entire faceplate is scanned approximately 30 frames per second. The European standard is 25 frames per second. Raster scanning is typically done by scanning the electron beam from left to right and top to bottom. The process is such that the system is designed to start the integration with zero accumulated charge. For the fixed scan time the charge accumulated is proportional to the intensity of that portion of the image being considered. The output of the camera is a continuous voltage signal for each line scanned. The voltage signal for each scan line is subsequently sampled and quantized resulting in a series of sampled voltages being stored in digital memory. This analog-to-digital conversion process for the complete screen (horizontal and vertical) results in a two-dimensional array of picture elements (pixels). Typically, a single pixel is quantized to between 6 and 8 bits by the A/D converter.

Another approach to obtaining a digitized image is by use of the charge-coupled device (CCD). In this technology, the image is projected by a video camera onto the CCD which detects, stores, and reads out the accumulated charge generated by the light on each portion of the image. Light detection occurs through the absorption of light on a photoconductive substrate (e.g., silicon). Charges accumulate under positive control electrodes in isolated wells due to voltages applied to the central electrodes. Each isolated well represents one pixel and can be transferred to output storage registers by varying the voltages on the metal control electrodes. This is illustrated in Figs. 7-3(a) and (b).

Figure 7-4 indicates one type of CCD imager. Charges are accumulated for the time it takes to complete a single image after which they are transferred line by line into a storage register. For example, register A in Fig. 7-4

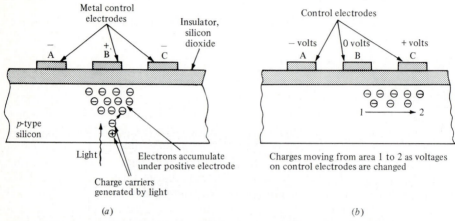

(a)

(b)

Figure 7-3 Basic principle of charge-coupled device. (a) accumulation of an electron charge in a pixel element (b) movement of accumulated charge through the silicon by changing the voltages on the electrodes A, B, and C. (Reprinted with permission of McGraw-Hill, Inc. [10].)

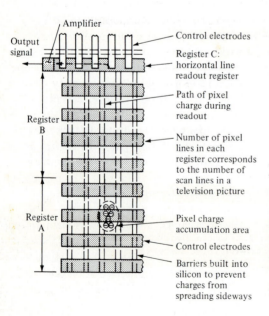

Figure 7-4 One type of charge-coupled-device imager. Register A accumulates the pixel charges produced by photoconductivity generated by the light image. The B register stores the lines of pixel charges and transfers each line in turn into register C. Register C reads out the charges laterally as shown into the amplifier. (Reprinted with permission of McGraw-Hill, Inc. [10].)

accumulates the pixel charge produced by the light image. Once accumulated for a single picture, the charges are transferred line by line to register B. The pixel charges are read out line by line through a horizontal register C to an output amplifier. During readout, register A is accumulating new pixel elements. The complete cycle is repeated approximately every $\frac{1}{60}$th of a second.

Lighting Techniques

An essential ingredient in the application of machine vision is proper lighting. Good illumination of the scene is important because of its effect on the level of complexity of image-processing algorithms required. Poor lighting makes the

Table 7-1 Illumination techniques

Technique	Function/use
A. Front light source	
1. Front illumination	Area flooded such that surface is defining feature of image
2. Specular illumination (dark field)	Used for surface defect recognition (background dark)
3. Specular illumination (light field)	Used for surface defect recognition; camera in-line with reflected rays (background light)
4. Front imager	Structured light applications; imaged light superimposed on object surface—light beam displaced as function of thickness
B. Back light source	
1. Rear illumination (lighted field)	Uses surface diffusor to silhouette features; used in parts inspection and basic measurements
2. Rear illumination (condenser)	Produces high-contrast images; useful for high magnification application
3. Rear illumination (collimator)	Produces parallel light ray source such that features of object do not lie in same plane
4. Rear offset illumination	Useful to produce feature highlights when feature is in transparent medium
C. Other miscellaneous devices	
1. Beam splitter	Transmits light along same optical axis as sensor; advantage is that it can illuminate difficult-to-view objects
2. Split mirror	Similar to beam splitter but more efficient with lower intensity requirements
3. Nonselective redirectors	Light source is redirected to provide proper illumination
4. Retroreflector	A device that redirects incident rays back to sensor; incident angle capable of being varied; provides high contrast for object between source and reflector
5. Double density	A technique used to increase illumination intensity at sensor; used with transparent media and retroreflector.

task of interpreting the scene more difficult. Proper lighting techniques should provide high contrast and minimize specular reflections and shadows unless specifically designed into the system.

The basic types of lighting devices used in machine vision may be grouped into the following categories:

1. Diffuse surface devices. Examples of diffuse surface illuminators are the typical fluorescent lamps and light tables.
2. Condenser projectors. A condenser projector transforms an expanding light source into a condensing light source. This is useful in imaging optics.
3. Flood or spot projectors. Flood lights and spot lights are used to illuminate surface areas.
4. Collimators. Collimators are used to provide a parallel beam of light on the subject.
5. Imagers. Imagers such as slide projectors and optical enlargers form an image of the target at the object plane.

Various illumination techniques have been developed to use these lighting devices. Many of these techniques are presented in Table 7-1. The purpose of these techniques is to direct the path of light from the lighting device to the camera so as to display the subject in a suitable manner to the camera.

As shown in Table 7-1, there are two basic illumination techniques used in machine vision: front lighting and back lighting. Front lighting simply means that the light source is on the same side of the scene as the camera. Accordingly, reflected light is used to create the image viewed by the camera. In back lighting the light source is directed at the camera and is located behind the objects of interest. The image seen by the camera is a silhouette of the object under study. Back lighting is suitable for applications in which a silhouette of the object is sufficient for recognition or where there is a need to obtain relevant measurements. The table also lists other miscellaneous techniques that may be used to provide illumination.

Analog-to-Digital Signal Conversion

For a camera utilizing the vidicon tube technology it is necessary to convert the analog signal for each pixel into digital form. The analog-to-digital (A/D) conversion process involves taking an analog input voltage signal and producing an output that represents the voltage signal in the digital memory of a computer. A/D conversion consists of three phases: sampling, quantization, and encoding.

Sampling A given analog signal is sampled periodically to obtain a series of discrete-time analog signals. This process is illustrated in Fig. 7-5. By setting a specified sampling rate, the analog signal can be approximated by the sampled digital outputs. How well we approximate the analog signal is determined by the sampling rate of the A/D converter. The sampling rate should be at least

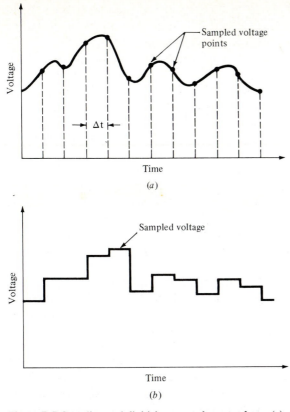

Figure 7-5 Sampling and digitizing an analog waveform. (a) analog waveform indicating sampling interval, t, and sampled voltage points. (b) digital approximation to analog signal.

twice the highest frequency in the video signal if we wish to reconstruct that signal exactly.[14]

Example 7-1 Consider a vision system using a vidicon tube. An analog video signal is generated for each line of the 512 lines comprising the faceplate. The sampling capability of the A/D converter is 100 nanoseconds (100×10^{-9} s). This is the cycle time required to complete the A/D conversion process for one pixel. Using the American standard of 33.33 milliseconds ($\frac{1}{30}$ s) to scan the entire faceplate consisting of 512 lines, determine the sampling rate and the number of pixels that can be processed per line.

The cycle time per pixel is limited by the A/D conversion process to
$$100 \times 10^{-9} \text{ s}$$

$$= 0.1 \times 10^{-6} \text{ s/pixel}$$

The scanning rate for the 512 lines in the faceplate is $\frac{1}{30}$ s

$$= 33.33 \times 10^{-3} \text{ s}$$

Accordingly, the scanning rate for each line is

$$= (33.33 \times 10^{-3} \text{ s})/512 \text{ lines}$$

$$= 65.1 \times 10^{-6} \text{ s/line}$$

The number of pixels that can be processed per line is therefore

$$= \frac{65.1 \times 10^{-6} \text{ s/line}}{0.1 \times 10^{-6} \text{ s/pixel}}$$

$$= 651 \text{ pixels/line}$$

In practice, an allowance would have to be made for the time required when the electron beam is shut off during its raster from one line to the next. This dead time would decrease the number of pixels used in the vidicon system. For example, the number of pixels that the system could handle might be reduced from 651 pixels per line to 512 pixels per line, which is similar to the number of pixels per line used in home television. For a 512×512 pixel image, there are a total of 262,144 pixels to consider during the faceplate scanning period. Each pixel must be processed and some type of image-processing function must be carried out during the sampling period. This results in substantial demand on the digital computer performing the function. The requirement on the capability of the vision system computer has been one of the limiting factors in the development of machine vision. Systems with a smaller number of pixels ($128 \times 128 = 16,384$ pixels or $256 \times 256 = 65,536$ pixels) impose lower computational requirements; however, the image resolution of these systems is much lower than for systems possessing a greater pixel density. For a given video signal representing one line, the number of samples taken determines the horizontal resolution of the imaging system. The total number of lines determines the vertical resolution.

Quantization Each sampled discrete-time voltage level is assigned to a finite number of defined amplitude levels. These amplitude levels correspond to the gray scale used in the system. The predefined amplitude levels are characteristic to a particular A/D converter and consist of a set of discrete values of voltage levels. The number of quantization levels is defined by

$$\text{number of quantization levels} = 2^n$$

where n is the number of bits of the A/D converter. A large number of bits enables a signal to be represented more precisely. For example, an 8-bit converter would allow us to quantize at $2^8 = 256$ different values whereas 4 bits would allow only $2^4 = 16$ different quantization levels.

Encoding The amplitude levels that are quantized must be changed into digital code. This process, termed encoding, involves representing an amplitude level by a binary digit sequence. The ability of the encoding process to distinguish between various amplitude levels is a function of the spacing of each quantization level. Given the full-scale range of an analog video signal, the spacing of each level would be defined by

$$\text{quantization level spacing} = \frac{\text{full-scale range}}{2^n}$$

The quantization error resulting from the quantization process can be defined as

$$\text{quantization error} = \pm\tfrac{1}{2}(\text{quantization level spacing})$$

Example 7-2 A continuous video voltage signal is to be converted into a discrete signal. The range of the signal after amplification is 0 to 5 V. The A/D converter has an 8-bit capacity. Determine the number of quantization levels, the quantization level spacing, the resolution, and the quantization error.

For an 8-bit capacity, the number of quantization levels is $2^8 = 256$. The A/D converter resolution $= \frac{1}{256} = 0.0039$ or 0.39 percent. For the 5-V range,

$$\text{quantization level spacing} = \frac{(5\text{ V})}{(2^8)} = 0.0195\text{ V}$$

$$\text{quantization error} = \pm\tfrac{1}{2}(0.0195\text{ V}) = 0.00975\text{ V}$$

To represent the voltage signal in binary form involves the process of encoding. This is accomplished by assigning the sequence of binary digits to represent increasing quantization levels.

Example 7-3 For the A/D converter of Example 7-2, indicate how the voltage signal might be indicated in binary form.

Of the 8 bits available, we can use these bits to represent increasing quantization levels as follows:

Voltage range, V	Binary number	Gray scale
0–0.0195	0000 0000	0 (black)
0.0195–0.0390	0000 0001	1 (dark gray)
0.0390–0.0585	0000 0010	2
⋮	⋮	⋮
4.9610–4.9805	1111 1110	254 (light gray)
4.9805–5.0	1111 1111	255 (white)

If we had a voltage signal which, for example, varied over a ±5-V range, we could then assign the first bit to indicate polarity and use the succeeding 7 bits to identify specific voltage ranges. The number of quantization levels would be $2^7 = 128$ and the quantization level spacing would be 5 V divided by 128 or 0.039 V.

Image Storage

Following A/D conversion, the image is stored in computer memory, typically called a frame buffer. This buffer may be part of the frame grabber or in the computer itself. Various techniques have been developed to acquire and access digital images. Ideally, one would want to acquire a single frame of data in real time. The frame grabber is one example of a video data acquisition device that will store a digitized picture and acquire it in $\frac{1}{30}$s. Digital frames are typically quantized to 8 bits per pixel. However a 6-bit buffer is adequate since the average camera system cannot produce 8 bits of noise-free data. Thus the lower-order bits are dropped as a means of noise cleaning. In addition the human eye can only resolve about $2^6 = 64$ gray levels. A combination of row and column counters are used in the frame grabber which are synchronized with the scanning of the electron beam in the camera. Thus, each position on the screen can be uniquely addressed. To read the information stored in the frame buffer, the data is "grabbed" via a signal sent from the computer to the address corresponding to a row–column combination. Such frame grabber techniques have become extremely popular and are used frequently in vision systems.

7-3 IMAGE PROCESSING AND ANALYSIS

The discussion in the preceding section described how images are obtained, digitized, and stored in a computer. For use of the stored image in industrial applications, the computer must be programmed to operate on the digitally stored image. This is a substantial task considering the large amount of data that must be analyzed. Consider an industrial vision system having a pixel density of 350 pixels per line and 280 lines (a total of 98,000 picture elements), and a 6-bit register for each picture element to represent various gray levels; this would require a total of $98,000 \times 6 = 588,000$ bits of data for each $\frac{1}{30}$ s. This is a formidable amount of data to be processed in a short period of time and has led to various techniques to reduce the magnitude of the image-processing problem. These techniques include:

1. Image data reduction
2. Segmentation
3. Feature extraction
4. Object recognition

We will discuss these techniques of image data analysis in the following subsections.

Image Data Reduction

In image data reduction, the objective is to reduce the volume of data. As a preliminary step in the data analysis, the following two schemes have found common usage for data reduction:

1. Digital conversion
2. Windowing

The function of both schemes is to eliminate the bottleneck that can occur from the large volume of data in image processing.

Digital conversion reduces the number of gray levels used by the machine vision system. For example, an 8-bit register used for each pixel would have $2^8 = 256$ gray levels. Depending on the requirements of the application, digital conversion can be used to reduce the number of gray levels by using fewer bits to represent the pixel light intensity. Four bits would reduce the number of gray levels to 16. This kind of conversion would significantly reduce the magnitude of the image-processing problem.

Example 7-4 For an image digitized at 128 points per line and 128 lines, determine (a) the total number of bits to represent the gray level values required if an 8 bit A/D converter is used to indicate various shades of gray, and (b) the reduction in data volume if only black and white values are digitized.

(*a*) For gray scale imaging with $2^8 = 256$ levels of gray

$$\text{number of bits} = 128 \times 128 \times 256 = 4,194,304 \text{ bits}$$

(*b*) For black and white (binary bit conversion)

$$\text{number of bits} = 128 \times 128 \times 2 = 16,384 \text{ bits}$$

$$\text{reduction in data volume} = 4,194,304 - 16,384$$

$$= 4,177,920 \text{ bits}$$

Windowing involves using only a portion of the total image stored in the frame buffer for image processing and analysis. This portion is called the window. For example, for inspection of printed circuit boards, one may wish to inspect and analyze only one component on the board. A rectangular window is selected to surround the component of interest and only pixels within the window are analyzed. The rationale for windowing is that proper recognition of an object requires only certain portions of the total scene.

Segmentation

Segmentation is a general term which applies to various methods of data reduction. In segmentation, the objective is to group areas of an image having similar characteristics or features into distinct entities representing parts of the image. For example, boundaries (edges) or regions (areas) represent two natural segments of an image. There are many ways to segment an image.

Three important techniques that we will discuss are:

1. Thresholding
2. Region growing
3. Edge detection

In its simplest form, *thresholding* is a binary conversion technique in which each pixel is converted into a binary value, either black or white. This is accomplished by utilizing a frequency histogram of the image and establishing what intensity (gray level) is to be the border between black and white. This is illustrated for an image of an object as shown in Fig. 7-6. Figure 7-6(*a*) shows a regular image with each pixel having a specific gray tone out of 256 possible gray levels. The histogram of Fig. 7-6(*b*) plots the frequency (number of pixels) versus the gray level for the image. For histograms that are bimodal in shape, each peak of the histogram represents either the object itself or the background upon which the object rests. Since we are trying to differentiate between the object and background, the procedure is to establish a threshold

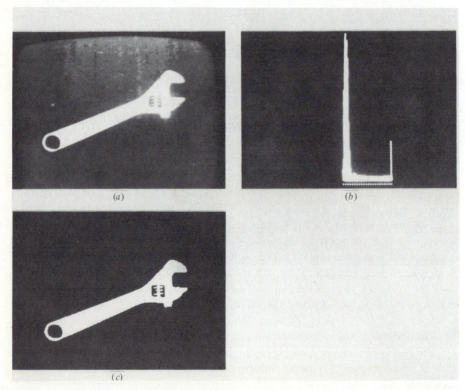

(*a*)

(*b*)

(*c*)

Figure 7-6 Obtaining a binary image by thresholding. (a) image of object with all gray-levels present, (b) histogram of image (c) binary image of object after thresholding. (Photos courtesy of Robotics Laboratory, Lehigh University.)

(typically between the two peaks) and assign, for example, a binary bit 1 for the object and 0 for the background. The outcome of this thresholding technique is illustrated in the binary-digitized image of Fig. 7-6(c). To improve the ability to differentiate, special lighting techniques must often be applied to generate a high contrast.

It should be pointed out that the above method of using a histogram to determine a threshold is only one of a large number of ways to threshold an image. It is however the method used by many of the commercially available robot vision systems today. Such a method is said to use a global threshold for the entire image. In some cases this is not possible and a local thresholding method as described below may be employed.

When it is not possible to find a single threshold for an entire image (for example, if many different objects occupy the same scene, each having different levels of intensity), one approach is to partition the total image into smaller rectangular areas and determine the threshold for each window being analyzed.

Thresholding is the most widely used technique for segmentation in industrial vision applications. The reasons are that it is fast and easily implemented and that the lighting is usually controllable in an industrial setting.

Once thresholding is established for a particular image, the next step is to identify particular areas associated with objects within the image. Such regions usually possess uniform pixel properties computed over the area. The pixel properties may be multidimensional; that is, there may be more than a single attribute that can be used to characterize the pixel (e.g., color and light intensity). We will avoid this complication and confine our discussion to single pixel attributes (light intensity) of a region.

Region growing is a collection of segmentation techniques in which pixels are grouped in regions called grid elements based on attribute similarities. Defined regions can then be examined as to whether they are independent or can be merged to other regions by means of an analysis of the difference in their average properties and spatial connectiveness. For instance, consider an image as depicted in Fig. 7-7(a). To differentiate between the objects and the background, assign 1 for any grid element occupied by an object and 0 for background elements. It is common practice to use a square sampling grid with pixels spaced equally along each side of the grid. For the two-dimensional image of a key as shown, this would give the pattern indicated in Fig. 7-7(b). This technique of creating "runs" of 1s and 0s is often used as a first-pass analysis to partition the image into identifiable segments or "blobs." Note that this simple procedure did not identify the hole in the key of Fig. 7-7(a). This could be resolved by decreasing the distance between grid points and increasing the accuracy with which the original image is represented.

For a simple image such as a dark blob on a light background, a runs technique can provide useful information. For more complex images, this technique may not provide an adequate partition of an image into a set of meaningful regions. Such regions might contain pixels that are connected to each other and have similar attributes, for example, gray level. A typical

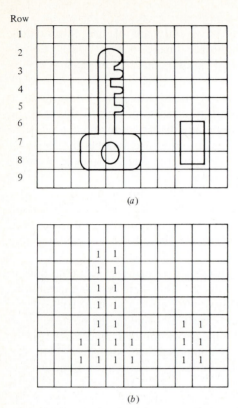

(a)

(b)

Figure 7-7 Image segmentation (a) image pattern with grid. (b) segmented image after runs test.

region-growing technique for complex images could have the following procedure:

1. Select a pixel that meets a criterion for inclusion in a region. In the simplest case this could mean select white pixel and assign a value of 1.
2. Compare the pixel selected with all adjacent pixels. Assign an equivalent value to adjacent pixels if an attribute match occurs.
3. Go to an equivalent adjacent pixel and repeat process until no equivalent pixels can be added to the region.

This simple procedure of "growing" regions around a pixel would be repeated until no new regions can be added for the image.

The region growing segmentation technique described here is applicable when images are not distinguishable from each other by straight thresholding or edge detection techniques. This sometimes occurs when lighting of the scene cannot be adequately controlled. In industrial robot vision systems, it is common practice to consider only edge detection or simple thresholding. This is due to the fact that lighting can be a controllable factor in an industrial setting and hardware/computational implementation is simpler.

Edge detection considers the intensity change that occurs in the pixels at

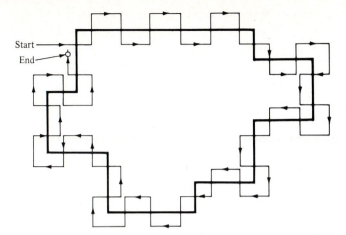

Figure 7-8 Edge following procedure to detect the edge of a binary image.

the boundary or edges of a part. Given that a region of similar attributes has been found but the boundary shape is unknown, the boundary can be determined by a simple edge following procedure. This can be illustrated by the schematic of a binary image as shown in Fig. 7-8. For the binary image, the procedure is to scan the image until a pixel within the region is encountered. For a pixel within the region, turn left and step, otherwise, turn right and step. The procedure is stopped when the boundary is traversed and the path has returned to the starting pixel. The contour-following procedure described can be extended to gray level images.

Feature Extraction

In machine vision applications, it is often necessary to distinguish one object from another. This is usually accomplished by means of features that uniquely characterize the object. Some features of objects that can be used in machine vision include area, diameter, and perimeter. A feature, in the context of vision systems, is a single parameter that permits ease of comparison and identification. A list of some of the features commonly used in vision applications is given in Table 7-2. The techniques available to extract feature values for two-dimensional cases can be roughly categorized as those that deal with boundary features and those that deal with area features. The various features can be used to identify the object or part and determine the part location and/or orientation.

The region-growing procedures described before can be used to determine the area of an object's image. The perimeter or boundary that encloses a specific area can be determined by noting the difference in pixel intensity at the boundary and simply counting all the pixels in the segmented region that are adjacent to pixels not in the region; that is, on the other side of the

Table 7-2 Basic features and measures for object identification for two-dimensional objects

Gray level (maximum, average, or minimum)

Area

Perimeter length

Diameter

Minimum enclosing rectangle

Center of gravity—For all pixels (n) in a region where each pixel is specified by (x, y) coordinates, the x and y coordinates of the center of gravity are defined as

$$\text{C.G.}_x = \frac{1}{n}\sum_x x$$

$$\text{C.G.}_y = \frac{1}{n}\sum_y y$$

Eccentricity—A measure of "elongation." Several measures exist of which the simplest is

$$\text{Eccentricity} = \frac{\text{maximum chord length } A}{\text{maximum chord length } B}$$

where maximum chord length B is chosen perpendicular to A.

Aspect ratio—The length-to-width ratio of a boundary rectangle which encloses the object. One objective is to find the rectangle which gives the minimum aspect ratio.

Thinness—This is a measure of how thin an object is. Two definitions are in use

$$(a)\ \text{Thinness} = \frac{(\text{perimeter})^2}{\text{area}}$$

This is also referred to as compactness.

$$(b)\ \text{Thinness} = \frac{\text{diameter}}{\text{area}}$$

The diameter of an object, regardless of its shape, is the maximum distance obtainable for two points on the boundary of an object.

Holes—Number of holes in the object.

Moments—Given a region, R, and coordinates of the points (x, y) in or on the boundary of the region, the pqth order moment of the image of the region is given as

$$M_{pq} = \sum_{x,y} x^p y^q$$

boundary. An important objective in selecting these features is that the features should not depend on position or orientation. The vision system should not be dependent on the object being presented in a known and fixed relationship to the camera.

The preceding measures provide some basic methods to analyze images in a two-dimensional plane. Various other measures exist for the three-dimensional case as well. To illustrate some of the two-dimensional definitions and measures, the following example is provided.

Example 7-5 Consider the schematic of the image in Fig. 7-9. Determine the area, the minimum aspect ratio, the diameter, the centroid, and the thinness measures of the image.

For the image, the minimum boundary rectangle is as shown. For ease of calculation, the origin is translated to O' with x', y' coordinates. The area may be determined from the moment, $M_{o'o'}$ as

$$M_{o'o'} = \sum_{x', y'} x'y' = 24 \text{ pixels}$$

The minimum aspect ratio would use the minimum boundary rectangle and would simply be

$$\text{minimum aspect ratio} = \frac{\text{length}}{\text{width}} = \frac{9}{4}$$

with the diameter = 9 pixels. The centroid position for the $n = 24$ pixels in the region would be calculated from

$$\text{C.G.}_{x'} = \frac{1}{n} \sum_{x'} x'$$

$$\text{C.G.}_{y'} = \frac{1}{n} \sum_{y'} y'$$

or

$$\text{C.G.}_{x'} = \tfrac{1}{24}[4(\tfrac{1}{2}) + 4(\tfrac{3}{2}) + 4(\tfrac{5}{2}) + 2(\tfrac{7}{2}) + \cdots + 2(\tfrac{17}{2})]$$

$$= \tfrac{1}{24}(90) = \tfrac{15}{4} \text{ units}$$

$$\text{C.G.}_{y'} = \tfrac{1}{24}[3(\tfrac{1}{2}) + 9(\tfrac{3}{2}) + 9(\tfrac{5}{2}) + 3(\tfrac{7}{2})]$$

$$= \tfrac{96}{48} = 2 \text{ units}$$

where pixel location is taken at the midpoint of each pixel cell. (C.G. =

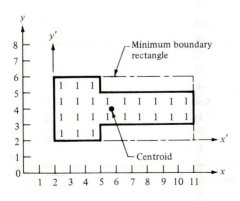

Figure 7-9 Schematic of pixel pattern for Example 7-5.

center of gravity.) The centroid is indicated in Fig. 7-9. Of the two thinness measures defined, the calculations would result in

$$\text{Compactness} = \frac{(\text{perimeter})^2}{\text{area}} = \frac{26^2}{24} = 28.17$$

$$\text{Thinness} = \frac{\text{diameter}}{\text{area}} = \frac{9}{24} = \frac{3}{8}$$

Object Recognition

The next step in image data processing is to identify the object the image represents. This identification problem is accomplished using the extracted feature information described in the previous subsection. The recognition algorithm must be powerful enough to uniquely identify the object. Object recognition techniques used in industry today may be classified into two major categories:

1. Template-matching techniques
2. Structural techniques

Template-matching techniques are a subset of the more general statistical pattern recognition techniques that serve to classify objects in an image into predetermined categories. The basic problem in template matching is to match the object with a stored pattern feature set defined as a model template. The model template is obtained during the training procedure in which the vision system is programmed for known prototype objects. These techniques are applicable if there is not a requirement for a large number of model templates. The procedure is based on the use of a sufficient number of features to minimize the frequency of errors in the classification process. The features of the object in the image (e.g., its area, diameter, aspect ratio, etc.) are compared to the corresponding stored values. These values constitute the stored template. When a match is found, allowing for certain statistical variations in the comparison process, then the object has been properly classified.

Structural techniques of pattern recognition consider relationships between features or edges of an object. For example, if the image of an object can be subdivided into four straight lines (the lines are called primitives) connected at their end points, and the connected lines are at right angles, then the object is a rectangle. This kind of technique, known as syntactic pattern recognition, is the most widely used structural technique. Structural techniques differ from decision-theoretic techniques in that the latter deals with a pattern on a quantitative basis and ignores for the most part interrelationships among object primitives. A detailed discussion of pattern recognition techniques is the subject of complete books and is beyond the scope of this text.

It can be computationally time consuming for complete pattern recognition. Accordingly, it is often more appropriate to search for simpler regions or edges within an image. These simpler regions can then be used to extract the required features. The majority of commercial robot vision systems make use of this approach to the recognition of two-dimensional objects. The recognition algorithms are used to identify each segmented object in an image and assign it to a classification (e.g., nut, bolt, flange, etc.)

7-4 TRAINING THE VISION SYSTEM

The purpose of vision system training is to program the vision system with known objects. The system stores these objects in the form of extracted feature values which can be subsequently compared against the corresponding feature values from images of unknown objects. Typical features used in machine vision were discussed in the previous section and summarized in Table 7-2.

Training of the vision system should be carried out under conditions as close to operating conditions as possible. Physical parameters such as camera placement, aperture setting, part position, and lighting are the critical conditions that should be simulated as closely as possible during the training session.

Vision system manufacturers have developed application software for each individual system marketed. The software is typically based on a high-level programming language. For example, Object Recognition Systems Inc. uses the "C" language, and Automatix Inc. uses their internally developed language called RAIL (for Robot Automatic Incorporated Language). There are two versions of RAIL, one for automated vision systems and the other for robot programming. We will present portions of the robot programming language in the Appendix to Chapter Nine.

7-5 ROBOTIC APPLICATIONS

Many of the current applications of machine vision are inspection tasks that do not involve the use of an industrial robot. A typical application is where the machine vision system is installed on a high-speed production line to accept or reject parts made on the line. Unacceptable parts are ejected from the line by some mechanical device that is communicating with the vision system.

Machine vision applications can be considered to have three levels of difficulty. These levels depend on whether the object to be viewed is controlled in position and/or appearance. Controlling the position of an object in a manufacturing environment usually requires precise fixturing. Controlling the appearance of an object is accomplished by lighting techniques. Also, appearance is influenced by the object's surface texture or coloration. The three levels of difficulty used to categorize machine vision applications in an industrial setting are:

1. The object can be controlled in both position and appearance.
2. Either position or appearance of the object can be controlled but not both.
3. Neither position nor appearance of the object can be controlled.

The third level of difficulty requires advanced vision capabilities. The objective in engineering the vision application is to lower the level of difficulty involved, thereby reducing the level of sophistication of the vision system required in the application. For example, one problem that occurs in object recognition is that the recognition process is facilitated if the object is in a known position and orientation. Parts in a factory are typically not positioned and oriented in this manner. This problem can be reduced from a third level to a first level of difficulty by fixturing the parts and using techniques such as structured lighting to control the appearance.

In this section we will emphasize the use of machine vision in robotic applications. Robotic applications of machine vision fall into the three broad categories listed below:

1. Inspection
2. Identification
3. Visual servoing and navigation

The first category is one in which the primary function is the inspection process. This is carried out by the machine vision system, and the robot is used in a secondary role to support the application. The objectives of machine vision inspection include checking for gross surface defects, discovery of flaws in labeling (during final inspection of the product package), verification of the presence of components in assembly, measuring for dimensional accuracy, and checking for the presence of holes and other features in a part. When these kinds of inspection operations are performed manually, there is a tendency for human error. Also, the time required in most manual inspection operations requires that the procedures be accomplished on a sampling basis. With machine vision, these procedures are carried out automatically, using 100 percent inspection, and usually in much less time. In Chap. Fifteen, we will return to robotic inspection in which many of the applications of robots make use of machine vision.

The second category, *identification*, is concerned with applications in which the purpose of the machine vision system is to recognize and classify an object rather than to inspect it. Inspection implies that the part must be either accepted or rejected. Identification involves a recognition process in which the part itself, or its position and/or orientation, is determined. This is usually followed by a subsequent decision and action taken by the robot. Identification applications of machine vision include part sorting, palletizing and depalletizing, and picking parts that are randomly oriented from a conveyor or bin.

In the third application category, *visual servoing and navigational control*, the purpose of the vision system is to direct the actions of the robot (and other devices in the robot cell) based on its visual input. The generic example of robot visual servoing is where the machine vision system is used to control the trajectory of the robot's end effector toward an object in the workspace. Industrial examples of this application include part positioning, retrieving parts moving along a conveyor, retrieving and reorienting parts moving along a conveyor, assembly, bin picking, and seam tracking in continuous arc welding. The use of machine vision in arc welding is described in Chap. Fourteen when we discuss the application of welding robots in detail. An example of navigational control would be in automatic robot path planning and collision avoidance using visual data. Clearly the visual data are just an important input in this type of task and a great deal of intelligence is required in the controller to use the data for navigation and collision avoidance. This and the visual servoing tasks remain important research topics and are not now viable applications of robot vision systems.

The bin-*picking* application is an interesting and complex application of machine vision in robotics which involves both identification and servoing. Bin picking involves the use of a robot to grasp and retrieve randomly oriented parts out of a bin or similar container. The application is complex because parts will be overlapping each other. The vision system must first recognize a target part and its orientation in the container, and then it must direct the end effector to a position to permit grasping and pickup. The difficulty is in the fact that the target part is jumbled together with many other parts (probably identical parts), and the conditions of contrast between the target and its surroundings are far from ideal for part recognition.

Solution of the bin-picking problem owes much to the pioneering work in vision research at the University of Rhode Island. At the time of writing, there are two commercially available bin-picking systems, one offered by Object Recognition Systems, Inc. (ORS) called the i-bot 1 system, and the other by General Electric Co. called BinVision. We will describe the ORS product to illustrate the operation of these systems. Figure 7-10 shows the i-bot system. It utilizes a Unimation PUMA robot to pick parts from the bin. The system is capable of identifying the position and orientation of three parts in a 5-s image-processing cycle. However, the parts must have a relatively large aspect ratio (length-to-width ratio) for the system to operate effectively. Also, the objects in the bin must be the same in terms of features such as size, shape, and texture.

Seam tracking in continuous arc welding is another example of visual servoing and navigation in robotics vision systems. Because the vision system must operate in the presence of an arc-welding torch, special problems in the interpretation of the vision image must be solved. A typical solution uses a form of structured light to project a known pattern on the parts to be welded and uses the observed geometric distortion to determine the path and other parameters required for a successful welding operation. We will discuss this and other sensor technologies used in robotic arc welding in Chap. Fourteen.

In addition to the three main categories of machine vision applications in robotics, there are other special purpose vision applications. An example of these special applications is one in which the vision system functions as a range

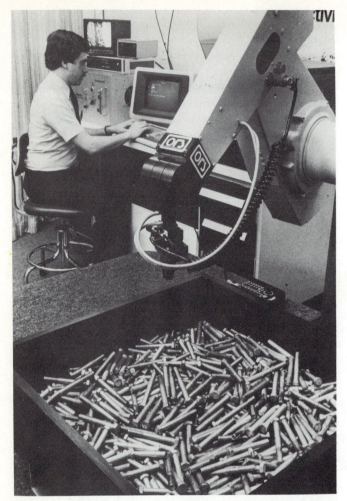

Figure 7-10 I-bot robot-vision system. (Photo courtesy of Object Recognition Systems, Inc.)

finder using triangulation or other techniques to determine the distance to the object.

Machine vision, coupled with the force and torque sensors discussed in Chap. Six, can immensely enhance a robot's applicability in manufacturing. Representative tasks in which such highly sensored systems could be applied include manipulation of parts during deburring, flash removal, and assembly. We will discuss the particular technical problems encountered in robotic assembly and the need for sensors in this application in Chap. Fifteen.

REFERENCES

1. I. Aleksander (ed.), *Artificial Vision for Robots*, Chapman and Hall, New York, 1983.
2. D. H. Ballard and C. M. Brown, *Computer Vision*, Prentice-Hall, Englewood Cliffs, NJ, 1982.

3. W. B. Gevarter, *An Overview of Machine Vision*, U.S. Department of Commerce, National Bureau of Standards, Document NBSIR 82-2582, September 1982.

4. W. B. Gevarter, "Machine Vision: The State of the Art," *Computers in Mechanical Engineering*, April 1982, pp. 25–30.

5. R. C. Gonzales and R. Safabaksh, "Computer Vision Techniques for Industrial Inspection and Robot Control: A Tutorial Overview," *Tutorial on Robotics*, C. S. G. Lees, R. C. Gonzales, and K. S. Fu (eds.), IEEE Computer Society Press, New York, 1984, pp. 299–324.

6. M. P. Groover and E. W. Zimmers, Jr., *CAD/CAM: Computer-Aided Design and Manufacturing*, Prentice-Hall, Englewood Cliffs, NJ, 1984, chaps. 17 and 19.

7. B. K. P. Horn, "Artificial Intelligence and the Science of Image Understanding," *Computer Vision and Sensor-Based Robotics*, G. G. Dodd and L. Rossel (eds.), Plenum Press, New York, 1979, pp. 68–77.

8. D. L. Hudson and J. E. Trombley, "Developing Industrial Applications for Machine Vision," *Computers in Mechanical Engineering*, April 1983, pp. 18–23.

9. T. Kanade, "Visual Sensing and Interpretation," *Computers in Mechanical Engineering*, April 1983, pp. 59–69.

10. S. P. Parker (ed.), *McGraw-Hill Encyclopedia of Electronics and Computers*, McGraw-Hill, New York, 1984.

11. C. A. Rosen, "Machine Vision and Robots," *Computer Vision and Sensor-Based Robotics*, G. G. Dodd and L. Rossel (eds.), Plenum Press, New York, 1979, pp. 3-22.

12. A. Rosenfeld, "Segmentation: Pixel-Based Methods," *Fundamentals in Computer Vision*, O. D. Faugeras (ed.), Cambridge University Press, 1983, pp. 225–237.

13. R. R. Schreiber, "Robot Vision: An Eye to the Future," *Robotics Today*, June 1983, pp. 53–57.

14. W. E. Snyder, *Industrial Robots: Computer Interfacing and Control*, Prentice-Hall, Englewood Cliffs, NJ, 1985.

15. R. Waldman, "Update on Research in Vision," *Robotics Today*, June 1983, pp. 63–67.

PROBLEMS

7-1 Consider a vision system which provides one frame of 256 lines every $\frac{1}{2}$ s. The system is a raster scan system. Assume that the time for the electron beam to move from one line to the next takes 15 percent of the time to scan a single line. Determine the sampling rate for the system if it is specified that there will be 320 pixels on each line.

7-2 A video voltage signal is to be digitized by an A/D converter. The maximum voltage range is +15 V. The A/D converter has a 6-bit capacity. Determine the number of quantizing levels, the quantization level spacing, and the quantization error.

7-3 For the A/D converter of Prob. 7-2, indicate how the voltage signal might be expressed in binary form. Compare the answer with one obtained for an A/D converter of only 4-bit capacity.

7-4 The accompanying figure represents a 8×10 array of pixels. Each element in the array

60	59	58	57	59	45	25	15
55	60	59	61	55	40	12	11
59	58	60	60	11	12	10	10
54	55	58	25	10	11	11	58
59	60	20	15	11	10	55	59
60	59	15	12	15	10	60	60
60	15	10	11	10	12	60	59
55	14	9	8	11	58	62	60
10	11	15	11	12	59	61	60
9	10	11	12	60	57	59	55

Figure P7-4

indicates the gray level value of the pixel. Construct a histogram for the array and choose an appropriate threshold value to use to separate black from white. A 6-bit converter was used to establish a gray level scale.

7-5 Consider the image representation of an object as given in the figure. White pixels are designated as 1s on a black background of 0s. Determine the boundary of the object by using the edge following procedure as discussed in the text. Start in the upper left-hand corner. Discuss any limitations of the technique.

0	0	0	0	0	0	0	0	0	0	0	0	0
0	0	0	0	1	1	1	1	1	1	1	1	0
0	0	1	1	1	1	1	1	1	1	1	1	0
0	1	1	1	1	0	0	0	1	1	1	1	0
0	1	1	1	1	0	0	0	1	1	1	0	0
0	1	1	1	1	1	1	1	1	0	0	0	0
0	1	1	1	1	1	1	1	1	1	1	1	0
0	1	0	0	1	1	1	1	0	0	1	1	0
0	1	0	0	1	1	1	1	1	1	1	1	0
0	1	0	0	0	0	0	0	0	0	1	1	0
0	0	0	0	0	0	0	0	0	0	1	1	0

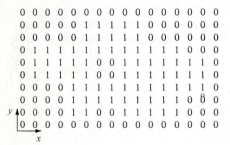

Figure P7-5

7-6 Consider a circle and an ellipse that might be viewed by a machine vision system. The circle has a 4-in. radius, whereas the ellipse has semiaxes a and b of 4 in. and 2 in., respectively. Apply the two definitions of thinness to both elements and compare the results.

7-7 For the image in the accompanying sketch in which white pixels are designated as 1s on a black background of 0s, determine (a) the area, (b) the minimum aspect ratio, (c) the eccentricity, and (d) the centroid of the image. Choose the coordinates of the frame in the lower left-hand corner as indicated.

	0	0	0	0	0	0	0	0	0	0	0	0	0	0	0	0
	0	0	0	0	0	1	1	1	1	1	0	0	0	0	0	0
	0	0	0	0	0	1	1	1	1	1	0	0	0	0	0	0
	0	1	1	1	1	1	1	1	1	1	1	1	1	0	0	0
	0	1	1	1	1	1	0	0	1	1	1	1	1	1	1	0
	0	1	1	1	1	1	0	0	1	1	1	1	1	1	1	0
	0	0	0	0	1	1	1	1	1	1	1	1	1	1	1	0
	0	0	0	0	1	1	1	1	1	1	1	1	1	0	0	0
y	0	0	0	0	1	1	0	0	1	1	1	1	1	0	0	0
	0	0	0	0	0	0	0	0	0	0	0	0	0	0	0	0

x

Figure P7-7

THREE

ROBOT PROGRAMMING AND LANGUAGES

Robot programming is concerned with teaching the robot its work cycle. A large portion of the program involves the motion path that the robot must execute in moving parts or tools from one location in the work space to another. These movements are often taught by showing the robot the motion and recording it into the robot's memory. However, there are other portions of the program that do not involve any movement of the arm. These other parts of the program include interpreting sensor data, actuating the end effector, sending signals to other pieces of equipment in the cell, receiving data from other devices, and making computations and decisions about the work cycle. Many of these other activities are best taught by programming the robot using a computer-like language.

Chapters Eight and Nine consider the two fundamental methods for programming today's industrial robots. Chapter Eight details the "teach-by-showing" methods of programming. Chapter Nine presents what we consider to be a comprehensive discussion of how robots are programmed with a computerlike robot language. There are several appendixes to Chap. Nine, which present summaries of some of the commercially available robot languages.

Advanced technology robots of the future, with versatile end effectors and sophisticated sensors, will be capable of responding to very high-level commands—higher, more general commands than we have in today's commercially available languages. The robots will have to interpret these high-level commands and act upon them. To do this, robots of the future must possess more intelligence than today's machines. In Chap. Ten, we survey the field of artificial intelligence to see what promise this technology holds for robotics.

EIGHT

ROBOT PROGRAMMING

In Chap. Two, we defined a robot program to be a path in space. It is really more than that. A robot today can do much more than merely move its arm through a series of points in space. Current technology robots can accept input from sensors and other devices. They can send signals to pieces of equipment operating with them in the cell. They can make decisions. They can communicate with computers to receive instructions and to report production data and problems. All of these capabilities require programming.

8-1 METHODS OF ROBOT PROGRAMMING

Robot programming is accomplished in several ways. Consistent with current industrial practice we divide the programming methods into two basic types:

1. Leadthrough methods
2. Textual robot languages

The leadthrough methods require the programmer to move the manipulator through the desired motion path and that the path be committed to memory by the robot controller. The leadthrough methods are sometimes referred to as "teach-by-showing" methods. Chronologically, the leadthrough methods represent the first real robot programming methods used in industry. They had their beginnings in the early 1960s when robots were first being used for industrial applications.

Robot programming with textual languages is accomplished somewhat like computer programming. The programmer types in the program on a CRT (cathode ray tube) monitor using a high-level English-like language. The procedure is usually augmented by using leadthrough techniques to teach the robot the locations of points in the workspace. The textual languages started to be developed in the 1970s, with the first commercial language appearing around 1979.

In addition to the leadthrough and textual language programming, another method of programming is used for simple, low-technology robots. We referred to these types of machines in Chap. Two as limited sequence robots which are controlled by means of mechanical stops and limit switches to define the end points of their joint motions. The setting of these stops and switches might be called a programming method. We prefer to think of this kind of programming as a manual setup procedure.

In this chapter we discuss the leadthrough methods. We also examine the basic features and capabilities of these programming methods. What functions must a typical robot be able to do, and how is it taught to do these functions using leadthrough programming? In the following chapter we examine the textual programming languages and their capabilities.

8-2 LEADTHROUGH PROGRAMMING METHODS

In leadthrough programming, the robot is moved through the desired motion path in order to record the path into the controller memory. There are two ways of accomplishing leadthrough programming:

1. Powered leadthrough
2. Manual leadthrough

The powered leadthrough method makes use of a teach pendant to control the various joint motors, and to power drive the robot arm and wrist through a series of points in space. Each point is recorded into memory for subsequent playback during the work cycle. The teach pendant is usually a small hand-held control box with combinations of toggle switches, dials, and buttons to regulate the robot's physical movements and programming capabilities. Among the various robot programming methods, the powered leadthrough method is probably the most common today. It is largely limited to point-to-point motions rather than continuous movement because of the difficulty in using the teach pendant to regulate complex geometric motions in space. A large number of industrial robot applications consist of point-to-point movements of the manipulator. These include part transfer tasks, machine loading and unloading, and spot welding.

The manual leadthrough method (also sometimes called the "walk-through" method) is more readily used for continuous-path programming where the motion cycle involves smooth complex curvilinear movements of the robot arm. The most common example of this kind of robot application is spray painting, in which the robot wrist, with the spray painting gun attached as the end effector, must execute a smooth, regular motion pattern in order to apply the paint evenly over the entire surface to be coated. Continuous arc welding is another example in which continuous-path programming is required and this is sometimes accomplished with the manual leadthrough method.

In the manual leadthrough method, the programmer physically grasps the robot arm (and end effector) and manually moves it through the desired motion cycle. If the robot is large and awkward to physically move, a special programming apparatus is often substituted for the actual robot. This apparatus has basically the same geometry as the robot, but it is easier to manipulate during programming. A teach button is often located near the wrist of the robot (or the special programming apparatus) which is depressed during those movements of the manipulator that will become part of the programmed cycle. This allows the programmer the ability to make extraneous moves of the arm without their being included in the final program. The motion cycle is divided into hundreds or even thousands of individual closely spaced points along the path and these points are recorded into the controller memory.

The control systems for both leadthrough procedures operate in either of two modes: teach mode or run mode. The teach mode is used to program the robot and the run mode is used to execute the program.

The two leadthrough methods are relatively simple procedures that have been developed and enhanced over the last 20 years to teach robots to perform simple, repetitive operations in factory environments. The skill requirements of the programmers are relatively modest and these procedures can be readily applied in the plant.

8-3 A ROBOT PROGRAM AS A PATH IN SPACE

This and the following sections of this chapter will examine the programming issues involved in the use of the leadthrough methods, with emphasis on the powered leadthrough approach. Let us begin our discussion with our previous definition of a robot program as a path in space. The locus of points along the path defines the sequence of positions through which the robot will move its wrist. In most applications, an end effector is attached to the wrist, and the program can be considered to be the path in space through which the end effector is to be moved by the robot.

Since the robot consists of several joints (axes) linked together, the definition of the path in space in effect requires that the robot move its axes through various positions in order to follow that path. For a robot with six axes, each point in the path consists of six coordinate values. Each coordinate value corresponds to the position of one joint. As discussed in Chap. Two, there are four basic robot anatomies: polar, cylindrical, cartesian, and jointed arm. Each one has three axes associated with the arm and body configuration and two or three additional joints associated with the wrist. The arm and body joints determine the general position in space of the end effector and the wrist determines its orientation. If we think of a point in space in the robot program as a position and orientation of the end effector, there is usually more than one

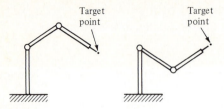

Figure 8-1 Two alternative axis configurations with end effector located at desired target point.

possible set of joint coordinate values that can be used for the robot to reach that point. For example, there are two alternative axis configurations that can be used by the jointed-arm robot shown in Fig. 8-1 to achieve the target point indicated. We see in the figure that although the target point has been reached by both of the alternative axis configurations, there is a difference in the orientation of the wrist with respect to the point. We must conclude from this that the specification of a point in space does not uniquely define the joint coordinates of the robot. Conversely, however, the specification of the joint coordinates of the robot does define only one point in space that corresponds to that set of coordinate values. Points specified in this fashion are said to be joint coordinates. Accordingly, let us refine our definition of a robot program by specifying it to be a sequence of joint coordinate positions, rather than merely a path in space. An advantage of defining a robot program in this way is that it simultaneously specifies the position and orientation of the end effector at each point in the path.

Let us consider the problem of defining a sequence of points in space. We will assume that these points are defined by specifying the joint coordinates as described above, although this method of specification will not affect the issues we are discussing here. For the sake of simplicity, let us assume that we are programming a point-to-point cartesian robot with only two axes, and only two addressable points for each axis. An addressable point is one of the available points (as determined by the control resolution) that can be specified in the program so that the robot can be commanded to go to that point. Figure 8-2 shows the four possible points in the robot's rectangular workspace. A program for this robot to start in the lower left-hand corner and traverse the perimeter of the rectangle could be written as follows:

Example 8-1

Step	Move	Comments
1	1,1	Move to lower left corner
2	2,1	Move to lower right corner
3	2,2	Move to upper right corner
4	1,2	Move to upper left corner
5	1,1	Move back to start position

The point designations correspond to the x, y coordinate positions in the cartesian axis system, as illustrated in the figure. In our example, using a robot

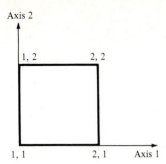

Axis 2

1, 2   2, 2

1, 1   2, 1   Axis 1   **Figure 8-2** Robot workspace for Example 8-1

with two orthogonal slides and only two addressable points per axis, the definition of points in space corresponds exactly with joint coordinate values.

Using the same robot, let us consider its behavior when performing the following program:

Example 8-2

Step	Move	Comments
1	1,1	Move to lower left corner
2	2,1	Move to lower right corner
3	1,2	Move to upper left corner
4	1,1	Move back to start position

The second program is the same as the first, except that the point in the upper right corner (2,2) has not been listed. Before explaining the implications of this missing point, let us recall that in Example 8-1, the move from one point to the next required only one joint to be moved while the other joint position remained unchanged. In this second program, the move from point 2,1 to point 1,2 requires both joints to be moved. The question that arises is: what path will the robot follow in getting from the first point to the second? Certainly one possibility is that both axes will move at the same time, and the robot will therefore trace a path along the diagonal line between the two points. The other possibility is that the robot will move only one axis at a time and trace out a path along the border of the rectangle, either through point 2,2 or through point 1,1.

The question of which path the robot will take between two programmed points is not a trivial one. It is important for the programmer to know the answer in order to plan out the motion path correctly. Unfortunately, there is no general rule that all robots follow. Limited-sequence nonservo robots, which are programmed using manual setup procedures rather than lead-through methods, can usually move both joints at the same time. Therefore, the path that is followed involves a slew motion (as described in Chap. Four), which is along the diagonal in our illustration. Other limited sequence robots

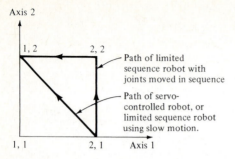

Axis 2

1, 2   2, 2

— Path of limited
sequence robot with
joints moved in sequence

— Path of servo-
controlled robot, or
limited sequence robot
using slow motion.

1, 1   2, 1   Axis 1

Figure 8-3 Likely path followed by the robot in Example 8-2 when operating at high speed. The arm trajectory misses the unprogrammed points.

move their joints in sequence rather than simultaneously. Usually, these robots that move one axis at a time do so by moving the lower numbered axes first. Thus, the path through point 2,2 is most likely in this example. However, there are no industry standards on this issue, and the programmer must make this kind of determination either from the user's manual or by experimentation with the actual robot. Servocontrolled robots, which are programmed by leadthrough and textual language methods, tend to actuate all axes simultaneously. Hence, with servocontrol, the robot would likely move approximately along the diagonal path between points 2,1 and 1,2. The differences between the paths for Example 8-2 are illustrated in Fig. 8-3.

As illustrated by the preceding discussion of Example 8-2, it is possible for the programmer to make certain types of robots pass through points without actually including the points in the program. The key phrase is "pass through." These are not addressable points in the program and the robot will not actually stop at them in the sense of an addressable point.

Methods of Defining Positions in Space

Irrespective of robot configuration, there are several methods that can be used by the programmer during the teach mode to actuate the robot arm and wrist. We list the following three methods:

1. Joint movements
2. x-y-z coordinate motions (also called world coordinates)
3. Tool coordinate motions

The first method is the most basic and involves the movement of each joint, usually by means of a teach pendant. The teach pendant has a set of toggle switches (or similar control devices) to operate each joint in either of its two directions until the end effector has been positioned to the desired point. This method of teaching points is often referred to as the joint mode. Successive positioning of the robot arm in this way to define a sequence of points can be a very tedious and time-consuming way of programming the robot.

To overcome this disadvantage, many robots can be controlled during the teach mode to move in *x-y-z* coordinate motions. This method, called the world coordinate system, allows the wrist location to be defined using the conventional cartesian coordinate system with origin at some location in the body of the robot. In the case of the cartesian coordinate robot, this method is virtually equivalent to the joint mode of programming. For polar, cylindrical, and jointed-arm robots, the controller must solve a set of mathematical equations to convert the rotational joint motions of the robot into the cartesian coordinate system. These conversions are carried out in such a way that the programmer does not have to be concerned with the substantial computations that are being performed by the controller. To the programmer, the wrist (or end effector) is being moved in motions that are parallel to the *x*, *y*, and *z* axes. The two or three additional joints which constitute the wrist assembly are almost always rotational, and while programming is being done in the *x-y-z* system to move the arm and body joints, the wrist is usually being maintained by the controller in a constant orientation. The *x-y-z* method of defining points in space is illustrated in Fig. 8-4 for a jointed-arm robot.

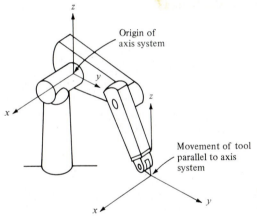

Figure 8-4 World mode or *X-Y-Z* method of defining points in space.

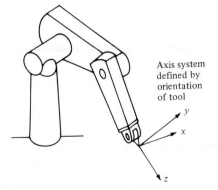

Figure 8-5 Tool mode of defining points in space.

Some robots have the capability for tool coordinate motions to be defined for the robot. This is a cartesian coordinate system in which the origin is located at some point on the wrist and the *xy* plane is oriented parallel to the faceplate of the wrist. Accordingly, the *z* axis is perpendicular to the faceplate and pointing in the same direction as a tool or other end effector attached to the faceplate. Hence, this method of moving the robot could be used to provide a driving motion of the tool. Again, a significant amount of computational effort must be accomplished by the controller in order to permit the programmer to use tool motions for defining motions and points. Figure 8-5 shows the tool coordinate system.

Reasons for Defining Points

The preceding examples and discussion are intended to argue that there are some good reasons for defining points in space in a robot program, rather than relying on the robot to pass through an undefined point or to actuate a certain axis before it actuates another axis. The two main reasons for defining points in a program are:

1. To define a working position for the end effector
2. To avoid obstacles

The first category is the most straightforward. This is the case where the robot is programmed to pick up a part at a given location or to perform a spot-welding operation at a specified location. Each location is a defined point in the program. This category also includes safe positions that are required in the work cycle. For example, it might be necessary to define a safe, remote point in the workspace from which the robot would start the work cycle.

The second category is used to define one or more points in space for the robot to follow which ensures that it will not collide with other objects located in the workcell. Machines, conveyors, and other pieces of equipment in the work volume are examples of these obstacles. By defining a path of points around these obstacles, the collisions can be prevented.

Speed Control

Most robots allow for their motion speed to be regulated during the program execution. A dial or group of dials on the teach pendant are used to set the speed for different portions of the program. It is considered good practice to operate the robot at a relatively slow speed when the end effector is operating close to obstacles in the workcell, and at higher speeds when moving over large distances where there are no obstacles. This gives rise to the notion of "freeways" within the cell. These are possible pathways in the robot cell which are free of obstructions and therefore permit operation at the higher velocities.

The speed is not typically given as a linear velocity at the tip of the end effector for robots programmed by leadthrough methods. There are several reasons for this. First, the robot's linear speed at the end effector depends on how many axes are moving at one time and which axes they are. Second, the speed of the robot depends on its current axis configuration. For example, the top speed of a polar coordinate robot will be much greater with its arm fully extended than with the arm in the fully retracted position. Finally, the speed of the robot will be affected by the load it is carrying due to the force of acceleration and deceleration. All of these reasons lead to considerable computational complexities when the control computer is programmed to determine wrist end velocity.

In the next chapter, we will be able to define the speed explicitly using the textual languages so that the wrist or even the end effector velocity can be programmed in more conventional units (e.g., millimeters per second or inches per second). This capability is not available with all computer-controlled robots because of the reasons mentioned above. However, we will assume that it is available for our purposes in Chap. Nine.

8-4 MOTION INTERPOLATION

Suppose we were programming a two-axis servocontrolled cartesian robot with eight addressable points for each axis. Accordingly, there would be a total of 64 addressable points that we can use in any program that might be written. The work volume is illustrated in Fig. 8-6. Assuming the axis sizes to be the same as our previous limited sequence robot, a program for the robot to perform the same work cycle as Example 8-1 would be as follows:

Example 8-3

Step	Move
1	1,1
2	8,1
3	8,8
4	1,8
5	1,1

If we were to remove step 3 in this program (similar to Example 8-2), our servocontrolled robot would execute step 4 by tracing a path along the diagonal line from point 8,1 to point 1,8. This process is referred to as interpolation. This internal algorithm followed by the robot controller to get between the two points is somewhat more complicated than it appears from our simple illustration. Also, as indicated in Chap. Four, there are different interpolation schemes that can be specified by the robot to get from one point to another. Before discussing these differences, let us describe the most basic interpolation process, called joint interpolation.

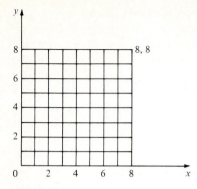

Figure 8-6 Robot work space with 8×8 addressable points.

In *joint interpolation*, the controller determines how far each joint must move to get from the first point defined in the program to the next. It then selects the joint that requires the longest time. This determines the time it will take to complete the move (at a specified speed). Based on the known move time, and the amount of the movement required for the other axes, the controller subdivides the move into smaller increments so that all joints start and stop their motions at the same time. Consider, for example, the move from point 1,1 to point 7,4 in the grid of Fig. 8-6. Linear joint 1 must move six increments (grid locations) and joint 2 must move three increments. To determine the joint interpolated path, the controller would determine a set of intermediate addressable points along the path between 1,1 and 7,4 which would be followed by the robot. The following program illustrates the process.

Example 8-4

Step	Move	Comments
1	1,1	User specified starting point
2	2,2	Internally generated interpolation point
3	3,2	Internally generated interpolation point
4	4,3	Internally generated interpolation point
5	5,3	Internally generated interpolation point
6	6,4	Internally generated interpolation point
7	7,4	User specified end point

The reader should note that the controller alternatively moves both axes, or just one axis. Also, for each move requiring actuation of both axes, the two axes start and stop together. This kind of actuation causes the robot to take a path as illustrated in Fig. 8-7. The controller does the equivalent of constructing a hypothetically perfect path between the two points specified in the program, and then generates the internal points as close to that line as possible. The resulting path is not a straight line, but is rather an approximation. The controller approximates the perfect path as best it can within the limitations imposed by the control resolution of the robot (the available

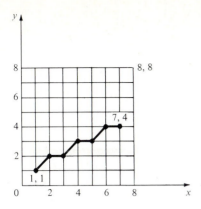

Figure 8-7 Interpolated path taken by robot in Example 8-4.

addressable points in the work volume). In our case, with only 64 addressable points in the grid, the approximation is very rough. With a much larger number of addressable points and a denser grid, the approximation would be better.

The reader might have noticed that the interpolation procedure used above created a straight line approximation. This is usually referred to as straight line interpolation, yet we have described it as joint interpolation. Because we are dealing with a cartesian robot in the above illustration which has only linear axes, joint interpolation and straight line interpolation are the same. For other robots with a combination of rotational and linear joints (cylindrical and polar configurations), or all rotational joints (jointed arm configuration), straight line interpolation produces a path that is different from joint interpolation. For joint interpolation, the algorithm is outlined in the preceding discussion. In *straight line interpolation*, the robot controller computes the straight line path between two points and develops the sequence of addressable points along the path for the robot to pass through. As indicated the procedure is identical to the example given in Example 8-4.

Consider a robot that has one rotational axis (axis 1) and one linear axis (axis 2), where each axis has eight addressable points. This creates a total of 64 addressable points which form the grid shown in Fig. 8-8. The grid is polar rather than rectilinear. During an interpolation procedure, this has the effect of creating moves of different lengths (from the viewpoint of euclidean geometry). For example, compare the move from 1,1 to 3,2 with the move from 1,7 to 3,8. The addressability of a robot with rotational axes is not uniform in euclidean space. Moves that are made close to the axis of rotation are significantly smaller than moves that are far from the rotation joint. This change in robot configuration has implications for the interpolation schemes used by the controller. Although the descriptions given above still apply for joint interpolation and straight line interpolation, it is clear that the paths taken by the robot will be affected by the change in anatomy. The incremental moves executed by the robot consist of combinations of rotational moves (along axis 1) and linear moves (along axis 2). We leave the visualization of

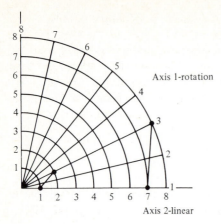

Figure 8-8 Grid for a robot with one rotational and one linear axis, with each axis divided into eight addressable locations.

these effects to the reader as exercises at the end of the chapter. For a jointed-arm robot, even one with only two joints, the problem of visualizing the two types of interpolation becomes even more difficult than for the rotational–linear joint combination under consideration here.

On many robots, the programmer can specify which type of interpolation scheme to use. The possibilities include:

1. Joint interpolation
2. Straight line interpolation
3. Circular interpolation
4. Irregular smooth motions (manual leadthrough programming)

For many commercially available robots, joint interpolation is the default procedure that is used by the controller. That is, the controller will follow a joint interpolated path between two points unless the programmer specifies straight line (or some other type of) interpolation.

Circular interpolation requires the programmer to define a circle in the robot's workspace. This is most conveniently done by specifying three points that lie along the circle. The controller then constructs an approximation of the circle by selecting a series of addressable points that lie closest to the defined circle. The movements that are made by the robot actually consist of short-straight-line segments. Circular interpolation therefore produces a linear approximation of the circle. If the gridwork of addressable points is dense enough, the linear approximation looks very much like a real circle. Circular interpolation is more readily programmed using a textual programming language than with leadthrough techniques.

In manual leadthrough programming, when the programmer moves the manipulator wrist to teach spray painting or arc welding, the movements typically consist of combinations of smooth motion segments. These segments are sometimes approximately straight, sometimes curved (but not necessarily circular), and sometimes back-and-forth motions. We are referring to these movements as *irregular smooth motions*, and an interpolation process is involved in order to achieve them. To approximate the irregular smooth pattern being taught by the programmer, the motion path is divided into a sequence of closely spaced points that are recorded into the controller memory. These positions constitute the nearest addressable points to the path followed during programming. The interpolated path may consist of thousands of individual points that the robot must play back during subsequent program execution.

8-5 WAIT, SIGNAL, AND DELAY COMMANDS

Robots usually work with something in their work space. In the simplest case, it may be a part that the robot will pick up, move, and drop off during execution of its work cycle. In more complex cases, the robot will work with other pieces of equipment in the workcell, and the activities of the various equipment must be coordinated. This situation introduces the subject of workcell control, which we shall discuss in more detail in Chap. Eleven. For the moment, let us introduce the kinds of basic programming commands that must be employed in workcell control.

Nearly all industrial robots can be instructed to send signals or wait for signals during execution of the program. These signals are sometimes called interlocks, and their various applications in workcell control will be discussed in Chap. Eleven. The most common form of interlock signal is to actuate the robot's end effector. In the case of a gripper, the signal is to open or close the gripper. Signals of this type are usually binary; that is, the signal is on–off or high-level–low-level. Binary signals are not readily capable of including any complex information such as force sensor measurements. The binary signals used for the robot gripper are typically implemented by using one or more dedicated lines. Air pressure is commonly used to actuate the gripper. A binary valve to actuate the gripper is controlled by means of two interlock signals, one to open the gripper and the other to close it. In some cases, feedback signals can be used to verify that the actuation of the gripper had occurred, and interlocks could be designed to provide this feedback data.

In addition to control of the gripper, robots are typically coordinated with other devices in the cell also. For example, let us consider a robot whose task is to unload a press. It is important to inhibit the robot from having its gripper enter the press before the press is open, and even more obvious, it is important that the robot remove its hand from the press before the press closes.

To accomplish this coordination, we introduce two commands that can be used during the program. The first command is

SIGNAL M

which instructs the robot controller to output a signal through line M (where M is one of several output lines available to the controller). The second command is

WAIT N

which indicates that the robot should wait at its current location until it receives a signal on line N (where N is one of several input lines available to the robot controller).

Let us suppose that the two-axis robot of Fig. 8-2 is to be used to perform the unloading of the press in our example. The layout of the workcell is illustrated in Fig. 8-9, which is similar to Fig. 8-6. The platten of the press (where the parts are to be picked up) is located at 8,8. The robot must drop the parts in a tote pan located at 1,8. One of the columns of the press is in the way of an easy straight line move from 8,8 to 1,8. Therefore, the robot must move its arm around the near side of the column in order to avoid colliding with it. This is accomplished by making use of points 8,1 and 1,1. Point 8,1 will be our position to wait for the press to open before entering the press to remove the part, and the the robot will be started from point 1,1, a point in space known to be safe in the application. We will use controller ports 1 to 10 as output (SIGNAL) lines and ports 11 through 20 as input (WAIT) lines. Specifically, output line 4 will be used to actuate (SIGNAL) the press, and output lines 5 and 6 will be used to close and open the gripper, respectively. Input line 11 will be used to receive the signal from the press indicating that it has opened (WAIT). The following is our program to accomplish the press unloading task (the sequence begins with the gripper in the open position).

Example 8-5

Step	Move or signal	Comments
0	1,1	Start at home position
1	8,1	Move to wait position
2	WAIT 11	Wait for press to open
3	8,8	Move to pickup point
4	SIGNAL 5	Signal gripper to close
5	8,1	Move to safe position
6	SIGNAL 4	Signal press to actuate
7	1,1	Move around press column
8	1,8	Move to tote pan
9	SIGNAL 6	Signal gripper to open
10	1,1	Move to safe position

Each step in the program is executed in sequence, which means that the SIGNAL and WAIT commands are not executed until the robot has moved to the point indicated in the previous step.

The operation of the gripper was assumed to take place instantaneously so that its actuation would be completed before the next step in the program was started. Some grippers use a feedback loop to ensure that the actuation has occurred before the program is permitted to execute the next step. A WAIT instruction can be programmed to accomplish this feedback. One of the exercises at the end of the chapter deals with this problem.

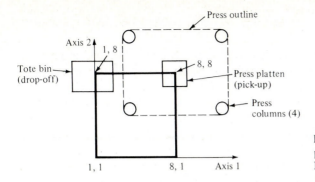

Figure 8-9 Robot work space for press unloading operation of Example 8-5.

An alternative way to address this problem is to cause the robot to delay before proceeding to the next step. In this case, the robot would be programmed to wait for a specified amount of time to ensure that the operation had taken place. The form of the command for this second case has a length of time as its argument rather than an input line. The command

DELAY X SEC

indicates that the robot should wait X seconds before proceeding to the next step in the program. Below, we show a modified version of Example 8-5, using time as the means for assuring that the gripper is either opened or closed.

Example 8-6

Step	Move or signal	Comments
0	1,1	Start at home position
1	8,1	Move to wait position
2	WAIT 11	Wait for press to open
3	8,8	Move to pickup point
4	SIGNAL 5	Signal gripper to close
5	DELAY 1 SEC	Wait for gripper to close
6	8,1	Move to safe position
7	SIGNAL 4	Signal press that hand is clear
8	1,1	Move around press column
9	1,8	Move to tote pan
10	SIGNAL 6	Signal hand to open
11	DELAY 1 SEC	Wait for gripper to open
12	1,1	Move to home position

The reader is cautioned that our programs above are written to look like computer programs. This is for convenience in our explanation of the programming principles. The actual teaching of the moves and signals is accomplished by leading the arm through the motion path and entering the nonmotion instructions at the control panel or with the teach pendant. In the majority of industrial applications today, robots are programmed using one of the leadthrough methods. Only with the textual language programming do the programs read like computer program listings.

8-6 BRANCHING

Most controllers for industrial robots provide a method of dividing a program into one or more branches. Branching allows the robot program to be subdivided into convenient segments that can be executed during the program. A branch can be thought of as a subroutine that is called one or more times during the program. The subroutine can be executed either by branching to it at a particular place in the program or by testing an input signal line to branch to it. The amount of decision logic that can be incorporated into the program varies widely with controllers. However, most controllers allow for a branch to be identified or assigned one of a preestablished group of names. They permit the use of an incoming signal to invoke a branch. Most controllers allow the user to specify whether the signal should interrupt the program branch currently being executed, or wait until the current branch completes. The interrupt capability is typically used for error branches. An error branch is invoked when an incoming signal indicates that some abnormal event (e.g., an unsafe operating condition) has occurred. Depending on the event and the design of the error branch, the robot will either take some corrective action or simply terminate the robot motion and signal for human assistance.

A frequent use of the branch capability is when the robot has been programmed to perform more than one task. In this case, separate branches are used for each individual task. Some means must be devised for indicating which branch of the program must be executed and when it must be executed. A common way of accomplishing this is to make use of external signals which are activated by sensors or other interlocks. The device recognizes which task must be performed, and provides the appropriate signal to call that branch. This method is frequently used on spray painting robots which have been programmed to paint a limited variety of parts moving past the workstation of a conveyor. Photoelectric cells are frequently employed to identify the part to be sprayed by distinguishing between the geometric features (e.g., size, shape, the presence of holes, etc.) of the different parts. The photoelectric cells are used to generate the signal to the robot to call the spray painting subroutine corresponding to the particular part.

Given the concept of a branch or subroutine that might be repeated in a program, an additional concept is readily introduced. Robot programs have thus far been discussed as consisting of a series of points in space, where each point is defined as a set of joint coordinates corresponding to the number of degrees of freedom of the robot. These points are specified in absolute coordinates. That is, when the robot executes the program or the branch of the program, each point is visited at exactly the same location every time. The new concept involves the use of a relocatable branch.

A relocatable branch allows the programmer to specify a branch involving a set of incremental points in space that are performed relative to some defined starting point for the branch. This would permit the same motion subroutine to be performed at various locations in the workspace of the robot. Many industrial robots have the capacity to accept relocatable branches as part of a program. The programmer indicates that a relocatable branch will be defined (the method of doing this varies from robot to robot), and the controller records relative or incremental motion points rather than absolute points.

Let us illustrate these branching concepts by developing two versions of a robot program to perform a palletizing operation. A pallet is a common type of container used in a factory to hold and move parts. Suppose that the operation required the robot to pick up parts from an input chute, and place them on a pallet with 24 positions as depicted in Fig. 8-10. When a start signal is given, the robot must begin picking up parts and loading them into the pallet, continuing until all 24 positions on the pallet are filled. The robot must then generate a signal to indicate that the pallet is full, and wait for the start signal to begin the next cycle.

For the purpose of discussion, several points will be defined that the robot must go to during the execution of the palletizing program. We will use names for convenience to identify these points, with the understanding that the names must be defined in terms of robot joint coordinates. When the robot is directed to go to the point name in the program, it goes to the associated joint coordinates.

Point name	Explanation
SAFE	Safe location to start and stop
PICKUP	Location of part pickup at end of chute
INTER	Intermediate point above chute to pass through
LOC1	Location of first pallet position
LOC2	Location of second pallet position
:	:
:	:
LOC24	Location of 24th pallet position
ABOVE1	Location above first pallet position
:	:
:	:
ABOVE24	Location above 24th pallet position

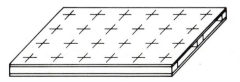

Figure 8-10 Pallet with 24 positions used to illustrate branching in robot programs.

In creating robot programs for palletizing operations of this type, the robot is programmed to approach a given part from a direction chosen to avoid interference with other parts. Accordingly, the various pallet locations would be approached from points at some distance directly above each location. The 24 points named ABOVE1 through ABOVE24 have been designated for this purpose. Similarly, the point named INTER has been defined above the chute for the same reason.

The speed at which the program is executed should be varied during the program. When the gripper is approaching a pickup or dropoff point, the speed setting should be at a relatively slow value. When the robot moves larger distances between the chute and the pallet, higher speeds would be programmed.

As our example programs in this chapter have become more complicated, progressing from a simple listing of points in space to the identification of points as variable names, and the use of commands such as WAIT and SIGNAL, we have seen the need to change the format of our program listing. Once again we are faced with the need to expand the way we express the program. In the programs below, we introduce the MOVE command followed by the name of the point to which the command refers. The instruction

MOVE SAFE

means that the robot should move its end effector to the point in space called SAFE. That point has been defined so the robot knows where to go when it comes to that instruction in the program.

The program is initiated upon receipt of a start signal on input line 11. The robot is interlocked so that it must await a signal on input line 12 indicating that a part has been delivered by the chute and is ready for pickup. To operate the gripper, our program will use output line 5 to open the gripper, and output line 6 to close the gripper. We assume that the gripper is fast acting, and that no feedback signals or waiting time are needed (as suggested by Examples 8-5 and 8-6) to ensure that the operation of the gripper has taken place. Finally, output line 7 will be used to indicate that the pallet is full.

We begin our program development by presenting a robot program in which no use is made of the branching capability. This will serve as a starting point to show how efficiencies can be obtained by use of branches in the subsequent programs.

The reader will note that our program contains a total of 243 commands and that portions of the program are very repetitious. Obviously, the repetition could be reduced by means of branches (subroutines) that can be written and called during the program. Let us consider the task of fetching and placing the parts that is repeated 24 times in Example 8-7. The task can be divided into two subtasks, one to fetch the part from the chute position, and the second to place the part in the correct pallet position. This division can be seen if we examine a portion of the program corresponding to the actions involved in fetching and placing any given part (we will use the second part to illustrate).

Example 8-7

Step	Command	Comments
1	MOVE SAFE	Move to the starting safe position
2	WAIT 11	Wait for start signal on line 11

(The following portion of the program directs the robot to pick up first part.)

Step	Command	Comments
3	MOVE INTER	Go to the intermediate point above chute
4	WAIT 12	Wait for next part from chute
5	SIGNAL 5	Open gripper
6	MOVE PICKUP	Move gripper to pick up part
7	SIGNAL 6	Close gripper
8	MOVE INTER	Depart to intermediate point above chute
9	MOVE ABOVE1	Move to point above first pallet location
10	MOVE LOC1	Position part in first pallet location
11	SIGNAL 5	Open gripper
12	MOVE ABOVE1	Depart slowly from pickup point

(The next portion of the program directs the robot to pick up second part.)

Step	Command	Comments
13	MOVE INTER	Go to the intermediate point above chute
14	WAIT 12	Wait for next part from chute
15	SIGNAL 5	Open gripper
16	MOVE PICKUP	Move gripper to pick up part
17	SIGNAL 6	Close gripper
18	MOVE INTER	Depart to intermediate point above chute
19	MOVE ABOVE2	Move to point above second pallet location
20	MOVE LOC2	Position part in second pallet location
21	SIGNAL 5	Open gripper
22	MOVE ABOVE2	Depart slowly from pickup point

(The preceding portions of the program are repeated for the next 21 parts.)

.
.
.

(The next portion of the program directs the robot to pick up 24th part.)

Step	Command	Comments
232	MOVE INTER	Go to the intermediate point above chute
233	WAIT 12	Wait for next part from chute
234	SIGNAL 5	Open gripper
235	MOVE PICKUP	Move gripper to pick up part
236	SIGNAL 6	Close gripper
237	MOVE INTER	Depart to intermediate point above chute
238	MOVE ABOVE2	Move to point above second pallet location
239	MOVE LOC2	Position part in second pallet location
240	SIGNAL 5	Open gripper
241	MOVE ABOVE2	Depart slowly from pickup point

(The pallet is now full.)

Step	Command	Comments
242	MOVE INTER	Go to the intermediate safe position
243	SIGNAL 7	Signal that pallet is full

This portion of the program is repeated below, indicating where the subdivision occurs. (The following is the "fetch" subtask.)

13	MOVE INTER	Go to the intermediate point above chute
14	WAIT 12	Wait for next part from chute
15	SIGNAL 5	Open gripper
16	MOVE PICKUP	Move gripper to pick up part
17	SIGNAL 6	Close gripper
18	MOVE INTER	Depart to intermediate point above chute

(The following is the "place" subtask.)

19	MOVE ABOVE2	Move to point above second pallet location
20	MOVE LOC2	Position part in second pallet location
21	SIGNAL 5	Open gripper
22	MOVE ABOVE2	Depart slowly from pickup point

Notice that the coding for the "fetch" portion of the task is identical regardless of which part number is being picked up. We will identify this portion of the program as a branch and we will name it "FETCH." It can be expressed as follows:

BRANCH FETCH	Indicates the following is branch FETCH
MOVE INTER	Go to the intermediate point above chute
WAIT 12	Wait for next part from chute
SIGNAL 5	Open gripper
MOVE PICKUP	Move gripper to pick up part
SIGNAL 6	Close gripper
MOVE INTER	Go to intermediate point above chute
END BRANCH	This is the end of the branch

The second subtask to place the part in a numbered pallet position is nearly the same for the 24 repetitions except for the fact that the location changes on each replication. It would therefore constitute a relocatable branch. We will name it "PLACE" and to use it for each part we must somehow move the end effector to the point above the pallet location corresponding to that part. One way to do this is to use incremental positioning in which the robot is directed to move a certain distance rather than to a particular point. The PLACE subroutine might be expressed as follows for this method:

BRANCH PLACE	Indicates the following is branch PLACE
MOVE Z(−50)	Position part in pallet
SIGNAL 5	Open gripper to release part
MOVE Z(+50)	Depart from pickup point
END BRANCH	This is the end of the branch

The MOVE commands indicate that the robot should move its end effector in the z-axis direction by a distance of 50 mm. MOVE Z(−50) directs the robot to move down 50 mm and the command MOVE Z(+50) directs it to move in a positive *z* direction by 50 mm. In this case the robot would first have to be moved to the point above the required pallet location. The sequence for a particular part placement (we will use part number 2 for illustration) would be as follows:

| MOVE ABOVE2 | Move to point above 2nd pallet location |
| PLACE | Calls branch PLACE for execution |

In the incremental positioning approach, it is very important that each ABOVE point (ABOVE1 through ABOVE24) be located very precisely relative to the respective pallet position. Otherwise, the PLACE branch will position the parts inaccurately on the pallet.

For the method described above, each of the 24 must be defined in the program, and this could require a great deal of tedious work by the programmer. The final program, benefiting from the use of the FETCH and PLACE branches defined above, might be listed as indicated below.

Example 8-8

Step	Command	Comments
1	BRANCH FETCH	Indicates the following is branch FETCH
2	MOVE INTER	Go to the intermediate point above chute
3	WAIT 12	Wait for next part from chute
4	SIGNAL 5	Open gripper
5	MOVE PICKUP	Move gripper to pick up part
6	SIGNAL 6	Close gripper
7	MOVE INTER	Depart to intermediate point above chute
8	END BRANCH	This is the end of the branch
9	BRANCH PLACE	Indicates the following is branch PLACE
10	MOVE Z(−50)	Position part in pallet
11	SIGNAL 5	Open gripper to release part
12	MOVE Z(+50)	Depart from pickup point
13	END BRANCH	This is the end of the branch
14	MOVE SAFE	Move robot to the starting safe position
15	WAIT 11	Wait for start signal on line 11
16	FETCH	Fetch first part
17	MOVE ABOVE1	Move to first position
18	PLACE	Place first part
19	FETCH	Fetch second part
20	MOVE ABOVE2	Move to second position
21	PLACE	Place second part
.	.	.
85	FETCH	Fetch 24th part
86	MOVE ABOVE24	Move to 24th position
87	PLACE	Place 24th part
88	MOVE INTER	Go to the intermediate safe position
89	SIGNAL 7	Signal that pallet is full

It is clear from these two examples (Examples 8-7 and 8-8) that there is significant efficiency in robot programming when branches are used. The lines of code have been reduced from 243 to 89 for the palletizing operation described above. This is a substantial reduction in the programming effort.

There are other ways in which the use of branches can make programming of robots easier. A good example is in the editing of a robot program. The problem of editing an existing program can be facilitated when the program is divided into branches, even if the branches are used only once during the work cycle. Instead of editing the entire program when an error is discovered, the individual branch can be edited. For instance, when a robot is programmed using the manual leadthrough method, the editing of the program is never easy and sometimes impossible. This is due to the fact that these programs are recorded into the robot control memory by sampling the location of the current robot joint configuration many times per second. Accordingly, to correct an individual point or series of points would require the user to know which of thousands of points must be changed. Rather than make the editorial corrections in the control memory, it is common practice to divide such a program into logical sections (branches). When a program is found to be in error, the particular branch containing the error is deleted from the program and retaught using the manual leadthrough method.

8-7 CAPABILITIES AND LIMITATIONS OF LEADTHROUGH METHODS

Some of the teach pendants used for commercially available robots possess a wide range of capabilities. Defining points in space or setting the speed can be easily done using the toggle switches and dials of a simple teach pendant. For WAIT, SIGNAL, and DELAY commands, special buttons must be added to the pendant or these functions must be defined using the console at the main controller. The programming of branches can also be accomplished with a teach pendant in various ways. For example, a branch can be executed by using a toggle switch mounted either on the teach pendant or on the controller itself. When this method is used, there are several toggle switches corresponding to a predetermined set of branches. For example, suppose the controller has the capability for three branches to be defined. This would allow the programmer to define each branch, and then call it into the regular program at the desired times by manipulating the toggle switches to indicate when each of the branches are to be executed. Some of the teach pendants are quite sophisticated in terms of the kinds of instructions that can be programmed into the controller. In Appendix A9 after Chap. Nine, we describe the operation of the teach pendant supplied by United States Robots for the MAKER 110 model. This is an example of a programming device with features that are somewhat beyond those of the usual teach pendant and more like a textual language. We must define some additional programming terms in Chap. Nine usually associated with robot programming languages before discussing the MAKER 110 system.

Although the leadthrough programming controls offer the above capabilities, there are certain inherent limitations with the leadthrough methods. These limitations can be summarized as follows:

1. The robot cannot be used in production while it is being programmed.
2. As the complexity of the program increases, it becomes more difficult to accomplish leadthrough programming using the currently available methods.
3. Leadthrough programming is not readily compatible with modern computer-based technologies such as CAD/CAM, data communications networking, and integrated manufacturing information systems.

The fact that leadthrough programming requires the presence of the robot precludes the use of the robot in production. This has important economic implications. Since programming takes time away from production, it means that the batch size of parts to be produced by the robot must be sufficiently large to minimize the contribution of the programming cost. If the lot size is too small, it might take longer to prepare the program than to run it.

The second limitation deals with the fact that robots are being employed in production applications of increasing complexity, and being called on to perform increasingly sophisticated functions. With the leadthrough methods, it is difficult to program these kinds of functions. As we have progressed through this chapter in our discussion of the various robot programming functions, it has become more and more difficult to describe the individual steps that are to be performed by the robot without resorting to some form of computerlike coding. It is more difficult to define many of the nonmotion activities with a teach pendant than with a textual programming language.

The third limitation is concerned with the problem of interfacing robots to other computer-based systems in the corporation. One of the important goals in manufacturing is to establish computer-integrated manufacturing (CIM) systems in which the data and information necessary to make a product is captured originally on a CAD/CAM data base, and downloaded through the various manufacturing planning steps ultimately to the shop floor where robots and other automated systems work. The various components of the CIM system need to be able to communicate with each other and with the central plant computer. The use of leadthrough programming procedures does not lend itself to the communications and data base requirements of this kind of computer-integrated factory. Textual robot languages are more suited to these needs. We continue our study of robot programming in the next chapter by examining the opportunities and capabilities offered by these languages.

REFERENCES

1. M. P. Groover and E. W. Zimmers, Jr., *CAD/CAM: Computer-Aided Design and Manufacturing*, Prentice-Hall, Englewood Cliffs, NJ, 1984, chap. 10.
2. E. Kafrissen and M. Stephens, *Industrial Robots and Robotics*, Reston, Reston, VA, 1984, chap. 9.

3. R. P. Paul, *Robot Manipulators: Mathematics, Programming, and Control*, The MIT Press, Cambridge, MA, 1981, chap. 10.
4. R. R. Schreiber, "How to Teach a Robot," *Robotics Today*, June 1984, pp. 51–56.
5. R. Thomas, "Programming Expands Limits of Robot Controllers," *Industrial Enginnering*, April 1983, pp. 34–40.
6. L. L. Toepperwein, M. T. Blackman, et al., "ICAM Robotics Application Guide," *Technical Report AFWAL-TR-80-4042*, Vol. II, Materials Laboratory, Air Force Wright Aeronautical Laboratories, Ohio, April 1980.

PROBLEMS

8-1 Using the 8×8 square grid illustrated in Fig. 8-6, show the path taken by a cartesian coordinate robot if it is directed to move between the following sets of points in the grid using linear interpolation:

(*a*) Point (1,1) and point (6,6).
(*b*) Point (2,1) and point (8,2).
(*c*) Point (2,2) and point (7,5).

8-2 Using the gridwork illustrated in Fig. 8-8 for a robot with one rotational axis and one linear axis, show the path taken by the robot if it is directed to move between the following sets of points in the grid using joint interpolation:

(*a*) Point (1,1) and point (6,6).
(*b*) Point (2,1) and point (8,2).
(*c*) Point (2,2) and point (7,5).

8-3 Using the gridwork illustrated in Fig. 8-8 for a robot with one rotational axis and one linear axis, show the path taken by the robot if it is directed to move between the following sets of points in the grid using linear interpolation:

(*a*) Point (1,1) and point (6,6).
(*b*) Point (2,1) and point (8,2).
(*c*) Point (2,2) and point (7,5).

8-4 Rewrite the program in Example 8-5 in the text so that it includes the use of WAIT instructions to make sure that the gripper has opened and closed properly before the next step in the program has been executed. Use the coding format we have adopted in this chapter to write the program.

The following programming exercises are intended to illustrate the programming features of a robot that is programmed using the powered leadthrough method. This type of robot includes many of the small teaching robots, such as the Microbot TeachMover and the Rhino robot.

8-5 Using a pen mounted in the robot's end effector, program the robot to write your first and last initials in large rectangular letters (about 6 in. high) on a sheet of paper attached to the surface of the work table.

8-6 The robot is to be programmed to pick up a part from point A and move it to point B, followed by a move to a neutral position. Points A and B are to be defined by the programmer within the robot's work volume. Then the robot should pick up the part at point B and move it back to point A, followed by a move to the previous neutral position. The robot can be operated continually in the "run" mode to repeat the motion pattern over and over.

8-7 As an enhancement to Problem 8-6, if the capability exists on the robot used (e.g., Microbot TeachMover), program the robot to check each time it attempts to pick up the part to determine whether or not it has closed on the part. If the part is in the gripper, then continue the program. If the part is not in the gripper, the robot should move to the neutral position, provide a signal of some kind (e.g., light or buzzer), and wait 5 s. It should then attempt the pickup again.

8-8 Program the robot to pick up two blocks (the blocks are of different sizes) from fixed positions on either side of a center position, and to stack the blocks in the center position. The larger block will always be on one side of the center and the smaller block will always be on the other side of the center position. The smaller block is to be placed on top of the larger block.

8-9 This exercise is similar to the Prob. 8-8 except that the positions of the two blocks can be exchanged at random. It is not known whether the larger block is on one side of the center or the other. The robot must be programmed to always pick up the larger block first, place it at the center position, and to then pick up the smaller block and place it on top of the larger block.

8-10 The robot is to be programmed to pick up a part from a known fixed position on a conveyor and to place it at an upstream location on the conveyor so that the conveyor will deliver it back to the pickup point. The fixed pickup position is established by means of a mechanical stop along the conveyor so that the part is always in the same orientation and location for the robot.

NINE

ROBOT LANGUAGES

The previous chapter on robot programming was devoted to the types of task commands that can be implemented on a robot that is taught using lead-through programming, specifically the teach pendant leadthrough method. The current chapter will be concerned with textual languages for robot programming. Most of the robot languages implemented today use a combination of textual programming and teach pendant programming. The textual language is used to define the logic and sequence of the program, while the specific point locations in the workspace are defined using teach pendant control. This can be thought of as an off-line–on-line programming method in the sense that the program itself can be written off-line with the textual language, while the points must be defined on-line with the teach pendant.

In our discussion a textual programming language will be described which represents a composite of the available commercial robot languages. We have attempted to extract features from the various languages to present the concepts and principles in textual programming. In addition, we have prepared several appendixes at the end of the chapter which describe a number of the programming systems used in robotics. These appendixes were prepared from the "user's manuals" and similar documents provided by the companies marketing these systems.

9-1 THE TEXTUAL ROBOT LANGUAGES

The first textual robot language was WAVE, developed in 1973 as an experimental language for research at the Stanford Artificial Intelligence Laboratory. Research involving a robot interfaced to a machine vision system was accomplished using the WAVE language. The research demonstrated the feasibility of robot hand–eye coordination. Development of a subsequent language began in 1974 at Stanford. The language was called AL and it could be used to control multiple arms in tasks requiring arm coordination.

Many of the concepts of WAVE and AL went into the development of the first commercially available robot textual language, VAL (for Victor's Assembly Language, after Victor Scheinman). VAL was introduced in 1979 by Unimation, Inc., for its PUMA robot series. This language was upgraded to VAL II and released in 1984.

Work in robot language development was also taking place at the T. J. Watson Research Labs of the IBM Corporation, starting around 1976. Two of the IBM languages were AUTOPASS and AML (A Manufacturing Language), the second of which has been commercially available since 1982 with IBM's robotic products. Both of these languages are directed at assembly and related tasks.

Some of the other textual languages for robots that should be mentioned include RAIL, introduced in 1981 by Automatix for robotic assembly and arc welding, as well as machine vision; MCL (Manufacturing Control Language), developed under U.S. Air Force sponsorship by McDonnell-Douglas as an enhancement of the APT (Automatically Programmed Tooling) numerical control part programming language; and HELP, available from the General Electric Company under license from the Italian firm DEA.

9-2 GENERATIONS OF ROBOT PROGRAMMING LANGUAGES

The textual robot languages possess a variety of structures and capabilities. These languages are still evolving. In this section we identify two generations of textual languages and speculate about what a future generation might be like.

First Generation Languages

The "first generation" languages use a combination of command statements and teach pendant procedures for developing robot programs. They were developed largely to implement motion control with a textual programming language, and are therefore sometimes referred to as "motion level" languages.[11] Typical features include the ability to define manipulator motions (using the statements to define the sequence of the motions and the teach pendant to define the point locations), straight line interpolation, branching, and elementary sensor commands involving binary (on–off) signals. In other words, the first generation languages possess capabilities similar to the advanced teach pendant methods used to accomplish the robot programming instructions described in Chap. Eight. They can be used to define the motion sequence of the manipulator (MOVE), they have input/output capabilities (WAIT, SIGNAL), and they can be used to write subroutines (BRANCH). For writing a program of low-to-medium complexity, a shop person would likely find the teach pendant methods of programming easier to use, whereas people with computer programming experience would probably find the first generation languages easier to use. The VAL language is an example of a first generation robot programming language.

Common limitations of first generation languages include inability to specify complex arithmetic computations for use during program execution, the inability to make use of complex sensors and sensor data, and a limited capacity to communicate with other computers. Also, these languages cannot be readily extended for future enhancements.

Second Generation Languages

The second generation languages overcome many of the limitations of the first generation languages and add to their capabilities by incorporating features that make the robot seem more intelligent. This enables the robot to accomplish more complex tasks. These languages have been called "structured" programming languages[11] because they possess structured control constructs used in computer programming languages. Commercially available second generation languages include AML, RAIL, MCL, and VAL II. Programming in these languages is very much like computer programming. This might be considered a disadvantage since a computer programmer's skills are required to accomplish the programming. The second generation languages commonly make use of a teach pendant to define locations in the work space.

The features and capabilities of these second generation languages can be listed as follows[13]:

1. Motion control. This feature is basically the same as for the first generation languages.
2. Advanced sensor capabilities. The enhancements in the second generation languages typically include the capacity to deal with more than simple binary (on–off) signals, and the capability to control devices by means of the sensory data.
3. Limited intelligence. This is the ability to utilize information received about the work environment to modify system behavior in a programmed manner.
4. Communications and data processing. Second generation languages generally have provisions for interacting with computers and computer data bases for the purpose of keeping records, generating reports, and controlling activities in the workcell.

The motion control capability in some of the second generation languages goes beyond the previous generation by including more complex geometry problems than straight line interpolation. The MCL language, for instance, is based on APT. Accordingly, MCL includes many of the geometry definition features contained in APT. For example, lines, circles, planes, cylinders, and other geometric elements can be defined in APT and MCL.

The advanced sensor capabilities include the use of analog signals in addition to binary signals, and the ability to communicate with devices that are controlled by these signals. Control of the gripper is an example of the enhanced sensor capabilities of the second generation languages. Typical control of the gripper using a first generation language involves commands to open or close the gripper. Second generation languages permit the control of sensored grippers which can measure forces (these grippers were described in Chap. Six). The sensor monitors the forces or pressures during closure against an object, and the robot controller is able to regulate the amount of gripping force that is applied to the object.

The third feature provided by the second generation languages is limited intelligence. Instead of merely repeating the same motion pattern over and over, with slight differences for different product configurations, the robot has the capacity to deal with irregular events that occur during the work cycle in a way that seems intelligent. The intelligence is limited in the sense that it must be programmed into the robot controller. The robot cannot figure out what to do on its own beyond what it has been programmed to do. The error recovery problem illustrates this intelligence feature that might be programmed into the robot controller. Suppose the holding fixture in a robotic machining cell malfunctions by failing to close properly against the workpart. The robot's intelligent response might be to open the fixture, grasp the part and lift it out, reinsert the part back into the fixture, and signal for closure. If this recovery procedure works, the activities in the cell resume under regular programmed control. If not, the procedure might be repeated once or twice or some other action might be taken. The robot gives the appearance of behaving in an intelligent way, but it is operating under algorithms that have been programmed into its controller. By contrast with this error recovery procedure, an "unintelligent" response of the robot would be to merely stop all work in the cell in the event of a malfunction.

First generation languages are quite limited in their ability to communicate with other computers. Typically, any communication with other controllers and similar external devices must be accomplished by means of the WAIT and SIGNAL commands through the input/output ports of the robot. Second generation languages possess a greater capacity to interact with other computer-based systems. The communications capability would be used for maintaining production records on each product, generating performance reports, and similar data processing functions.

A related feature of some of the second generation languages is extensibility. This means that the language can be extended or enhanced by the user to handle the requirements of future applications, future sensing devices, and future robots, all of which may be more sophisticated than at the time the language is initially released. It also means that the language can be expanded by developing commands, subroutines, and macro statements (with a mechanism for passing parameter values from the main program) that are not included in the initial instruction set.

Future Generation Languages

Future generation robot languages will involve a concept called "world modeling." Other terms sometimes used instead of world modeling include model-based languages and task-object languages. In a programming scheme based on world modeling, the robot possesses knowledge of the three-dimensional world and is capable of developing its own step-by-step procedure to perform a task based on a stated objective of what is to be accomplished.

According to this definition, there are two basic ingredients of a programming language based on a world-modeling system. The first is that the robot system has in its control memory a three-dimensional model of its work environment. This model includes the robot manipulator itself, the worktable, fixtures, tools, parts, and so on. The model might be generated either by inputing three-dimensional geometric data into the control memory or by providing the robot with the capacity to see the work environment and properly interpret what it sees. In this latter case, the robot develops its own three-dimensional model of the workspace.

The second ingredient is the capacity for automatic self-programming. In effect, the human programmer gives the system an objective, and the system develops its own program of actions required to accomplish the objective. To use an extreme example of world modeling, borrowed from Grossman,[4] the programmer might write the following program for the robot:

ASSEMBLE TYPEWRITER

The robot would have to obey this command by developing its own step-by-step plan for selecting the separate components and assembling them in the proper order to make the typewriter. The level of intelligence implied by this example is many years away, and some might argue that robots will never be intelligent enough to deal with such a high-level command. What is more likely than the above macro command is that the task of assembling the typewriter would be divided into smaller subtasks that can be more readily managed by the robot. Nevertheless, it is clear that the kind of intelligence required in world modeling is of a different type and higher level than the preprogrammed intelligence represented by the error recovery example described above under second generation languages. The robot intelligence needed in world modeling includes the ability to solve problems and make decisions relying on other than preprogrammed instructions. One general approach to this problem involves the use of artificial intelligence, and in the following chapter we discuss this topic and its potential applications in robotics. Another approach is one that has been adopted in research on hierarchical control systems for robotics and other programmable automation at the National Bureau of Standards.[1] In Chap. Nineteen, we describe this work at NBS.

In principle, it should be possible with future generation languages to accomplish robot programming completely off-line without requiring the need for a teach pendant to physically show each point in the program to the robot.

In world modeling, the robot possesses a geometric model of its workspace by which it knows the desired locations without being taught each point. The obvious disadvantage of the first and second generation languages is the interruption of regular production work while the teach pendant is used to define the locations named in the program. This disadvantage would be avoided with off-line programming. In Chap. Eleven, we return to the topic of off-line programming when we discuss work cell simulation by computer graphics. Future applications of off-line programming are likely to require some form of simulation to verify the correctness of the program in advance of its downloading to the robot.

There are problems to be solved before future generation languages using off-line programming become a reality. Certainly one of these problems deals with the accuracy of the world model contained in the robot's memory. The model is not the same as the real world. There will always be some positional error between the actual physical objects in the work environment and the computer model used by the robot. For intricate tasks (e.g., assembly operations), the positional errors may mean the difference between success and failure in performing the task.

A second problem with off-line programming is concerned with the technology of artificial intelligence, hierarchical control, and similar approaches that would permit the robot to accept a high-level objective-oriented command (e.g., ASSEMBLE TYPEWRITER) and translate that instruction into a series of actions required to accomplish it. World modeling for robot programming is an area of significant interest in a number of research and development laboratories, both in academia and industry.

9-3 ROBOT LANGUAGE STRUCTURE

The second generation languages represent the current state of the art in textual languages. In this and the following sections of this chapter we describe the features that characterize the second generation languages in use today for robotics. Some of these features, of course, are applicable to first generation languages as well, but our discussion will include the more advanced capabilities that usually go beyond the first generation.

The language must be designed to operate with a robot system as illustrated in Fig. 9-1. It must be able to support the programming of the robot, control of the robot manipulator, and interfacing with peripherals in the work cell (e.g., sensors, and equipment). It should also support data communications with other computer systems in the factory.

Operating Systems

In using the textual languages, the programmer has available a CRT monitor, an alphanumeric keyboard, and a teach pendant. There should also be some means of storing the programs, either on magnetic tape or disk. Using the

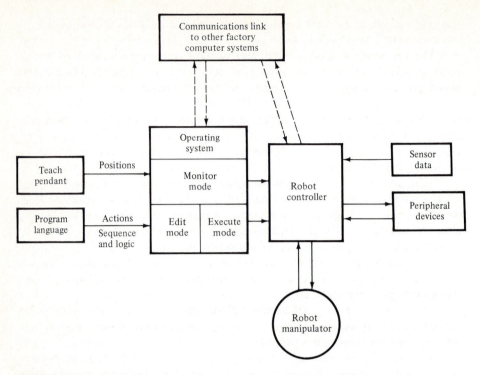

Figure 9-1 Diagram of robot system showing various components of the system that must be coordinated by means of the language.

language requires that there be some mechanism that permits the user to determine whether to write a new program, edit an existing program, execute (run) a program, or perform some other function. This mechanism is called an operating system, a term used in computer science to describe the software that supports the internal operation of the computer system. The purpose of the operating system is to facilitate the operation of the computer by the user and to maximize the performance and efficiency of the system and associated peripheral devices. The definition and purpose of the operating system for a robot language are similar.

A robot language operating system contains the following three basic modes of operation:

1. Monitor mode
2. Run mode
3. Edit mode

The monitor mode is used to accomplish overall supervisory control of the system. It is sometimes referred to as the supervisory mode. In this mode, the user can define locations in space using the teach pendant, set the speed control for the robot, store programs, transfer programs from storage back into control memory, or move back and forth between the other modes of operation such as edit or run.

The run mode is used for executing a robot program. In other words, the robot is performing the sequence of instructions in the program during the run mode. When testing a new program in the run mode, the user can typically employ debugging procedures built into the language to help in developing a correct program. For example, the program may call for the manipulator to exceed its joint limits in moving from one point named in the program to the next. Since the robot cannot do this, an error message would print out on the monitor and the robot would stop. This condition can be corrected by returning to the edit mode and adjusting the program or by redefining the point. Most modern robot languages permit the user to cross back into the monitor or edit modes while the program is being executed, so that another program can be written. In some cases, it is even possible to edit the current program although there are inherent dangers in doing this.

The edit mode provides an instruction set that allows the user to write new programs or to edit existing programs. Although the operation of the editing mode differs from one language system to another, the kinds of editing operations that can be performed include the writing of new lines of instructions in sequence, deleting or making changes to existing instructions, and inserting new lines in a program.

As with a computer programming language, the robot language program is processed by the operating system using either an interpreter or a compiler. An interpreter is a program in the operating system that executes each instruction of the source program (in our case, the source program is the user's robot language program) one at a time. VAL is an example of a robot language that is processed by an interpreter. A compiler is a program in the operating system that passes through the entire source program and pretranslates all of the instructions into machine level code that can be read and executed by the robot controller. MCL is an example of a robot language that is processed by a compiler. Compiled programs usually result in faster execution times.[2] On the other hand, a source program that is processed by an interpreter can be edited more readily since recompilation of the entire program is not required.

Robot Language Elements and Functions

Let us examine some of the basic elements and functions that should be incorporated into the language to enable the robot to perform tasks of medium to high complexity. We will organize our discussion into the following headings:

Constants, variables, and other data objects
Motion commands
End effector and sensor commands
Computations and operations

Program control and subroutines
Communications and data processing
Monitor mode commands

The following sections will discuss these basic elements of second generation robot languages. Where appropriate we will illustrate elements using vocabulary statements from actual robot languages. In those cases we will usually identify the language. In other cases, we will invent vocabulary words and statements that are composites of more than a single language. There is a risk in these sections that the text will read like the programmer's manual for some robot language. We ask the reader's indulgence if our descriptions seem like this.

9-4 CONSTANTS, VARIABLES, AND OTHER DATA OBJECTS

Like other computer programming languages, a robot language should permit the programmer to specify constants, variables, and other forms of data for use in the program. The second generation robot languages possess a rich set of capabilities for accomplishing the definition of these data objects.

Constants and Variables

A constant is a value used in the program that does not change during execution of the program. A variable in computer programming is a symbol or symbolic name that can change in value during execution of the program. Constants and variables can be integers (whole numbers), real numbers containing a decimal point, or strings that are enclosed in quotes. The range of values for numerical constants and variables depends on the computer system on which the language is implemented. The bit capacity of the central processing unit establishes a limit on the range of values that the computer can handle. Integers and real numbers can have positive or negative values, indicated by a + or − sign. A string is a sequence of 8-bit alphanumeric characters or symbols (with 8 bits, there are up to 256 possible characters or symbols), usually surrounded by a marker such as '. An example of a string is a 'robot'.

The exact syntax for the above constants might vary from one language to the next. We shall attempt in our discussion to reflect the most prevalent practice and not to concern ourselves with the syntax issue.

Variable symbols and names are created from the alphanumeric character set (letters a through z and digits 0 through 9). There are rules in the various robot languages for establishing variable names. Typical rules are that the variable name must begin with an alphabetic letter, and that it must not be identical to a language vocabulary word. In the second generation languages, variables can be specified for integers, real numbers, or strings. First generation languages were generally limited to the specification of integer variables that might be used, for example, for counting routines in the program logic.

Aggregates and Location Variables

An aggregate is an ordered set of constants or variables. AML, for example, permits the specification of an aggregate by enclosing it with the bracketing symbols ⟨ and ⟩, and by separating the elements in the aggregate by commas. Two examples of aggregates in the AML language are

⟨50.526, 236.003, 14.581, 25.090, 125.750⟩
⟨'we', 'they'⟩

The first example is an aggregate consisting of five real numbers. The second example consists of two strings. It is not necessary that the aggregate contain elements that are all of the same type. Any combination of integers, real numbers, and strings can be contained in the same aggregate.

An aggregate can be used to specify the joint coordinate values of a robot's joints. The first example above could be used to define a five-axis robot's joint coordinate values for a point in space. The method of specification might be done as follows:

DEFINE A1 = POINT ⟨50.526, 236.003, 14.581, 25.090, 125.750⟩

The usual interpretation of the aggregate in this statement is that the first three values (50.526, 236.003, 14.581) define the position of the wrist, or the tool attached to the wrist, in world space (x-y-z coordinates). The remaining values (25.090, 125.750) refer to the rotations of the wrist joints in degrees relative to some neutral reference position.

9-5 MOTION COMMANDS

Among the most important functions in a robot language are those which control the movement of the manipulator arm. This section describes how the textual languages accomplish these functions.

MOVE and Related Statements

One of the most important functions of the language, and the principal feature that distinguishes robot languages from computer programming languages, is manipulator motion control. In Chap. Eight, we defined the basic motion command, the MOVE statement

MOVE A1

This causes the end of the arm (end effector) to move from its present position to the point (previously defined), named A1. Recall from the previous chapter that a point is defined in terms of the robot's joint positions, and so A1 defines the position and orientation of the end effector. This MOVE statement generally causes the arm to move with a joint-interpolated motion. There are variations on the MOVE statement. For example, the VAL II language provides for a straight line move with the statement:

MOVES A1

The suffix S stands for straight line interpolation. The controller computes a straight line trajectory from the current position to the point A1 and causes the robot arm to follow that trajectory.

In some cases, the trajectory must be controlled so that the end effector passes through some intermediate point as it moves from the present position to the next point defined in the statement. This intermediate point is referred to as a via point. The need for the via point arises in applications in which there are obstacles and clearances to be considered along the motion path. For example, in removing a part from a production machine, the arm trajectory would have to be planned so that no interference occurs with the machine. The move statement for this situation might read like the following:

MOVE A1 VIA A2

This command tells the robot to move its arm to point A1, but to pass through via point A2 in making the move.

A related move sequence involves an approach to a point and departure from the point. The situation arises in many material-handling applications, in which it is necessary for the gripper to be moved to some intermediate location above the part before proceeding to it for the pickup. This is what is called an approach, and the robot languages permit this motion sequence to be done in several different ways. We will use VAL II to illustrate. Suppose the robot's task is to pick up a part from a container. We assume that the gripper is initially open. The following sequence might be used:

APPRO A1, 50
MOVES A1
SIGNAL (to close gripper)
DEPART 50

The APPRO command causes the end effector to be moved to the vicinity of point A1, but offset from the point along the tool z axis in the negative direction (above the part) by a distance of 50 mm. From this location the end effector is moved straight to the point A1 and closes its gripper around the part. The DEPART statement causes the robot to move away from the pickup point along the tool z axis to a distance of 50 mm. The provision is available in VAL II for the APPRO and DEPART statements to be performed using straight line interpolation rather than joint interpolation. These commands are APPROS and DEPARTS, respectively.

In addition to absolute moves to a defined point in the workspace, incremental moves are also available to the programmer. In the incremental move, the direction and distance of the move must be defined. This is commonly done by specifying the particular joint(s) to be moved and the distance of the move. Move distances for linear joints are defined in inches or millimeters, while rotational joint moves are specified in degrees of rotation. The following examples from AML illustrate the possibilities:

DMOVE(1, 10)
DMOVE(⟨4,5,6⟩, ⟨30,−60,90⟩)

DMOVE is the command for an incremental or "Delta" move. In parenthesis, the joint and the distance of the incremental move are specified. The first example moves joint 1 (assumed to be a linear joint) by 10 in. The second example commands an incremental move of axes 4, 5, and 6 by 30°, −60°, and 90°, respectively.

In the AL language, which is designed for multiple arm control, the MOVE statement can be used to identify which arm is to be moved. Robots of the future might possess more than a single arm, and we present the AL statement to illustrate how this might be done.

MOVE ARM2 TO A1

The robot is instructed to move its arm number 2 from the current position to point A1.

SPEED Control

The SPEED command is used to define the velocity with which the robot's arm is moved. When the SPEED command is given in the monitor mode (preparatory to executing a program), it indicates some absolute measure of velocity available for the robot. This might be specified as

SPEED 60 IPS

which indicates that the speed of the end effector during program execution shall be 60 in./sec unless it is altered to some other value during the program. If no units are given, the speed command usually indicates some value relative to the robot designer's concept of "normal" speed. For instance,

SPEED 75

indicates that the robot should operate at 75 percent of normal speed during program execution (unless altered during the program).

When the speed command is included as a statement in the robot program, it can be used either to specify the actual speed (e.g., 60 IPS), or it can indicate that the robot should operate at a certain percent of the speed that was specified under monitor mode before program execution. For example, if SPEED 60 IPS was specified under monitor command, and the following statement appeared in the program

SPEED 75

it would mean that the subsequent statements should be performed at a speed that is 75 percent of 60 IPS (45 in./sec).

Definition of Points in the Workspace

Our motion control programs have made use of points in the workspace. The locations of these points must be defined for the program. As indicated earlier, the definition of point locations is usually done by means of a teach pendant. The pendant is used to drive the robot arm to the desired position and orientation. Then, with a command typed into the keyboard such as

HERE A1

that point location is named A1. (The HERE statement is used in the VAL language.) The position and orientation of each joint are captured in control memory as an aggregate such as

⟨50.526, 236.003, 14.581, 25.090, 125.750⟩

As indicated previously, the first three values are the x-y-z coordinates in world space and the remaining values are wrist rotation angles.
An alternative way of specifying points in space is to name the point and designate its coordinate values by typing them into control memory directly without using the teach pendant. We have used the following method for specifying these coordinate values

DEFINE A1 = POINT ⟨50.526, 236.003, 14.581, 25.090, 125.750⟩

There are, of course, problems in defining points in this way because of the programmer's difficulty in knowing the coordinates in the work cell of the desired position for the robot end effector.

Paths and Frames

Several points can be connected together to define a path in the workspace. The following statement might be used to specify a path

DEFINE PATH1 = PATH(A1,A2,A3,A4)

Accordingly, the path PATH1 consists of the connected series of points A1, A2, A3, and A4, defined relative to the robot's world space. The beginning of the path is A1 and the end of the path is the last point that is specified in the series. A path can consist of two or more points. All points specified in the path statement must have been previously defined. The manner in which the robot moves between the points in the path is determined by the motion statement. For example, the statement

MOVE PATH1

would indicate that the robot arm would move through the sequence of positions defined in PATH1 using a joint-interpolated motion between the points. If the statement

MOVES PATH1

is used, this indicates that straight line interpolation must be used to move between the points in the path.

A reference frame is a cartesian coordinate system that may have other points or paths defined relative to it. For example, suppose a part has several identical features replicated in its design, each with a different position and orientation. Furthermore, suppose that the robot must be programmed to process each of the identical features on the part. An example of this kind of situation might be the routing of the same pattern at several locations around the contour of a curved plate. The pattern would consist of a path which contains several moves, and although all paths are identical, their positions and orientations in space vary around the part. In this situation, it would be convenient to program the routing path relative to a reference frame and then redefine the reference frame for each location on the part.

The following statement conveys the concept of the frame definition in robot programming

DEFINE FRAME1 = FRAME(A1,A2,A3)

The variable name given to the frame is FRAME1. Its position in space is defined using the three points, A1, A2, and A3. A1 becomes the origin of the frame, A2 is a point along the x axis, and A3 is a point in the xy plane. Accuracy is improved in the internal calculations as the separation between points is increased. The three points uniquely define the cartesian coordinate system of the new frame. The z axis is perpendicular to the xy plane, with its positive direction pointing to form a right-hand coordinate system.

In our illustration of the series of routing operations around the contoured surface of a curved plate, let us assume that there are nine identical patterns to be made, each one with a different reference frame. We can define the nine frames as FRAME1, FRAME2, ..., FRAME9. A routing path, called ROUTE, can now be defined relative to one of these frames (any frame will do) by means of a statement such as the following

DEFINE ROUTE:FRAME1 = PATH(P1,P2,P3,P4,P5,P6,P7)

where the series of points P1 through P7 defines the routing pattern at the first position on the part identified by FRAME1. Instead of the seven points being defined relative to the world space coordinate system, they are defined relative to the new coordinate system FRAME1. When the robot is commanded to follow the path, the statement must include the definition of the reference frame, as follows

MOVES ROUTE:FRAME1

To repeat the same path, only relocating and reorienting it relative to successive reference frames, we would use the following commands as required to execute the sequence of routing operations

$$\text{MOVES ROUTE:FRAME2}$$
$$.$$
$$\text{MOVES ROUTE:FRAME3}$$
$$.$$
$$.$$
$$.$$
$$\text{MOVES ROUTE:FRAME9}$$

Using the computational methods for compound transformations described in Chap. Four, the path ROUTE is transformed in the robot space into each new frame that is specified in the move command. Each of the points in ROUTE is transformed into the new frame, and the straight line segment path is executed accordingly.

If the programmer were to use the statement MOVES ROUTE, that is, without specifying the particular frame, then the default condition would require the robot to interpret the command so that the world space coodinate system were used as the reference frame. The path would be transformed into the robot's world cartesian coordinate system.

9-6 END EFFECTOR AND SENSOR COMMANDS

In Chap. Eight, we made use of the SIGNAL and WAIT statements to initiate output signals or await input signals. The signals were binary (on–off) which imposed limitations on the level of control that could be exercised. The second generation languages have more advanced input–output capabilities.

End Effector Operation

One of the uses of the SIGNAL commands in the previous chapter was to operate the gripper: SIGNAL 5 to close the gripper and SIGNAL 6 to open the gripper. In most robot languages, there are better ways of exercising control over the end effector operation. The most elementary commands are

$$\text{OPEN} \quad \text{and} \quad \text{CLOSE}$$

VAL II distinguishes between differences in the timing of the gripper action. The two commands OPEN and CLOSE cause the action to occur during execution of the next motion, while the statements

$$\text{OPENI} \quad \text{and} \quad \text{CLOSEI}$$

cause the action to occur immediately, without waiting for the next motion to begin. This latter case results in a small time delay which can be defined by a parameter setting in VAL II.

The preceding statements accomplish the obvious actions for a non-servoed gripper. Greater control over a servoed gripper operation can be achieved in several ways. For instance, the command

CLOSE 40 MM   or   **CLOSE 1.575 IN**

when applied to a gripper that has servocontrol over the width of the finger opening would close the gripper to an opening of 40 mm (1.575 in.). Similar commands would control the opening of the gripper.

Some grippers also have tactile and/or force sensors built into the fingers. These permit the robot to sense the presence of the object and to apply a measured force to the object during grasping. For example, a gripper servoed for force measurement can be controlled to apply a certain force against the part being grasped. The command

CLOSE 3.0 LB

indicates the type of command that might be used to apply a 3-lb gripping force against the part.

Force control of the gripper can be substantially more refined than the preceding command. For a properly instrumented hand, the AL language statement

CENTER

provides a fairly sophisticated level of control for tactile sensing. Invoking this command causes the gripper to slowly close until contact is made with the object by one of the fingers. Then, while that finger continues to maintain contact with the object, the robot arm shifts position while the opposite finger is gradually closed until it also makes contact with the object. The CENTER statement allows the robot to center its arm around the object rather than causing the object to be moved by the gripper closure. This could be useful in determining the position of an object whose location is only approximately known by the robot.

For end effectors that are powered tools rather than grippers, the robot must be able to position the tool and operate it. An OPERATE statement (based roughly on a command available in the AL language) might be used to control the powered tool. For example, consider the following sequence of commands:

OPERATE TOOL (SPEED = 125 RPM)
OPERATE TOOL (TORQUE = 5 IN LB)
OPERATE TOOL (TIME = 10 SEC)

We are assuming a powered rotational tool such as a powered screwdriver. All three statements apply to the operation. However, the first two statements are mutually exclusive; either the tool can be operated at 125 r/min or it can be operated with a torque of 5 in.-lb. The driver would be operated at 125 r/min until the screw began to tighten, at which point the torque statement would take precedence. The third statement indicates that after 10 sec the operation will terminate.

Sensor Operation

Let us consider some additional control features of the SIGNAL, WAIT, and similar statements beyond those described in Chap. Eight. The SIGNAL command can be used both for turning on or off an output signal. The statements

SIGNAL 3, ON

.
.
.

SIGNAL 3, OFF

would allow the signal from output port 3 to be turned on at one point in the program and turned off at another point in the program. The signal in this illustration is assumed to be binary. An analog output could also be controlled with the SIGNAL command. We will reserve for analog signals the input/output ports numbered greater than 100. The statement could be written as follows

SIGNAL 105, 4.5

This would provide an output of 4.5 units (probably volts) within the allowable range of the output signal.

The on–off conditions can also be applied with the WAIT command. In the following sequence, the robot provides power to some external device. The WAIT is used to verify that the device has been turned on before permitting the program to continue. Later in the program, the robot turns off the device and the device signals back that it has been turned off before the program continues. The relevant commands are as follows:

```
SIGNAL 5, ON      Robot turns on the device
WAIT 15, ON       Device signals back that it is on
    .
    .
    .
SIGNAL 5, OFF     Robot turns off the device
WAIT 15, OFF      Device signals back that it is off
```

The WAIT statement can be used for analog signals as well as binary digital signals in the same manner as the SIGNAL command.

Instead of identifying input and output signals by their I/O port number, it is often more convenient to define a variable name for the signal. This is usually easier for the programmer to remember. It also permits the variable to be used in the program, so that its value might be changed within the program by means of some logical operators (to be covered in Sec. 9-7). The variables could be defined as follows

```
DEFINE MOTOR1 = OUTPORT 5
DEFINE SENSR3 = INPORT 15
```

This would permit the preceding input output statements to be written in the following way

SIGNAL MOTOR1, ON
WAIT SENSR3, ON

.

.

.

SIGNAL MOTOR1, OFF
WAIT SENSR3, OFF

It is also possible to define an analog signal, either input or output, as a variable that is used during program execution. The statement

DEFINE VOLT1 = OUTPORT 105

specifies that the variable VOLT1 will be used with output port 105. At some point in the program, that variable could be computed to be a particular value (e.g., 4.5 V), and that value could be sent to the designed device in the cell by the statement

SIGNAL VOLT1

The value of VOLT1 would be signaled to the external device through output port 105. Similar use can be made of the WAIT command for an analog input signal. Specification of the variable name and associated input port is done by

DEFINE VOLTS3 = INPORT 115

Subsequent use of the variable can be made in a WAIT statement as follows

WAIT VOLT3

which indicates that the program execution should wait for the value of the signal on input port 115 to have a value that is greater than or equal to VOLT3. The programmer must keep in mind what the normal signal level is likely to be (whether it is normally greater than or less than VOLT3) since this may influence the logic of the program.

The REACT Statement

The REACT statement is a statement in VAL and VAL II used to continuously monitor an incoming signal and to respond to a change in the signal in some manner. It will serve to illustrate the types of commands used to interrupt regular execution of the robot program in response to some higher priority event. A typical use of this kind of command is when some error or safety hazard has occurred in the workcell and the condition is detected by one of the sensors.

In our discussion, we will expand the scope of the REACT statement slightly beyond its use in VAL or VAL II in order to be consistent with our other language statements. (Readers should refer to the VAL II Manual[15] for clarification of the exact use of the VAL statement.) A typical form of the statement would be as follows

REACT 17, SAFETY

The statement is interpreted as follows. Input line 17 is to be continuously monitored, and when a change in its signal value occurs, branch to a subroutine called SAFETY. (We will discuss subroutines in Sec. 9-8.) The input signal on line 17 is a binary signal which is at either of two levels (e.g., on or off). Input line 17 could be identified by some variable name if the programmer wishes (e.g., DEFINE VAR17 = INPORT 17). The change in the signal value which invokes the REACT statement is a change from one level to the other. If the signal is normally off, and it suddenly comes on, then REACT transfers program control to the subroutine identified in the statement.

The normal use of the REACT or equivalent types of statements in textual languages is to complete the current motion command before interrupting. In some cases, an immediate reaction is required. This is accomplished by the REACTI statement

REACTI 17, SAFETY

This causes immediate suspension of the regular program execution, so that a transfer to SAFETY is done at once. Manipulator motion stops immediately.

It is possible to define various levels of priority which would govern whether one REACT command takes precedence over another. This could be accomplished in the following way

REACT 17, SAFETY (PRIORITY 1)

If a REACT command with a lower priority, say (PRIORITY 4), were encountered while executing a higher priority REACT, say (PRIORITY 1), the current subroutine would be completed before transferring control to the new subroutine. On the other hand, if a subroutine were being executed at PRIORITY 4, a REACT at PRIORITY 1 would interrupt the current subroutine.

Analog I/O signals can be used with the REACT or REACTI commands (but not in VAL or VAL II). The analog signal level at which the reaction occurs must be specified in the statement. This might be done as follows

REACT 117, 4.5, SAFETY (PRIORITY 3)

This indicates that the transfer to subroutine SAFETY must occur if and when the input signal on port 117 becomes greater than or equal to 4.5 V. When variable names are used, the relevant statements are

DEFINE VAR17 = INPORT 17

.

.

.

REACT VAR17, SAFETY (PRIORITY 3)

The REACT command facilitates the design of an "interrupt" system in the ROBOT program. The purpose of an interrupt system is to transfer control from one part of the program to another (or to a subroutine) in response to conditions that take priority over regular program execution. In our case, the interrupt would be invoked upon receipt of an incoming signal as a result of the REACT instruction. Perhaps the incoming signal is expected to stay within specified limits for safe operation of the cell. If it did not, the interrupt system would halt the regular program and branch to a special subroutine. We shall illustrate the use of this command as part of an interrupt system in Example 9-3.

9-7 COMPUTATIONS AND OPERATIONS

The need arises in many robot programs to perform arithmetic computations and other types of operations on constants, variables, and other data objects. The standard set of mathematical operators in second generation languages are

+	addition
−	subtraction
*	multiplication
/	division
**	exponentiation
=	equal to

Precedence rules are established that evaluate an expression from left to right, with parentheses used to indicate that expressions within parentheses should be evaluated first.

Some of the languages also have the capability to calculate the common trigonometric, logarithmic, exponential, and similar functions. The following is a list of these functions that we can make use of in some of the problem exercises:

SIN(A)	Sine of an angle A
COS(A)	Cosine of an angle A
TAN(A)	Tangent of an angle A
COT(A)	Cotangent of an angle A
ASIN(A)	Arc sine of an angle A
ACOS(A)	Arc cosine of an angle A
ATAN(A)	Arc tangent of an angle A
ACOT(A)	Arc cotangent of an angle A
LOG(X)	Natural logarithm of X
EXP(X)	Exponential function e**X
ABS(X)	Absolute value of X
INT(X)	Largest integer less than or = X
SQRT(X)	Square root of X

In addition to the arithmetic and trigonometric operators, relational operators are also used to evaluate and compare expressions. The common relational operators are listed below

EQ	Equal to
NE	Not equal to
GT	Greater than
GE	Greater than or equal to
LT	Less than
LE	Less than or equal to

These relational operators are used to evaluate a relational expression to be true or false. They typically evaluate to -1 if the expression is true and 0 if the relation is false. Some of the languages allow more complex comparisons to be made in a single statement. For example, the AML language includes the following relational operator

$$VS$$

The VS relational operator evaluates to -1, 0, or $+1$, based on whether the first argument in the expression is less than, equal to, or greater than the second argument.

Logical operators are also sometimes useful in robot programming; however, not all of the second generation languages possess this capability. The common logical operators include;

AND	Logical AND operator
OR	Logical OR
NOT	Logical complement

These operators are sometimes limited in their use to integer constants and variables. The AND and OR operators work on two values while the NOT operator works with only one value.

Examples of expressions which make use of these various operators include the following

$3+(5*N)$	Arithmetic expression
$N=N+1$	Arithmetic expression
TAN(30)	Trigonometric function
COUNT GT 10	Relational expression

9-8 PROGRAM CONTROL AND SUBROUTINES

In writing a robot program it is often necessary to exercise control over the sequence in which the statements are executed by means of branches and subroutines. A variety of instructions are available in the textual robot languages to control the logical flow of the program and to name and use subroutines in the program.

Program Sequence Control

Before considering subroutines, let us examine some of the possible instructions that can be used to control logic flow in the program. The following
types of statements are available in the second generation languages.

GOTO 10

This indicates an unconditional branch to statement 10. The GOTO statement
can be used with a logical expression as follows

IF (logical expression) GOTO 10

This indicates that if the logical expression is true, then the program branches
to statement 10. Otherwise, it continues to the next statement in the program.
An example will serve to illustrate some of the arithmetic expressions and
program control commands.

> **Example 9-1** This program is for the same palletizing operation as in
> Examples 8-7 and 8-8 of the previous chapter. To review, the robot must
> pick up parts from an incoming chute and deposit them onto a pallet. The
> pallet has four rows that are 50 mm apart and six columns that are 40 mm
> apart. The plane of the pallet is assumed to be parallel to the xy plane.
> The rows of the pallet are parallel to the x axis and the columns of the
> pallet are parallel to the y axis. Figure 9-2 shows the arrangement of the
> pallet. The objects to be picked up are about 25 mm tall (approximately
> 1.0 in.). We will use the following constants and variables in our program:

Variables:

ROW	The row number (integer value)
COLUMN	The column number (integer value)
X	An x-coordinate value
Y	A y-coordinate value

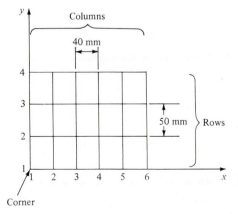

Corner

Figure 9-2 Pallet configuration for Examples
9-1 and 9-2.

Location constants:

PICKUP   The pickup point on the chute
CORNER   The corner starting point on the pallet

Location variables:

DROP   The dropoff point

The program to perform the palletizing operation is as follows (we provide commentary about some of the statements in the right margin):

```
      PROGRAM PALLETIZE
      DEFINE PICKUP = JOINTS(1,2,3,4,5)
      DEFINE CORNER = JOINTS(1,2,3,4,5)
      DEFINE DROP = COORDINATES(X,Y)
      OPENI
      ROW = 0                          Initialize ROW
  10  Y = ROW * 50.0                   Compute y for dropoff point
      COLUMN = 0                       Initialize COLUMN
  20  X = COLUMN * 40.0                Compute x for dropoff point
      DROP = CORNER + ⟨X,Y⟩            Define DROP for each iteration
      APPRO PICKUP, 50                 Pickup sequence
      MOVES PICKUP                     Pickup sequence
      CLOSEI                           Pickup sequence
      DEPART 50                        Pickup sequence
      APPRO DROP, 50                   Dropoff sequence
      MOVES DROP                       Dropoff sequence
      OPENI                            Dropoff sequence
      DEPART 50                        Dropoff sequence
      COLUMN = COLUMN + 1              Increment COLUMN variable
      IF COLUMN LT 6 GOTO 20           Check if COLUMN limit reached
      ROW = ROW + 1                    Increment ROW variable
      If ROW LT 4 GOTO 10              Check if ROW limit reached
      END PROGRAM
```

We see a further reduction in the number of logic steps that must be programmed. In our program above, the number of steps has been reduced to 26. There is also a reduction in the number of points in the workspace that must be taught using the teach pendant. In Example 8-8, the number of taught points was 27. In the above program the number of points that must be taught is two; the remaining positions are calculated in the program.

The IF statement provides the opportunity for a more complicated logical structure in the program, in the form of an IF...THEN...ELSE...END statement. This might be written in the following way:

IF (logical expression) THEN
.
(group of instructions)
.
ELSE
.
(group of instructions)
.
END

If the logical expression is true then the group of statements between THEN and ELSE are executed. If the logical expression is false then the group of statements between ELSE and END are executed. The program continues after the END statement.

DO loops can be accomplished in robot programs. The VAL II language will be used to illustrate the concept. The statement sequence would read as follows:

DO
.
(group of instructions)
.
UNTIL (logical expression)

The DO statement indicates the beginning of the DO loop. The group of instructions following the DO statement form a logical set whose variable value(s) would affect the logical expression associated with the UNTIL statement. On each execution of the group of instructions, the logical expression is evaluated. If the result is false the DO loop is executed again; if the result is true then the program continues at the first statement following the UNTIL step.

There are occasions during the program when it is necessary to delay or stop the execution of the program. The commands can take several different forms, depending on the circumstances that are foreseen by the programmer. The DELAY command can be used to delay the continuation of the program by a specified time, as indicated in Chap. Eight.

DELAY .5 SEC

The reason for the use of DELAY might be to ensure that some action has taken place before proceeding with the execution of the program.

The typical reasons for invoking a STOP command are related to unsafe conditions that have developed during workcell operation. The simplest form of stop command tells the controller to immediately stop execution of the program and the manipulator movement. We shall refer to this as a

STOP1

command. The next form of STOP command is one that stops the motion of the robot immediately, but permits continued execution of the program with the exception that no further moves of the manipulator are permitted. We shall call this a

STOP2

This permits other instructions following the STOP2 command to be executed. For example, in an emergency stop situation, it might be necessary to turn off other pieces of equipment in the workcell.

The third stop level permits the robot to complete the current motion instruction and then stop. This is a

STOP3

It permits continued execution of the program except for motion commands.

Finally, a fourth level of stop would be a temporary pause in the execution of the program. We shall differentiate this from the previous stop commands by use of the instruction word

PAUSE

which stops the program execution and reverts control back to the monitor mode. The program can be resumed by the operator using a

RESUME

command, which causes the program execution (including motion commands) to continue with the statement immediately following PAUSE. For convenience, the RESUME instruction could be abbreviated with an "R" or actuated with a special switch set up for the purpose. The PAUSE and RESUME statements might be utilized when a robot cell includes a human worker whose actions are variable and there is need to regulate the robot's cycle.

Subroutines

Subroutines were discussed in Chap. Eight under the heading of branching, but only elementary subroutines were considered. We used the statements

BRANCH FETCH

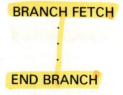

END BRANCH

to indicate the start and the ending of a branch or subroutine and to name the subroutine (FETCH). To call the branch during program execution, we simply used the name of the branch

FETCH

In the present chapter we will refer to branches by the more appropriate name SUBROUTINE. An advantage of the textual robot languages is that they generally permit subroutines to have arguments which can be named in a CALL statement and used within the subroutine. A subroutine with a single argument would be shown as follows

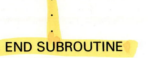

SUBROUTINE PLACE(N)

.
.
.

END SUBROUTINE

The argument N would be used somehow during the subroutine. We will demonstrate this capability in Example 9-2. The subroutine would be called using a statement that would identify the value of the argument; for example,

CALL PLACE(5)

Another statement used with subroutines is

RETURN

This statement results in termination of the subroutine and returns control of the program back to the statement following the CALL statement for the subroutine. The RETURN statement has a similar effect as the END BRANCH statement in the subroutine. The reason of its use is that there might be more than one place in the subroutine that the programmer might want to use to exit back to the main program.

The subroutine (branch) capability demonstrated in Chap. Eight was relatively simple with a minimum of logic. Subroutines of greater complexity are possible with the instruction set we have discussed in this section. Let us reconsider the previous example, this time using a subroutine with several arguments that can be called to perform the palletizing sequence.

Example 9-2 The palletizing operation is the same as in Example 9-1. We will use the same variable names as before except that the program will be more general so that it might be used for other pallet sizes. This can be done by defining the number of positions on the pallet in each direction and the pallet location spacing. Accordingly, we will need to define the following additional parameters:

MAXCOL	The number of columns on the pallet
MAXROW	The number of rows on the pallet
YSPACE	The spacing in the y direction (spacing between rows)
XSPACE	The spacing in the x direction (spacing between columns)

The main program is used with a PALLET subroutine. The subroutine uses the above parameters as arguments that must be identified when the subroutine is called. We organize the program as if it were used in production with empty pallets being brought into position in front of the robot and removed when they have been filled. The SIGNAL and WAIT commands are used for this purpose. SIGNAL 1 (output line 1) is used to initiate the delivery of an empty pallet into the loading position, and WAIT 11 (input line 11) is used in conjunction with a sensor device to ensure that the pallet delivery has been accomplished. After loading, SIGNAL 2 (output line 2) is used to initiate the removal of the filled pallet from the loading position, and WAIT 12 (input line 12) tests to make sure that it has been removed.

```
     PROGRAM PALLETIZE
     DEFINE PICKUP = JOINTS(1,2,3,4,5)
     DEFINE CORNER = JOINTS(1,2,3,4,5)
     DEFINE DROP = COORDINATES(X,Y)
     OPENI
5    SIGNAL 1
     WAIT 11
     CALL PALLET (MAXCOL = 6, MAXROW = 4, XSPACE = 40,
        YSPACE = 50)
     SIGNAL 2
     WAIT 12
     GOTO 5
     END PROGRAM

     SUBROUTINE PALLET (MAXCOL, MAXROW, XSPACE, YSPACE)
     ROW = 0
10   Y = ROW * YSPACE
     COLUMN = 0
20   X = COLUMN * XSPACE
     DROP = CORNER + ⟨X,Y⟩
     APPRO PICKUP, 50
     MOVES PICKUP
     CLOSEI
     DEPART 50
     APPRO DROP, 50
     MOVES DROP
     OPENI
     DEPART 50
     COLUMN = COLUMN + 1
     IF COLUMN LT MAXCOL GOTO 20
     ROW = ROW + 1
     IF ROW LT MAXROW GOTO 10
     END SUBROUTINE
```

Example 9-3 This program is not complete, but it contains the statements needed to illustrate a subroutine that might be used in an emergency stop situation. The pertinent statements are

.
.
.

REACT 113, 5.0, SAFETY

.
.
.

SUBROUTINE SAFETY
STOP2
SIGNAL 1, ON
SIGNAL 2, OFF
END SUBROUTINE

The REACT command indicates that the controller must measure the value of the signal on input line 113 and transfer to SUBROUTINE SAFETY if the SIGNAL ever exceeds a value of 5.0. Otherwise, the program execution continues without ever using this subroutine. The transfer of control operates as a program interrupt.

If SUBROUTINE SAFETY is invoked, it stops the robot arm immediately, but continues to step through the subsequent instructions in the subroutine. SIGNAL 1, ON might be used to turn on an alarm, and SIGNAL 2, OFF could be utilized to turn off a piece of machinery that works with the robot in the cell.

9-9 COMMUNICATIONS AND DATA PROCESSING

Communications in this section refers to the communication between the robot system and the operator or the robot system and other computer-based systems and their peripherals (e.g., storage devices, printers).

The capability of the robot system to communicate with the operator allows the robot to request information, to indicate the source of a malfunction, or to inform the operator about what it is presently doing. The most common device used by robot systems to communicate with a human operator is the CRT (cathode ray tube). Other possible devices include printers, character display devices other than a CRT, plotters, buzzers, and speech synthesizers. The programmer or operator must also interact with the system in order to prepare and edit programs, enter data requested by the system, and control the operation of the robot cell. Much of the communication is controlled by the programmer or operator by invoking the monitor mode of the system. The following section will describe some of the typical features and capabilities of the monitor mode. The common devices used by the operator to communicate with the system are the teach pendant and the alphanumeric keyboard. Other possible devices include joysticks, light pens (and other devices used to interact with a computer graphics system), and speech recognition systems.

The robot textual language would have one or more forms of READ and WRITE statements that could be employed by the programmer to design the system-to-operator and operator-to-system communication mechanism. The WRITE statement would be used to write messages (or files) to the operator on the CRT, and the READ statement would be used to read input from the operator to the system. The following is representative of a typical exchange that might occur during the system operation

```
WRITE 'ENTER part name of part placed on pallet'
READ (PARTNAME)
```

The dialog shows that the system has requested the operator to indicate which part has been loaded onto the next pallet that is to be transferred into the workcell. The READ statement is used to establish that the data entered by the operator at the console are to be stored as the variable name PART-NAME.

In Sec. 9-7, we developed several instructions that permitted the robot controller to send or receive signals from external devices (e.g., sensors and other equipment). The SIGNAL, WAIT, and REACT commands were all devoted to this purpose. The form of the signals in these transmissions is either binary or continuous analog. Similarly, advanced robot languages should offer commands that permit the transfer of data in digital form so that the robot system can communicate with other computer systems and digital-type devices. These commands should be designed to transmit or receive a file (consisting of anything from a single data value to an entire program) in the proper format so that the different devices can understand each other. Statements that satisfy these conditions could be cast in the following form

```
SEND DATA1 (CRT)
RECEIVE SUBROUTINE 6 (COMPUTER1)
```

The first statement is used to transmit a file called DATA1 to the operator's console. The console is the CRT and the argument in parentheses indicates the designated recipient and the format in which the data should be sent in order to be communicated to that device. In our example of sending data to the CRT, this represents an alternative to the WRITE statement and would be more convenient for certain data transmissions (e.g., where large amounts of data are involved). Methods of establishing files within the program are available and involve identifying the beginning and end of the file in some manner (e.g., BEGINFILE DATA1 .... ENDFILE DATA1). The system then saves DATA1 as a separate file for use as indicated in subsequent statements in the program, such as SEND DATA1 (CRT).

The second statement is used to receive a file called SUBROUTINE6 from an external device which is designated as COMPUTER1. Again, the data transmission must take place according to a particular format, and the argument (COMPUTER1) establishes that required format for the exchange. It is assumed that the data transmissions would be serial, that RS232 connectors would be used, and that other communication protocol standards would be adopted. Other characteristics of the transmission, such as baud rate, would be defined in an initializing statement for (CRT), (COMPUTER1), and the other devices which would exchange data in the system. The above examples are relatively simple, and tend to understate the steps required to set up the communications exchange between devices. Several of the operator's manuals listed at the end of the chapter go into more detail concerning the communications and data transfer issues, and the interested reader should consult these references.[2,6,18]

We anticipate that this field of digital communications will develop rapidly in the future and that future robots will be interfaced to CAD/CAM systems and plant level mainframe computers for the purpose of off-line programming, production scheduling, quality control, and performance reporting.

9-10 MONITOR MODE COMMANDS

As previously indicated, the monitor mode is used for functions such as entering locations by means of the teach pendant (HERE command) and setting the starting speed for execution of a robot program (SPEED command). Various functions relating to system supervision, data processing, and communications are also performed with the monitor commands. In this section we review some of the basic functions that can be accomplished in the monitor mode. We will use instruction statements that reflect typical language commands.

To open a file for the purpose of writing a new program or editing an existing program the EDIT command is used, followed by the name of the program. This command permits the operator to use the operating system's edit mode. When programming is finished, an EXIT command is given to store the program into the controller memory and to return from the edit mode to the monitor mode. If it is desired to store a program on a more permanent storage device (disk or tape) than control memory, the STORE (program name) command is used. The reverse procedure is READ (program name) which reads a file from storage into the robot controller memory. Files can also be displayed on the monitor using a LIST (program name) statement or printed to obtain a hard copy using a PRINT (program name). A statement such as DIRECTORY can be used to obtain a listing of the program names that are stored either in the controller memory or on the disk or tape. Programs can be deleted from memory or storage by means of a command like DELETE (program name) or ERASE (program name).

To have the robot execute a program, the user indicates this with an EXECUTE (program name) command. This is sometimes abbreviated by EX or EXEC. During execution the user may want to stop the program from continuing. A command such as ABORT or STOP is available to discontinue execution and stop the robot motion. This command in monitor mode is different from the various STOP commands that are used within the program to cause the program to be discontinued.

The various monitor commands are usually specific to a particular robot control operating system. In this section, we have only sampled the basic functions that are possible. Our exercise problems at the end of the chapter are concerned with robot programming inside the EDIT mode and will not make use of any of the monitor commands.

REFERENCES

1. J. S. Albus, A. J. Barbera, and R. N. Nagel, "Theory and Practice of Hierarchical Control," *Proceedings*, Twenty Third IEEE Computer Society International Conference, September 1981, pp. 18–39.
2. Automatix Incorporated, *RAIL Software Reference Manual*, Document No. MN-RB-07, Rev. 5.00, October 1983.
3. G. Carayannis, "Programming Languages for Intelligent Robots," (written presentation) Quebec, September 1983.
4. D. D. Grossman, "Robotic Software," *Mechanical Engineering*, August 1982, pp. 46–47.
5. W. A. Gruver, B. I. Soroka, J. J. Craig, and T. L. Turner, "Evaluation of Commercially Available Robot Programming Languages," *Conference Proceedings*, vol. 2, 13th International Symposium on Industrial Robots and Robots 7, Chicago, IL, April 1983.
6. IBM Corporation, *IBM 7565 Manufacturing System, A Manufacturing Language Concepts and User's Guide*, Base Publication 8509012, December 1982, and *A Manufacturing Language Reference*, Base Publication 8509016, October 1983.
7. S. Motiwalla, "Development of a Software System for Industrial Robots," *Mechanical Engineering*, August 1982, pp. 36–39.
8. M. S. Mujtaba, R. Goldman, and T. Binford, "Stanford's AL Robot Programming Language," *Computers in Mechanical Engineering*, August 1982, pp. 50–57.
9. SRL Working Group, "The Current Environment of Robot Languages," *Final Report*, R-83-RSP-01, March 1983.
10. R. R. Schreiber, "How to Teach a Robot," *Robotics Today*, June 1984, pp. 51–56.
11. W. E. Snyder, *Industrial Robots: Computer Interfacing and Control*, Prentice-Hall, Englewood Cliffs, NJ, 1985, chap. 15.
12. B. I. Soroka, "What Can't Robot Languages Do?" *Conference Proceedings*, 13th International Symposium on Industrial Robots and Robots 7, vol. 2, Chicago, IL, April 1983.
13. R. H. Taylor, P. D. Summers, and J. M. Meyer, "AML: A Manufacturing Language," *The International Journal of Robotics Research*, Fall 1982, pp. 19–41.
14. R. Thomas, "Programming Expands Limits of Robot Controllers," *Industrial Engineering*, April 1983, pp. 34–40.
15. R. B. Thornhill, "Application Software for Robotics—Current Applications and Future Industry Needs," *Proceedings*, CAM-I 12th Annual Meeting, Dallas, TX, November 1983.
16. R. B. Thornhill, "Generic Programming of Robots—The CAM-I Robotics Software Project," *Conference Papers*, Robotics Research, The Next Five Years and Beyond, Society of Manufacturing Engineers, Lehigh University, August 1984.
17. L. L. Toepperwein, M. T. Blackman, et al., "ICAM Robotics Application Guide," *Technical Report AFWAL-TR-80-4042*, vol. II, Materials Laboratory, Air Force Wright Aeronautical Laboratories, Ohio, April 1980.
18. Unimation, Inc., *Programming Manual—User's Guide to VAL II* (398T1), Version 1.1, August 1984.
19. B. O. Wood and M. A. Fugelson, "MCL, the Manufacturing Control Language," *Conference Proceedings*, 13th International Symposium on Industrial Robots and Robots 7, vol. 2, Chicago, IL, April 1983.

PROBLEMS

The following problems are textual robot programming exercises. The programs can be written using either the language statements developed in the text of the chapter or one of the commercially available robot languages. In some cases, owing to equipment limitations, it may be necessary to use the language instruction set proposed in this chapter to write the robot programs. The preferred procedure is to use one of the languages in conjunction with an actual robot and to test the program out on the robot. When used as laboratory teaching exercises, some of the exercises can be facilitated with physical containers, pallets, or fixtures, as indicated in the problems.

9-1 Write a program to instruct the robot to pick up bottles from a fixed location on a conveyor and insert them into a cardboard carton. A mechanical stop along the conveyor is used to locate the parts in a known position. The bottles are to be loaded into the carton about 12 in. away from the pickup point. Each carton holds two parts. Open cartons are presented to the robot and then subsequently closed and sealed at a different location. The open cartons are 4.0 in. tall and measure 5.0 by 10.0 in. The bottles to be loaded are 4.5 in. in diameter. Make a sketch of the workstation before programming and identify the various points used in the program. This sketch would be used by a technician to subsequently set up the workstation.

9-2 This problem is intended to simulate a continuous arc-welding operation. The problem is to weld together three rectangular steel plates into a partial box. The three steel plates are fixtured together in preparation for the welding operation. The continuous welds must be made along the three joints as indicated in Fig. P9-2(*a*). All three joints can be assumed to involve straight line welding. The welding torch must be held at an angle of 45° as illustrated in Fig. P9-2(*b*) in order to provide an even deposition of the welding rod into the seam.

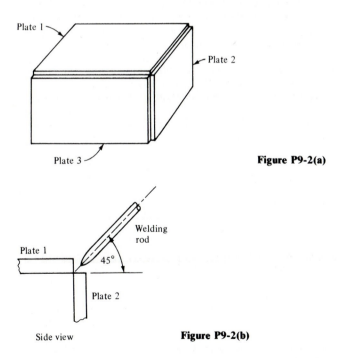

Plate 1

Plate 2

Plate 3

Figure P9-2(a)

Plate 1

Welding rod

45°

Plate 2

Side view

Figure P9-2(b)

9-3 A robot is to be programmed to unload parts from one pallet and load them onto another pallet. The parts are located on the unload pallet (pallet 1) in a 3 by 4 pattern in known fixed positions, 40 mm apart in both directions. The two directions of the pallet are assumed to be parallel to the *x* and y world coordinate axes of the robot. The parts are to be placed on the load pallet (pallet 2) in a 2 by 6 pattern, 40 mm apart in both directions. The two directions of the pallet are again assumed to be parallel to the x and y world coordinate axes of the robot. Make a sketch of the workstation setup before you begin programming.

9-4 As an additional exercise, the robot is to be programmed to repeat the above task only in reverse order, so that the objects are returned from pallet 2 back to pallet 1. Combining this program with the program of the preceding problem would permit the robot to continue moving parts perpetually between the two trays.

9-5 A robot is to be programmed to unload parts from one pallet and load them onto another pallet. The parts are located on the unload pallet in a 3 by 4 pattern in known fixed positions, 40 mm apart in both directions. The two directions of the pallet are assumed to be parallel to the x and y world coordinate axes of the robot. The parts are to be placed on the load pallet in a 2 by 6 pattern, 40 mm apart in both directions. The two directions of the load pallet are at an orientation of 30° with respect to the x and y world coordinate axis system of the robot. Make a sketch of the workstation setup before you begin programming.

9-6 Write a robot program to pick parts off a conveyor and load them into a pallet that is about 12 in. from the pickup point. A mechanical stop on the conveyor is used to locate the parts in a known position for the pickup. The parts are to be arranged in a 3 by 4 pattern, 40 mm apart in both directions. The two directions of the pallet are assumed to be parallel to the x and y world coordinate axes of the robot, respectively.

9-7 Repeat Prob. 9-2 but make use of the PATH statement, as described in the text.

9-8 With a felt-tipped pen as the robot's tool (either attached to the wrist or held in the gripper), program the robot to write two or three different letters or numbers on a piece of paper lying flat on the table in front of the robot. The programmer has the choice of which numbers and/or letters to use. Define each of the letters as a separate path using the PATH statement. Then provide a mechanism in your program for a user of the robot to enter the letter at the terminal in order for the robot to execute writing of that character.

9-9 A cube, 4 in. on each side, is to have five holes drilled in five of its faces in the pattern illustrated in Fig. P9-9(a). The five faces to be drilled are shown in Fig. P9-9(b). (Instead of an actual drilling operation, the student may want to substitute a marking operation on a block of a different size using a felt-tipped pen in the robot's end effector.) Use the PATH and FRAME commands to program this operation.

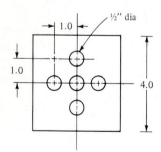

Figure P9-9(a)

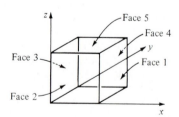

Figure P9-9(b)

9-10 This programming exercise is designed to have the robot make use of interlocks with other pieces of hardware in its workcell. The instructions demonstrated are WAIT and SIGNAL. The other pieces of hardware in the cell consist of: (1) a part rest with a limit switch designed to sense that a part is in place in the rest; (2) a rotating table that can be turned on and off by means of an external signal. The program should be written so that the robot waits for a part to be placed in the rest by a human operator. The input interlock is used for this purpose. When the part is placed in the rest, the robot picks up the part and places it in the center of the table which is presently turned off. The robot signals the table to rotate for 15 s (perhaps symbolizing some production process). At the end of the 15 s, the robot turns off the rotation, retrieves the part, and places it in some neutral location near the part rest.

9-11 Two different style workparts travel down a conveyor and are positioned in a robot workstation for a certain processing operation. (These may be two different car body styles coming down an automobile spot-welding line.) Two photocells are used to indicate that workparts have arrived in the workstation and to distinguish between the two styles. The operation of the photocells is as follows. The first photocell is actuated (ON) when the part has arrived and is registered in place. The second photocell is actuated (ON) when part style A is located in place in the workstation, and not actuated (OFF) when part style B is in place in the station. Using the WAIT commands with the incoming photocell signals, write a robot program that will call either of two processing subroutines depending on which part is presented to the workstation. The two subroutines are called by the statements: CALL ASTYLE, and CALL BSTYLE. In your preparation of the program, identify any other interlock and sensor requirements and incorporate them into the program. The program should be written so that it continues to repeat indefinitely.

9-12 A robotic workcell is to be programmed to perform the following cycle. Three different parts are to be processed through a production machine. The parts enter the workcell on a conveyor and are positioned at a fixed pickup point (call it PICKUP). A limit switch indicates that the part is in place (ON) or not in place (OFF). An identification system is used to determine which part is ready for pickup. The system indicates which part style is in place by means of three signal levels that can be fed into the robot controller. The three levels are: 1.0 V for style A, 3.0 V for style B, and 5.0 V for style C. The REACT command is to be used to test the incoming signal. A servocontrolled gripper is used to pick up the parts. Styles A and B require an opening width of 1.25 in. and style C requires an opening of 1.75 in. The part is then placed into the fixture of the production machine (call the load position LOAD) and the machine is signaled to begin its cycle. The fixture and the production machine must be told by the robot controller which part is being loaded because the holding and processing requirements are different for the three styles. When the machine cycle is completed, the machine signals back to the robot controller. The robot then removes the part from the machine and places it on an outgoing conveyor (call the position DROP). All pickup and dropoff steps must be done with APPRO and DEPART statements from above by 3.0 in. Write the program using the language statements in the chapter.

9-13 A robot loads and unloads part in an inspection machine which grades the parts into three categories: excellent, acceptable, and unacceptable. The robot must SIGNAL the machine that the part has been loaded, and the machine must signal back that the inspection process is completed. It must also indicate to the robot which of the three quality categories applies for each part inspected. Write a robot program to accomplish this procedure. Also, include in the program a computation algorithm to count the number of and proportion of parts and in each category. These values should be printed out after each production run of 10,000 parts.

9-14 Consider a cartesian coordinate robot with a workspace of 1.0 by 2.0 m in the xy plane and 0.75 m in the z direction. Somewhere in the workspace there is a pin projecting up from the xy table surface that is about 60 mm high. The pin is known to be 3 mm in diameter. It is used to identify the location of a reference frame from which location a number of assembly tasks will be done. The locations in the workspace of these assembly tasks are all known and defined from the pin. However, the coordinates of the pin are not known. Write a program using the REACTI statement and other language features described in the chapter that will scan the work table about 40 mm above the surface, to find the pin, and define its xy coordinates.

NINE-A

PROGRAMMING THE MAKER ROBOT

The MAKER robot system is produced by United States Robots, Incorporated, a subsidiary of Square D Company. The appendix is based on the *MAKER Robot System Operation Manual*, copyright 1984 by United States Robots. Portions of the manual are reprinted with permission of United States Robots, Inc.

The MAKER robot system is somewhat different to program than the

Table 9A-1 Teach pendant functions

Item	Function
Display	Error messages, status, operator prompts
Function 1–5	Buttons with associated displays for menu choices
Numeric	To enter numeric data
Prev Lev	To move to the previous menu level
Status	Displays status of overrides
Halt	Stops arm motion and puts the system in "halt" mode
Vert Align	Aligns the tool Z axis with the world Z axis
JT W/T	Selects joint or world/tool motions in manual mode
Relax	Relaxes the selected joint
Speed	Used to set monitor speed of arm
Grip 1,2	Used to operate the gripper controls
Motion switches	Used to move the joints or the arm in world or tool mode depending on JT W/T

other systems presented in this chapter. The MAKER is designed to be operated with a minimum of training of the operator. To accomplish this the MAKER is programmed using only the teach pendant. The teach pendant displays menus which allow the operator to select the necessary program commands. By responding to the menu options available, the operator is able to define points or to develop programs.

9A-1 THE TEACH PENDANT

The teach pendant is shown in Figure 9A-1. Table 9A-1 describes the function of each of the major elements of the teach pendant.

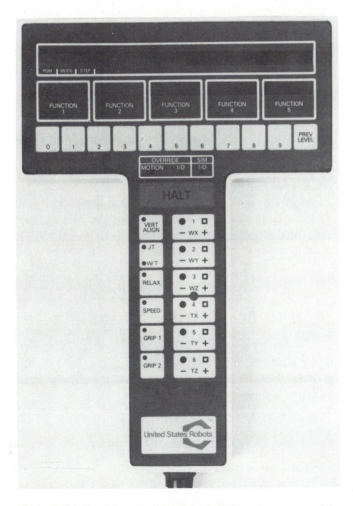

Figure 9A-1 Teach pendant for the MAKER robot system. (Photo courtesy United States Robots, Inc.)

9A-2 MOVING THE ROBOT

When the robot is in manual mode it is possible to move the robot using the teach pendant. The robot is in manual mode when the display reads MAN in the mode area of the display.

The robot can be moved in three ways: by individual joint, in world space, or in tool space.

In order to select joint space the JT W/T button should be pressed until the JT indicator is lit. When the JT indicator is lit the motion buttons will move the selected joint in the direction of the square or circle painted on the joints corresponding to the square or circle being selected on the teach pendant. For example, pushing the joint 3 button on the square will cause joint 3 on the robot to extend outward as long as the button is depressed.

If W/T is selected, the top three motions switches can be used to move the robot in the World X, Y and Z (WX, WY, WZ) directions and the bottom three switches can be used to move the robot in the Tool X, Y and Z (TX, TY, TZ) directions.

Pushing the VERT ALIGN (for vertical align) button aligns the tool Z axis with the world Z axis.

To adjust the speed with which the manipulator is moving the SPEED button is pushed and the display changes to:

MAN MONITOR SPEED: ⟨x⟩ ENTER NEW SPEED ⟨x⟩

The displays over the five function keys also change. Functions 1–3 are blank, 4 reads CLEAR and 5 reads ENTER.

A number between 1 and 100 is selected using the numeric keys and, if the number is correct, it is entered by pressing ENTER (function 5). The display then changes back to the manual display, including the new speed number.

9A-3 TEACHING POINTS

When in manual mode the five function keys display

TCH MODE   STP MODE   TRANS PTS   STORAGE   OTHER

In order to teach points we must get into teach/edit mode. This is done by selecting the TCH MODE function. The mode segment of the display now reads TCH and the rest of the display reads

SELECT TEACH/EDIT TYPE

Three choices are offered by the function key menu

PROGRAM   PALLET   POINT

Since we are concerned with teaching points we would select FUNCTION 3 for teaching POINTs.

If the POINT selection is made, the teach pendant will request a number between 1 and 9999. This number is the "name" of the point. If no point of that number has been previously taught then the display will read

POINT ⟨xxxx⟩ IS NOT TAUGHT GRP ⟨yyy⟩

The function keys offer the following choices

ACCEPT STEP RECORD GROUP OTHER

Selecting OTHER presents the following additional options

CONTROL COPY TRANS PTS DELETE OTHER

To define a point with the number displayed the operator moves the arm until it is in the desired position and pushes the RECORD button. The point is now recorded as that number. If desired the position may have been copied from another point by using the COPY option.

If the point has already been taught and is being edited, the options are the same as for teaching a new point, but the display will read

POINT ⟨xxxx⟩ IS TAUGHT GRP ⟨yyy⟩

In order to replace the point with new data the operator must first press RECORD and then the ACCEPT function.

Each point is assigned to a GROUP. The group name is a number between 1 and 255 and is identified by the display. Points are grouped in order so that they may be transformed as groups if necessary. The group number may be assigned or adjusted by using the GROUP function. We will not discuss the rest of the options in detail but we will describe their functions.

STEP is used to step the robot to different named points.

CONTROL is used to turn on and off the various system overides available during stepping. For example, it may disable the outputs from the controller to the peripheral equipment.

TRANS PTS allows the operator to transform a group of points as a unit. This allow the operator to move a set of points within the workcell while maintaining their relationship to each other.

DELETE is used to delete points.

9A-4 TEACHING PROGRAMS

Programs, like points, are taught while the system is in TCH mode as indicated by the display. In order to teach programs, the operator must select PROGRAM from the main TCH menu described earlier. When the PROGRAM selection is made the display will request a program number; this is the program ID. When the ID is selected, the display will read

⟨ID#⟩ TCH ⟨step#⟩ ⟨description of program step⟩

If no steps have been taught, the description will read, END OF PROGRAM.

The function keys will display

MOVE   STEP   GRIPPER   GOSUB   OTHER

Pressing OTHER causes the function displays to read

REPLACE   STEP   INSERT   DELETE   OTHER

Pressing OTHER again, causes

MODIFY   STEP   COPY   CONTROL   OTHER

Pressing OTHER again, returns us to the first of the three program menus. Obviously, there are a lot of options available for the types of steps to be programmed, and this is only the first level of the program menu! Notice that STEP and OTHER are always in the same position in the menu. This simplifies operation of the system.

We will begin with a description of the most fundamental type of program step, a MOVE.

When MOVE is selected the following is displayed (in addition to the program and step numbers)

MOVE PT ⟨xxxx⟩ SPD ⟨yy⟩ ⟨ABC⟩

where xxxx represents the point number, yy the speed, and ABC the motion type. The functions display

ACCEPT   POINT   SPEED   PTP/CP   OTHER

Pushing OTHER presents

ACCEPT   JI/ST   TOUCH   blank   OTHER

When all of the parameters are correct the operator pushes ACCEPT to teach that program step. The motion type ABC consists of three motion descriptors. These determine whether the motion will be

1. Point-to-point or continuous-path (PTP/CP)
2. Joint-interpolated or straight line (JI/ST)
3. Normal or touch-terminated (TOUCH)

A touch-terminated point stops the robot motion when a specified input sensor signal is true. This feature allows the robot to move in a direction until an external signal is generated, such as by a touch sensor. The three motion descriptors are selected using the appropriate function. For example, a move step directing the robot to move to point 342, in a straight line, as a continuous-path motion, with a normal termination at speed 75 would appear on the teach pendant display as

MOVE PT 342 SPD 75 SCN

If the specified point has not been taught then the system will ask the operator if he wishes to teach the point.

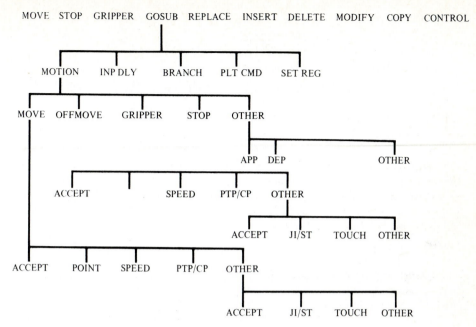

Figure 9A-2 Motion function menu.

MOVE is only one type of motion. Other types may be selected by pressing the REPLACE function, followed by the MOTION function. Figure 9A-2 illustrates the MOTION function and the resulting menu structure. The rest of this section will briefly describe the features available from the PROGRAM menus. Remember, that some selections may invoke additional menus.

STEP is used to step to a specific step in the program.

GRIPPER is used to teach a gripper operation step.

GOSUB is used for calling another program as a subroutine. It invokes a menu which allows choices such as unconditional subroutine calls, conditional subroutine calls (IF, THEN, ELSE), and loops.

REPLACE is used to invoke numerous step types.

MOTION invokes the various types of motion steps.

OFFMOVE is used to move the joints or the tool relative to their present position (at program execution time) by specified offsets.

APP is used to teach an approach to a point.

DEP is used to define a departure path from a point.

INP DLY is used for teaching various responses to input signals, such as delays, interlocks, reacts (jump to subroutines).

BRANCH invokes the same options as selecting GOSUB.

PLT CMD selects the special functions for operating on pallets.

SET REG allows the register operations such as setting, incrementing, and decrementing register values.

INSERT allows the insertion of steps between other steps.
DELETE is used to delete steps.
MODIFY allows simple modifications to taught steps.
COPY allows the copying of steps.
CONTROL allows the operator to set certain system control parameters.

9A-5 TEACHING PALLETS

A common task for a robot is to load or unload a fixed array of parts. Sometimes that array is a regular geometric pattern and sometimes it is a random collection of points. The MAKER system allows some special functions for dealing with pallets. In this case we will consider only "regular" pallets, that is, pallets whose points are arranged in a regular geometric pattern.

Consider a rectangular array of parts arranged in a 5×6 matrix. The task is for the robot to unload the parts in order from the array and to load them into a single position, perhaps a machine tool. Rather than consider the intricacies of the loading task, we will treat it as a subroutine which we will call program 43. The task facing us is to unload the parts in the pallet and to place them in the machine by calling subroutine 43.

We could teach the pallet by defining each of the 30 points, or we could define the pallet using the special features available. From the main TCH menu select the function PALLET. The display requests the pallet number (let us call this pallet 67) and then displays

67 TCH SELECT PALLET TYPE

and the functions read

REG PLT   IRR PLT

Since we are dealing with an array of parts we will select a REG PLT (regular pallet). The display will now read

67 TCH REG PLT DIM1$\langle x \rangle$ DIM2$\langle y \rangle$ DIM3$\langle z \rangle$

and the functions display

ORIGIN   DIM1   DIM2   DIM3   OTHER

Pressing OTHER causes the functions to display

COUNT   GROUP   CONTROL   blank   OTHER

We will describe each function in sequence.

ORIGIN is used to define the physical position (the point) at which we wish to begin unloading the pallet. It must also be the intersection point of the pallet axes, in our case at one of the corners.

DIM1 is used to define the number of points along the axis to be unloaded first.

DIM2 is used to define the points along the second axis (if it is a two- or three-dimensional pallet).

DIM3 is used to define the points along the third (if necessary) axis.

Using the DIMn function causes the display to read

REG PLT DIM ⟨n⟩ SIZE ⟨x⟩ IND ⟨q⟩

and the functions to display

ACCEPT   SIZE   END POS   MOVE END   INDEX

SIZE is used to define the number of points along that axis, in our example 5 or 6. END POS is used to define the position of the last point along that axis. The robot is moved to that point or the point can be (COPY) copied. MOVE END will move the arm to the endpoint if it has already been taught. INDEX is used to set the number which will be loaded or unloaded in the row first. This can be used to define which end is loaded or unloaded first.

COUNT is used to set the pallet counter. When unloading a pallet it is set to the number of points in the array (in our case, 30), when loading it is set to 0. It may be set to some other number if a pallet must begin in the middle of a run.

GROUP is used to assign the pallet points to a group number.

CONTROL is used to set control parameters of the system.

Having taught the points in the array as a pallet, there are special motion functions available for operating on that pallet.

Selecting PLT CMD from the REPLACE menu will cause the display to read

SELECT PALLET STEP TYPE

and the functions will display

PMOVE   PAPP   PUT   TAKE   RESET

PMOVE is a move to a pallet point defined by the present value of the pallet counter.

PAPP is an approach to the pallet point defined by the present pallet counter.

PUT adds 1 to the value of the pallet counter.

TAKE subtracts 1 from the pallet counter.

RESET resets the pallet counter to the empty or full value (as selected by the operator).

By comparing the value of the pallet counter to a value in a register we can tell if a pallet operation is complete and can stop or start again on a fresh pallet. Typically this would be done by comparing the pallet counter to a value in a register and using the IF-THEN-ELSE command.

9A-6 SUMMARY

The MAKER provides the same types of programming commands as many other systems but in a different environment. Rather than presenting the system as a programming language such as VAL and AML, the MAKER makes extensive use of menus in an attempt to simplify the man–machine interface.

REFERENCE

1. *MAKER Robot System Operation Manual*, United States Robots, Inc., King of Prussia, PA, 1984.

VAL II

This appendix presents a summary of the VAL II robot programming language, developed by Unimation, Inc., and used for its industrial robots. VAL II is an enhanced version of the VAL robot programming language, released by Unimation in 1979 for its PUMA series industrial robots. The appendix is based on the *User's Guide to VAL II*,[1] copyrighted by Unimation, Inc. Major portions of the appendix are reprinted from the manual by permission of Unimation Inc. Unimation makes no representation regarding the accuracy of the information presented in this appendix, and is not responsible for results obtained in using the information.

It is not the purpose of this appendix to provide a complete description of the VAL II programming language. Readers should not attempt to accomplish any actual programming of Unimation robot systems using this appendix as the sole reference. The available user manuals and other operating instructions should be consulted before progamming is attempted in VAL II. The objective of this summary is to acquaint the reader with some of the basic concepts and statements of the language, in order to build a general understanding of robot textual languages.

9B-1 GENERAL DESCRIPTION

VAL II is a computer-based control system and language designed for the Unimation industrial robots. It provides the capability to easily define the task a robot is to perform, since the tasks are defined by user-written programs. Other benefits include the ability to respond to information from sensor systems such as machine vision, improved performance in terms of arm trajectory generation, and working in unpredictable situations or using reference frames.

9B-2 MONITOR COMMANDS

The VAL II monitor mode is used for functions such as defining point locations in the work volume, editing programs, executing programs, calibrating the robot arm, and similar purposes. We will sample some of the commands in this section.

Defining Locations

There are several commands in VAL II for defining locations and determining the current location of the robot. In the following commands all distances and coordinate values are in millimeters. One of the methods of defining point locations is the HERE command. The operator uses the teach pendant to move the robot manipulator to the position to be defined and uses the command as follows

HERE P1

This command, given in the monitor mode, defines the variable P1 to be the current robot arm location. A related command is

WHERE

This command queries the system to display the current location of the robot in cartesian world coordinates and wrist joint variables. It also displays the current gripper opening if the robot is equipped with a position-servoed hand.

The TEACH command is used to record a series of location values under the control of the record button on the teach pendant. Each time the record button is pressed, a location variable is defined and given the value corresponding to the location of the robot at the instant the record button is pressed. Each successive location variable is automatically assigned a new name. The assigned name is derived from the name specified in the command. For example, if the command

TEACH P1

is typed into the monitor, the first recorded location variable is P1, the next is P2, the third is P3, and so forth. It is possible to teach a complete motion path by successively positioning the robot using powered leadthrough with the teach pendant and pressing the record button.

A third command for defining positions is the POINT command. For example, the command

POINT PA = P1

sets the value of PA (a location variable) equal to the value of P1.

Program Editing and Control

Commands for entering and exiting the program editing mode are the following (shown at the beginning and end of a program)

EDIT ASSEMBLY1

.

.

.

.

E

The EDIT command opens a program which is to be named ASSEMBLY1. The E command (E for EXIT) exits the programming mode and returns to the monitor mode. A variety of commands are available during editing for deleting programming commands, inserting commands, and so on.

Execution of the program requires specification of speed. The SPEED command in the VAL II monitor mode is given prior to program execution in the following way

SPEED 50

This specifies that the robot execution speed is 50 units on a scale which ranges between 0.39 (very slow) and 12800 (very fast), where 100 is the "normal" speed. Thus a setting of 50 would be below normal speed. The actual speed at which the motion sequence is executed is related to the product of the speed setting defined by this monitor command and the speed specified within the program by the SPEED instruction. For example, if the monitor speed setting is 50 (as above), and the SPEED is specified in the program as SPEED 60, then the actual speed of the motions following the command will be 30 percent of normal speed.

The command

EXECUTE ASSEMBLY1

causes the system to execute the program named ASSEMBLY1. VAL II provides for several arguments to be used with this command. For example,

EXECUTE ASSEMBLY1, 5

causes the program to be executed exactly five times.

Other Monitor Commands

Additional monitor commands are available in VAL II for storing programs onto disks (STORE), copying programs (COPY), loading files from disk to computer memory (LOAD), listing the file names contained on a disk (FLIST), renaming (RENAME) or deleting (DELETE) files, and so on.

One final monitor command that will be mentioned is the DO command. The DO command causes the robot to execute a specified programming instruction. For example,

DO (programming instruction)

causes the robot to carry out the particular programming instruction indicated. We will give examples of the use of this command in the following section.

9B-3 MOTION COMMANDS

There are a number of programming instructions used to define the motion path in VAL II. The basic instruction is

MOVE P1

which causes the robot arm to move by a joint-interpolated motion to the point P1. A related motion command is

MOVES P1

which causes the movement to be along a straight line path from the current location to the point P1.

VAL II has approach and depart commands which are written as follows

APPRO P1, 50
MOVE P1
DEPART 50

The first command moves the tool to position and orientation specified by P1, but offset from the point along the tool z axis by a distance of 50 mm. The MOVE command moves the tool to point P1. The DEPART instruction moves the tool away from P1 by 50 mm along the tool z axis. In this sequence the tool z axis direction is defined by the orientation of the tool in the P1 specification. The APPRO and DEPART statements are executed in joint-interpolated motions. Both commands can be carried out with straight line motions by specifying APPROS and DEPARTS statements, respectively.

The speed setting can be changed during the program by means of the SPEED command. The command applies to all subsequent motions until the speed is again altered. There are two ways to specify speed in a VAL II program. The first method is to specify a numerical value without a units argument.

SPEED 60

This is interpreted as a percentage of the speed specified in the monitor SPEED command, as explained above. The alternative way is to specify

SPEED 15 IPS

which indicates that the subsequent motion commands are to be executed at a speed of 15 in./s (IPS). Units can also be specified as millimeters per second (MMPS).

A single joint can be changed by a certain amount with the statement

DRIVE 4, −62.5, 75

This command changes the angle of joint 4 (assumed to be a rotational joint) by driving it 62.5° in the negative direction at a speed of 75 percent of the monitor speed.

Another command in VAL II used for motion control is to align the tool or end effector for subsequent moves. The statement is simply

ALIGN

This causes the tool to be rotated so that its z axis is aligned parallel to the nearest axis of the world coordinate system. It is useful for lining up the tool before a series of locations are taught in which it is important that the tool be properly oriented. This is done in the monitor mode with the DO command as follows

DO ALIGN

The DO command can also be used with other motion commands, as in DO MOVE P1 and DO DRIVE 4, −62.5, 75.

9B-4 HAND CONTROL

VAL II has provisions for controlling a pneumatically operated gripper or an electrically driven servoed hand. To operate the latter, certain conditions must be satisfied, which we shall not be concerned with here.

The basic commands are OPEN and CLOSE. With a pneumatic binary gripper, these commands cause the gripper to assume fully open and fully closed positions. These commands are executed during the next robot motion. The commands OPENI and CLOSEI cause the commands to be executed immediately rather than during the next motion.

With a servocontrolled gripper, the finger opening can be specified as follows

OPEN 75

.

.

.

CLOSEI 50

The first statement causes the gripper to open to 75 mm during the next motion of the robot, while the second statement causes the gripper to close immediately to 50 mm.

A related statement is GRASP. This allows a check to be made to determine if the gripper has closed by an expected amount. The command would work as follows

GRASP 12.7, 120

The command causes the gripper to close immediately, and checks if the final opening is less than the specified amount of 12.7 mm. If it is, the program branches to the statement 120 in the program. If the statement number is not specified, for example,

GRASP 12.7

then an error message is displayed at the operator's console. The GRASP statement provides a convenient method of grasping an object and testing to make sure that contact has been made.

The control of the gripper and simultaneous movement of the robot arm can be accomplished with a single statement rather than two separate commands as described above. The required statement is

MOVET P1, 75

This command generates a joint-interpolated motion from the previous position to the point P1, and during the motion, the hand opening is changed to 75 mm. The corresponding straight line motion command is

MOVEST P1, 75

These statements assume a servocontrolled gripper capable of responding to the 75-mm opening specification. For a pneumatically operated hand, the command is interpreted to mean "open" if the value specified is greater than zero, and "close" if the value is otherwise.

9B-5 CONFIGURATION CONTROL

For a jointed-arm robot with six joints, most points in the workspace can be reached by assuming one of eight possible spatial configurations. Normally, the robot would remain in the same configuration throughout program execution. The need for using one or more of the commands in this section is to permit the robot to achieve a position in one configuration which it would be unable to achieve in the alternative configuration. The statements

RIGHTY   or   LEFTY

provide for a change in the jointed-arm robot configuration so that the first three joints (base rotation, shoulder, and elbow) resemble a human's right or left arm, respectively. The statements

ABOVE   or   BELOW

request a change in robot configuration so that the elbow of the robot is pointed up (ABOVE) or down (BELOW). Finally, the statements

FLIP   or   NOFLIP

provide for a change in the range of operation of joint 5 (on a six-jointed robot) to positive (NOFLIP) or negative (FLIP) angles. The last two statements would be invoked to keep joints 4 and 6 within stop limits.

9B-6 INTERLOCK COMMANDS

The output and input signals of the robot controller can be determined in the VAL II program for communicating with other equipment in the workcell. Some of the possible commands are explained in this section.

The RESET command turns off all external output signals. This is useful in the initialization portion of a program to ensure that all external output signals are in a known state (off).

The SIGNAL command is used to turn output signals on or off. Positive signal numbers turn the corresponding signal on and negative numbers turn the corresponding signals off. For example, the statement

SIGNAL −1, 4

would turn output signal 1 off and output signal 4 on.

The WAIT command can be used to control program execution based on input signals. For example,

WAIT SIG(1001,−1003)

stops program execution until external (input) signal 1001 is turned on and input signal 1003 is turned off.

VAL II contains several versions of the REACT command for interrupting regular program execution due to external signals. The basic command is

REACT VAR1, SUBR2

This results in continuous monitoring of the external binary signal identified as variable VAR1 and looks for a transition of the signal from its current state to the alternative state. If the variable value is positive (as in the example above), VAL II looks for a transition from off to on. If the variable is negative (e.g., −VAR1), then VAL II looks for a transition from on to off. The reactions are triggered by signal transitions as indicated in the statement. For example, if a signal is to be monitored for a transition from off to on as in the above command, and the signal is already on, the reaction does not occur until the signal goes off and then on again. Another condition on the signal is that the change must remain stable for at least 30 ms to assure detection of a transition. When the specified reaction occurs, the program control is transferred to the subroutine named SUBR2.

Since the REACT statement is often used for safety interrupts, it is desirable to specify different levels of priority for different REACT statements. A priority argument can be added to the statement as follows

REACT VAR1, SUBR2, 5

The value of the priority argument must be in the range 1 to 127 to indicate the relative importance of the reaction. Higher values indicate higher priorities. A default condition is priority 1. When the specified signal transition is detected, VAL II reacts by checking the priority specified with the react instruction against the program priority setting at that time. If the REACT priority is greater, the normal program sequence is interrupted and the equivalent of a CALL SUBR2 is executed. Also, the program priority is raised to the REACT priority, thus locking out any reactions of lower or equal priority. When a program control is returned to the previous location in the program, the priority is restored to the value it had before the REACT was invoked.

If the REACT priority is less than or equal to the program priority when the signal transition is detected, the reaction is queued and is not invoked until program priority has been reduced. Accordingly, depending on the relative priorities, there can be a considerable delay between when the signal transition is noticed by VAL II and when the corresponding subroutine is executed.

The REACTI statement operates in a similar manner as REACT except that it stops the current robot motion immediately when the specified signal transition is detected. In REACT, the current motion command is completed before the reaction is invoked.

9B-7 INPUT/OUTPUT CONTROLS

Instructions are available in VAL II for communicating information to the operator and to other digital and analog devices. For communicating with the operator's terminal, the PROMPT and TYPE commands are available. The first of these statements is used to obtain data from the operator. The construction of the command is

PROMPT "Enter the number of parts: ", PIECES

The text string indicated in quotations is displayed on the operator's monitor, after which the system waits for the operator to respond by typing in the value requested and pressing the return key. This value is assigned to the variable name PIECES. Program execution then resumes.

A related statement is the TYPE command. This causes the specified output information to be displayed on the terminal. During processing of the TYPE statement, program execution normally waits for the output to be completed before continuing. This delay can be avoided, depending on output specifications by the programmer. An example of the TYPE statement is the following

TYPE /B, "The value of length is ", /F5.2, LENGTH

Suppose at the time of execution of this command, that the variable LENGTH has the value of 12.75 as determined somehow in the program. The /B causes a beep to be sounded at the system terminal. The /F5.2 in the statement specifies the format for displaying the value of the variable LENGTH. The following message is displayed on the CRT

The value of length is 12.75

VAL II contains provisions for communicating with other digital devices with IOPUT and IOGET commands. These instructions either send or receive (respectively) output to a digital input/output module. For example, the command

IOPUT OUTBUF = LENGTH

causes the system to write the current value of LENGTH to the memory location identified by the value of OUTBUF.

Finally, the system can handle analog signals if the robot controller system is configured with the optional analog I/O module. There are two functions, ADC and DAC. The ADC function is used as follows

VAR5 = ADC(5)

This statement evaluates the current input at analog input channel number 5 and returns it as an integer value in the range -2048 to 2047. VAL II assumes that the A/D converter accepts input signals in the maximum range from -10 to $+10$ V dc. Otherwise, a gain factor can be applied to compensate. The DAC command is used to output an analog voltage signal from either of two channels provided with the analog I/O option. The two channels are identified as 0 or 1. An example of the command is

DAC 1 = VAR1

The analog output voltage is proportional to the value on the right-hand side in the statement, VAR1. The range of the output signal is between -10 and $+10$ V, and can be specified as a constant, a real variable (as above), or an arithmetic expression. The digital-to-analog converter operates on 12 bits where the voltage value from -10 to $+10$ V is determined as a proportion of the binary value in the range from -2048 to 2047, respectively.

9B-8 PROGRAM CONTROL

The VAL II language contains a number of instruction sequences that can be used to logically organize sections of user programs and to control the order in which they are executed. These program control statements include IF...THEN...ELSE...END, WHILE...DO...END, DO...UNTIL, and others.

The format of the IF...THEN...ELSE...END is conventional except that the keyword END is used to signify the end of the sequence of statements. The same keyword is used to indicate the termination of the WHILE...DO sequence. The DO...UNTIL operates in the reverse sequence with respect to the WHILE...DO statement since the logical condition is tested after the statements or group of statements are executed.

Another program structure available in VAL II is the CASE... OF...VALUE...END sequence. It provides a means for executing one group of instructions from among any number of groups. The following example illustrates the use of the sequence:

```
10   PROMPT "Enter test value: ", X
     IF X < 0 GOTO 20
     CASE INT(X) OF
     VALUE 0, 2, 4, 6, 8, 10:
     TYPE "The number ", X, "is even."
     VALUE 1, 3, 5, 7, 9:
     TYPE "The number ", X, "is odd."
     ANY
     TYPE "The number ", X, "is larger than 10."
     END
     GOTO 10
20   TYPE "Stopping because of negative value."
     STOP
```

The program asks the operator to enter a test value. If the value is negative, then the program exits after displaying the message, "Stopping because of negative value." Otherwise, a CASE structure is used to classify the test value into one of three categories, based on its integer value: (1) even numbers from zero to 10, (2) odd numbers from one to nine, and (3) all other positive numbers.

9B-9 SUMMARY

Additional statements are available in the VAL II language. The purpose of this appendix is to provide a sampling of the important features and statements that are contained in VAL II. The interested reader is referred to the user's manual[1] available from Unimation, Inc.

REFERENCE

1. Unimation, Inc., a Westinghouse Company, *Programming Manual—User's Guide to VAL II*, Document 398T1 (Preliminary), Version 1.1, Danbury, Connecticut, August 1984.

NINE-C

RAIL

This appendix presents a summary of the RAIL robot programming language, developed by Automatix, Inc., and used for its Robovision and Cybervision systems. The appendix is based on the *RAIL Software Reference Manual*,[1] copyrighted by Automatix, Inc. Major portions of the appendix are reprinted from the manual by permission of Automatix.

It is not the purpose of this appendix to provide a complete description of the RAIL language. Readers should not attempt to accomplish any actual programming of Automatix robot systems using this appendix as the sole reference. The available User Manuals and other operating instructions should be consulted before programming is attempted in RAIL. The objective of this summary is to acquaint the reader with some of the basic concepts and statements of the language, in order to build a general understanding of robot textual languages.

9C-1 GENERAL DESCRIPTION

RAIL is Automatix's language for computer-aided manufacturing. It is designed to control the company's Robovision, Cybervision, and Autovision systems, and general manufacturing equipment. Robovision is a system for robotic arc welding. There are special-purpose RAIL commands for interfacing between the robot and the welding equipment. Cybervision is the system for assembly. RAIL contains commands for controlling the robot in an assembly operation, and for interfacing with such external devices as parts feeders, conveyors, sensors, and material-handling equipment. RAIL also contains commands for interfacing with Autovision (machine vision) systems, which are designed for inspection and identification in a manufacturing operation. This appendix describes some of the features specific to Robovision and Cybervision systems.

9C-2 LANGUAGE FEATURES

The RAIL language provides the following robot programming features, most of which have been described in general terms in the text of Chap. Nine:

Commands for moving the robot, including approaching or departing from locations.

Robot welding commands, and facilities for setting welding parameters, such as voltage and wire feed rate.

A simple method to access input or output lines connected to equipment such as fixtures, part detector switches, conveyors, grippers, or welding positioners.

Commands for loading programs from the file system, for displaying user variables or programs, for editing programs, and for storing programs.

Data types, including integers, real numbers, character strings, logical data, arrays, points, paths, and reference frames.

User-selected variable names (up to 20 characters long) for referencing I/O channels, locations, weld paths, and so on.

Program control structures similar to those provided by the Pascal language, including IF..THEN..ELSE, WHILE..DO, WAIT, and other functions that can be inserted in an application program to make logical decisions at run-time, to change the sequence of execution of instructions, or to repeat the execution of any instructions.

Arithmetic, comparative, and logical expressions.

A built-in function library, including functions for square root, sine, cosine, arc-tangent, absolute value, time-of-day, and interval timing.

9C-3 LOCATIONS

The RAIL language has three types of data for robot locations: points, paths, and reference frames.

Definition of Points

A point specifies the position and orientation of the tool tip, relative to a world reference frame. The world space frame is commonly located at the robot base. The point would be specified by a variable name as follows

$$P1 = (300.0, 200.0, 100.0, 50.0, 25.0, 35.0)$$

The first three values are linear dimensions, expressed in millimeters or inches, indicating the *x*, *y*, and *z* coordinates of the tool tip in the world space. The last three values are orientation angles, expressed in degrees, indicating the orientation of the tool tip relative to the world reference frame.

Paths

A path is a connected series of points. When the robot is commanded to move along a path in RAIL, it will move smoothly through each point in the path without stopping or slowing down at any of the intermediate points. A path is specified by indicating the points which define it, as in the following PATH statement.

$$SEAM1 = PATH (P1, P2, P3, P4)$$

Reference Frames

A reference frame is a coordinate system that has other locations defined relative to it. For example, if a pallet has several holes within it (for part insertion), the holes can be defined relative to the pallet since their location within the pallet does not vary. Therefore, when the pallet is moved, the user would only have to redefine the new location of the pallet in space, and not the location of each hole in the pallet. A frame is specified in RAIL by defining the origin, a point along the *x* axis, and a point lying in the *xy* plane.

9C-4 ROBOT MOTION STATEMENTS

There are a variety of statements to control the motion of the robot in the RAIL language. The basic command is the MOVE statement. RAIL supports three types of robot motion: straight line, slew, and circular. By using specific key words (described below), the user program can select the type of motion to be used for a given motion statement.

Straight Line Motion

Straight line motion of the tool tip is accomplished by means of linear interpolation. This motion is the default condition in RAIL; if the user does not specify one of the other motion types, the robot will move in a straight line motion. The command for a straight line motion is

MOVE P1

Slew Motion

This corresponds to joint-interpolated movement of the robot arm. The arm is moved from point to point using the least possible motion for each joint, and in a coordinated fashion so that all joints start and stop moving at the same time. The advantage of slew motion is that it usually permits faster execution of a move. The command for a slew motion is defined by using the key word SLEW as follows

MOVE SLEW P1

Circular Motion

This motion specification causes the robot arm to move so that the tool tip travels along a circular arc formed by the points indicated in the path. The key word CIRCLE is used to specify a circular interpolation in a motion statement as follows

MOVE CIRCLE PATH (P1, P2, P3, P4)

The arm will move the tool tip to the first point from its current location using straight line motion. It will then move from point P1 to P3 using a path consisting of a circular arc through P2. For paths containing more than three points (as in the example statement above), a new circular path would be followed between points P3 and P4 in which P2, P3, and P4 are used to define the new circle. For each additional point included in the path, a new circular arc is defined using the last two points combined with the next destination point in the sequence.

Other types of robot motion statements in RAIL include approach and depart statements, rotation of the gripper, actuation of the gripper, and actuation of a welding cycle.

Approach and Depart

These commands (similar to commands explained in the text of Chap. Nine) permit the robot to approach the specified point along the tool's z axis. The commands are

APPROACH 50 FROM P1
.
.
.
DEPART 50

The first statement causes the tool to move to a position 50 units (millimeters in this case) from the point P1 along the tool z axis. The last statement causes the tool to depart from the point by 50 mm.

Rotate the Tool

This command is used to rotate the end effector a specified number of degrees about the roll axis. For example, the command

ROTATE HAND 180

specifies that the end effector should be rotated 180° in the clockwise direction from its current position.

Open and Close Gripper

These statements command the opening and closing of a gripper-type end effector. The RAIL statements are

OPEN   and   CLOSE

Welding

This is a RAIL command that permits a welding operation to be carried out at a point along a path. The motion is executed by the special WELD command, explained in Sec. 9C-6.

9C-5 LEARN STATEMENT

The LEARN statement is used to define location data in conjunction with the robot teach pendant. The statement can be used to program points in world space or relative to other predefined locations. It can also be used to teach paths and frames. In the statement

LEARN P1, P2, P3, P4

the four points are learned by moving the arm to the desired locations using the motion buttons on the teach pendant, and pressing the record button to store each point location in the proper sequence.

Paths are learned by depressing the path button on the teach pendant to indicate the start of a path, and then moving the robot through the desired path by means of the teach pendant, depressing the record button at each successive point in the path. Depressing the path button a second time signals the end of the learn routine.

Frames are learned by means of the statement

LEARN FRAME COORD1

The LEARN FRAME statement prompts the user to teach three locations. These locations correspond to the origin of the frame, a point along the x axis, and a point in the xy plane of the frame coordinate system.

9C-6 WELDING

RAIL provides a number of features for controlling a welding robot. These features include the capability to do the following: move the robot along a path while controlling the welding process parameters; stepping the robot through the program execution, with the option of adjusting variables or process parameters at each step; and modifying the welding parameters using the teach pendant or the alphanumeric keyboard.

The basic weld command in RAIL is the statement WELD, followed by an identification of the path to be taken by the robot and the welding parameters to be used. The statement has the following format

WELD SEAM1 WITH WELDSCHED[3]

In this command, the weld path is called SEAM1, defined as a path in RAIL. The clause WITH WELDSCHED[3] identifies that the welding parameters to be used during this welding path are defined in welding schedule number 3. The welding parameters available for specification in RAIL include wire feed rate (inches per minute), voltage (volts), predwell and crater fill times (seconds), gas preflow and postflow times (seconds), and several parameters for specifying weaving motions across the weld gap. Welding schedules are defined using the format given below:

WELDSCHED[3, WIREFEED] = 50
WELDSCHED[3, VOLTAGE] = 18
WELDSCHED[3, PREDWELL] = 1.5
WELDSCHED[3, FILLDELAY] = 0
WELDSCHED[3, PREFLOW] = 1.0
WELDSCHED[3, POSTFLOW] = 0
WELDSCHED[3, WEAVE] = OFF

RAIL contains provisions for controlling weave patterns in arc welding. The parameters that must be specified include weave amplitude (inches or millimeters) and cycle time (seconds), as well as weave delay values. Each of these parameters is specified as part of the welding schedule using the above format.

9C-7 INPUT/OUTPUT

The controller unit used with the Automatix robots is the AI 32. This controller is equipped with the following I/O devices to control or communicate with other equipment:

48 solid state relays for I/O
4 digital potentiometers for output
16 12-bit A/D converters for input (optional)

RAIL statements can be used to associate user-selected variable names to interlock the controller with external equipment through these I/O devices. The user can then read from or write to the external equipment by using the variable name in the program. To declare an input or output port, the following type of statement is used

INPUT PORT SWITCH1 7

This defines a variable named SWITCH1 to be an input to the controller through input port number 7. The status of that variable can be checked in the program as needed. Output statements are defined in a similar manner

OUTPUT PORT MOTOR3 15

This defines the variable MOTOR3 as an output from the controller through port 15. During the program this output can be turned on or off by means of the statements

MOTOR3 = ON

MOTOR3 = OFF

The first statement assigns the value of output port 15 to be on (which has the value 1 in RAIL). The second statement assigns the value of output port 15 to be off (which has the value of 0 in RAIL). In effect these two statements close and open the relay for port 15.

There are two WAIT commands in RAIL, one of which is used in conjunction with an interlock condition. This is the WAIT UNTIL command. It is used to synchronize the execution of a RAIL program with some condition. The test condition must be some relational expression, such as one that tests the status of an input port variable. When the WAIT UNTIL statement is encountered, program execution is stopped until the test condition has been satisfied. For example,

WAIT UNTIL MOTOR3 = ON

WAIT UNTIL CONSPEED > 5.0

would require program execution to stop until the two input channel conditions (MOTOR3 = ON and CONSPEED > 5.0) are both true.

The second WAIT command is used to delay the program execution for a specific time interval. The time interval is specified in either seconds (SEC) or milliseconds (MSEC). If not specified, the time units are assumed to be seconds. The following statements illustrate the WAIT command

WAIT 2 SEC

WAIT 2

WAIT 2000 MSEC

Each of the above three statements would result in an identical time delay.

9C-8 OPERATOR I/O AND FILE SYSTEM

The RAIL language has a variety of commands for the programmer or operator to communicate with the robot cell. The standard output device used with RAIL is the CRT, and the teach pendant used with RAIL is called the interactive command module, or ICM. This section will sample the sections of the RAIL manual pertaining to these types of instructions.

The READ statement reads data supplied from outside the program, and assigns the data values to RAIL variables in the program. There are two forms of READ statement: READ and READS. The READ statement is used to read numerical input from the operator. When a READ statement is encountered, execution c₁ the program is halted until the operator has entered a number for each variable in the variable list. The numbers must be integers or real numbers. The READS statement is used to read character string input (i.e., text containing any ASCII character, upper or lower case alphabetic characters, numbers, punctuation, etc.) from the operator. The operator must enter a line of text for each variable in the variable list.

The WRITE statement takes RAIL expressions (e.g., text) in the program, formats them into an output line, and prints them externally. There are two forms of WRITE statement: WRITE and WRITEICM. The WRITE statement is used to write messages to the operator on the CRT. Each WRITE statement prints out a single line on the CRT. The following example will illustrate the READS, READ, and WRITE statements.

> WRITE ('Enter Part Name:')
> READS (PARTNAME)
>
> WRITE ('Enter part length:')
> READ (LENGTH)

The first WRITE statement causes the message to be printed on the CRT, and the READS statement causes the operator's input to be stored in the program as the variable PARTNAME. The second WRITE statement requests the operator to input a numerical value which will be used in the program as the value of the variable called LENGTH.

The WRITEICM statement is a RAIL function that writes a character string to the ICM display. For example,

> **WRITEICM ('OVER TEMP CONDITION')**

will print the message out to the ICM rather than the CRT. The limitation on the message is that it must be no longer than 24 characters in length, which is the length of the display on the interactive command module. If the character string is longer than this limit, then only the first 24 characters of the string are printed.

Other commands are concerned with the control of the RAIL file system. Two cartridge tapes are available in the Automatix robot controller to store RAIL system software and user files. Some of the commands are explained briefly here.

The SAVE command is used to save the current value of RAIL variables (and/or the definitions of user-written functions) into a file on tape. For example, the statement

> **SAVE 'PARTFILE' PARTCOUNT, GOODPARTS, BADPARTS**

provides a means in the program to save the current values of the variables (PARTCOUNT, GOODPARTS, BADPARTS) in the variable list on tape storage. The file name is PARTFILE. Once the file has been created using the SAVE command, it can be retrieved at any time using the LOAD statement. The statement

LOAD 'PARTFILE'

loads the file from the cartridge tape into computer memory.

In addition to the SAVE and LOAD commands, there is another command in RAIL available for file maintenance. This is the FILER command. This is used for operations such as listing the contents of a tape, copying or deleting files, and so on. When the FILER command is entered, the screen is cleared, and the following line is printed out at the top of the screen:

Filer: L)ist, D)elete, P)rint, M)ake, V)olume, I)nit tape, C)opy, Q)uit?

The line shows all of the available FILER commands. To run any of the commands, the user simply enters the first letter of the command name. L)ist is used to print a list of the names of the files stored on the tape. D)elete is used to delete a file. P)rint permits the user to print a listing of the file onto the CRT. M)ake is used to create a file from the keyboard. The V)olume command permits the user to list the volume names of the tapes currently in the two tape drives of the system. The I)nit tape command is used to initialize a tape by creating an empty file directory on the tape. All new tapes must be initialized in this way before they can be used to store files. C)opy allows the user to copy the contents of one file onto another file. Finally, the Q)uit command is used to quit the FILER and return to RAIL.

9C-9 PROGRAM CONTROL

The RAIL language contains an assortment of the typical program control statements such as IF...THEN, IF...THEN...ELSE, and others.

The WHILE...DO statement permits the programmer to continue executing a given statement so long as a specified test condition remains true. The following example illustrates its use:

```
WHILE CYCLESTOP = OFF DO
BEGIN
MOVE SLEW HOME
WAIT UNTIL SWITCH1 = ON
CLAMP = ON
APPROACH 50 FROM SEAM1
WELD SEAM1 WITH WELDSCHED[3]
DEPART 50
CLAMP = OFF
END
```

This example makes use of several statements discussed previously in the appendix. The keywords BEGIN and END are used here to enclose a compound statement (a group of statements to be executed in sequence).

9C-10 SUMMARY

Additional statements are available in the RAIL language. The purpose of this appendix is to provide a sampling of the important features and statements that are contained in RAIL. The interested reader is referred to the user's manual[1] available from Automatix, Inc.

REFERENCE

1. Automatix Inc., *RAIL Software Reference Manual*, Document No. MN-RB-07, Rev. 5.00, October 1983.

NINE-D

AML†

This appendix presents a summary of the AML robot programming language, developed by IBM Corporation, and used for its IBM 7565 Manufacturing System. The appendix is based on two AML publications cited in Refs. 1 and 2, copyrighted by IBM Corporation. Major portions of the appendix are reprinted from these manuals by permission of IBM.

It is not the purpose of this appendix to provide a complete description of the AML programming language. Readers should not attempt to accomplish any actual programming of IBM robot systems using this appendix as the sole reference. The available user manuals and other operating instructions should be consulted before programming is attempted in AML. The objective of this summary is to acquaint the reader with some of the basic concepts and statements of the language, in order to build a general understanding of robot textual languages.

9D-1 GENERAL DESCRIPTION

A Manufacturing Language (AML) is a high-level, interactive programming language designed for robotic programming. The language provides a variety of data types and operators, system subroutines, language control structures, display station and file I/O controls, interactive editing and debugging facilities, and system identifiers to help improve program readability.

AML is subroutine oriented. Programs are written by structuring the application as a set of calls to both user-written and system subroutines. The

†Portions of this appendix have been reprinted by permission from *A Manufacturing Language Concepts and User's Guide* (850912), © 1982, and *A Manufacturing Language Reference* (850915), © 1983, by International Business Machines Corporation.

system makes no distinction between user-written subroutines and system subroutines, and the system treats both types similarly when they are called. Consistent with our terminology in the text and in the other appendixes, we shall refer to the AML system subroutines as commands.

9D-2 AML STATEMENTS

An AML statement is analogous to an English sentence. Its meaning is interpreted by first interpreting the words and expressions in it. There are three statement forms in the AML language:

1. Executable statement
2. Variable declaration statement
3. Subroutine declaration statement

With the exception of certain special cases, all AML statements end in a semicolon (;).

Executable Statement

Executable statements perform calculations, comparisons, and other similar functions. They constitute the logic which the interpreter is to execute. Executable statements differ from the other two types because they do not reserve storage or provide names for variables or subroutines. An executable statement has the following form

label: expression;

The statement consists of an optional label and the expression to be interpreted. A semicolon ends the statement. The label is separated from the expression by a colon. An example of an executable statement is

CALC: 5 − 1/2*10;

CALC is the user-selected label, and the expression consists of a simple arithmetic computation.

Compound executable statements are a sequence of statements contained on several lines. They are denoted by BEGIN...END words which enclose the sequence.

Variable Declaration Statement

Variable declaration statements have the following form

id: NEW expression;

or

id: STATIC expression;

The components of the variable declaration statement are the following: the name of the variable (called the "id" in AML), a keyword (either NEW or STATIC) which identifies the statement as a declaration statement, and an expression which the interpreter evaluates to determine the variable type and initial value. A semicolon ends the statement. Two examples of variable declaration statements are given below:

$$X1: NEW\ 12.0;$$
$$X2: NEW\ (2.5*X1);$$

The first statement declares the variable X1 to have an initial value of 12.0. The second statement declares the variable X2 to be given by an expression which evaluates to the product of 2.5 times the current value of X1.

The keywords NEW and STATIC are used by the interpreter to determine how to treat a declared variable. For variables outside of a subroutine, there is no difference in the way the keyword is interpreted. The difference applies only for their use in a subroutine. NEW variables are interpreted each time the subroutine within which they reside is called. The interpreter interprets the expression on the right side of the colon every time the subroutine is called, and storage is reserved for the new variable each time. When the subroutine execution is completed, and control is returned to the main program, the storage for the new variable is cleared. STATIC variable declarations are only interpreted once after they are declared, and this occurs at the time of the first call to the subroutine in which they reside. Unlike NEW variables, STATIC variables are not cleared from storage when the subroutine finishes executing. STATIC variables retain their storage and the values stored there even when the program is not executing. The last value saved for any STATIC variable will be available for further use in calculations in the main program.

Subroutine Declaration Statement

A subroutine consists of a collection of AML statements. The subroutine declaration statements serve to reserve space in storage for the subroutine, and names the subroutine for later reference in the program. Subroutine declaration statements have the following form:

```
id: SUBR(p1, p2, p3, ...pn);
subrstatement1;
subrstatement2;
        .
        .
        .
subrstatementn;
END;
```

The components of the subroutine declaration are the following:

The id or name of the subroutine, which identifies the subroutine for subsequent reference, followed by a colon.

The keyword SUBR, which serves to declare the subroutine in AML.

Parameters (p1, p2, p3, ...pn) which are transferred from the main program for use in the subroutine call. If specified, these must be enclosed in parentheses.

The body of the subroutine, consisting of AML statements each ending with a semicolon:

> subrstatement1;
> subrstatement2;
> .
> .
> .
> subrstatementn;

These statements have the same form and follow the same rules as other AML statements. The semicolon after the last statement (subrstatementn) and before the keyword END is optional

The keyword END, indicating the end of the subroutine.

A semicolon follows the END word. Special rules regarding the semicolon apply for the END statement for subroutines nested within other subroutines.

9D-3 CONSTANTS AND VARIABLES

AML uses a wide assortment of constant and variable types. Constants and variables can be integer, real, or string. Aggregate constants and variables are also possible in AML, including aggregates containing a combination of real, integer, and string elements. Aggregates are enclosed within angle brackets, ⟨ ⟩, and the elements are separated by commas.

AML permits the definition of two-dimensional aggregates in addition to the conventional one-dimensional aggregates. A two-dimensional aggregate contains both rows and columns of elements. A typical use of a two-dimensional aggregate would be to define an array of coordinates in a robot workspace. An example given in Ref. 1 involves the definition of six points (x, y, and z coordinates) for the IBM 7565 robot. The variable declaration statement is

> R: NEW ⟨⟨−6.0, −25.0, 3.1⟩, ⟨−6.0, −2.2, 3.5⟩,
> ⟨1.3, .5, 3.2⟩, ⟨5.2, −25.1, 3.0⟩,
> ⟨8.1, −2.2, 3.3⟩, ⟨8.4, 15.5, 3.5⟩⟩;

The individual elements of the two-dimensional aggregate are contained within angle brackets and can be more than two dimensions. Hence, in this example, a two-dimensional aggregate is used to represent a three-dimensional array. The value of R can be depicted graphically for the six points as shown in Table 9D-1. The individual elements of the aggregate can be accessed in

Table 9D-1 Coordinate values corresponding to the variable declaration of a series of points in space

Declaration statement
R: NEW ⟨⟨−6.0, −25.0, 3.1⟩, ⟨−6.0, −2.2, 3.5⟩, ⟨1.3, .5, 3.2⟩, ⟨5.2, −25.1, 3.0⟩, ⟨8.1, −2.2, 3.3⟩, ⟨8.4, 15.5, 3.5⟩⟩;

Corresponding coordinate values

R	X	Y	Z
1	−6.0	−25.0	3.1
2	−6.0	−2.2	3.5
3	1.3	.5	3.2
4	5.2	−25.1	3.0
5	8.1	−2.2	3.3
6	8.4	15.5	3.5

AML by means of the symbols R(1), R(2), and so on. In other words, R(4) represents ⟨5.2, −25.1, 3.0⟩. This might be useful for identifying a point in space to which the robot would be commanded to move.

9D-4 PROGRAM CONTROL STATEMENTS

The AML language contains a number of program control statements for conditional execution of the program. These conditional expressions include IF...THEN, IF...THEN...ELSE, WHILE...DO, and REPEAT...UNTIL statements. The IF...THEN and IF...THEN...ELSE statements have the conventional interpretation.

The WHILE...DO statement allows the programmer to repetitively execute an expression until a condition is satisfied. The form of the statement is

WHILE condition DO expression

An example of its use is the following

```
COUNT: SUBR;
X: NEW 1;
WHILE X LE 3 DO
    BEGIN
        DISPLAY (X, EOL);
        X = X + 1;
        END;
    DISPLAY ('DONE', EOL);
    END;
```

The subroutine is called COUNT; it defines the new variable X, counts up to 3, and displays the value of each iteration, finally displaying the word "DONE". The WHILE...DO statement controls the iteration process. A compound statement is contained within the BEGIN...END words. The EOL word contained in the DISPLAY statements indicate "end-of-line" to the device, corresponding to a carriage return. The final END; terminates the subroutine. When called the subroutine creates the following output

```
1
2
3
DONE
```

The REPEAT...UNTIL statement is similar to the WHILE...DO statement except that the order of condition and expression is reversed. Its basic form is therefore

REPEAT expression UNTIL condition

For example, a subroutine similar to the above, but written using the REPEAT...UNTIL statement, is the following

```
COUNT: SUBR;
X: NEW 1;
REPEAT DISPLAY (X, EOL);
UNTIL (X = X + 1) EQ 4;
DISPLAY ('DONE', EOL);
END;
```

9D-5 MOTION COMMANDS

The AML language is designed for the IBM 7565 robot, which is a cartesian coordinate robot with a maximum configuration of six joints plus gripper, pictured in Fig. 9D-1. As shown in the figure, there are names for the individual joints and gripper. Table 9D-2 provides a listing of these joints together with information about their limits of travel. Joints 1, 2, and 3 are linear, while joints 4, 5, and 6 are rotational. The gripper is also considered linear in terms of the opening between its fingers.

Motion commands for the IBM 7565 require the specification of the joint and the distance to be moved. The distance can be specified either by naming the absolute coordinates in the workspace to move to, or by specifying the incremental distance to be moved. The two associated commands are MOVE and DMOVE. For example, the command

MOVE(⟨JX,JY⟩, ⟨8.5,1.5⟩);

indicates that the two joints JX and JY should be moved to their corresponding coordinate values of 8.5 and 1.5 in., respectively. It is also possible to specify the point to be moved to as a previous defined location variable, as in

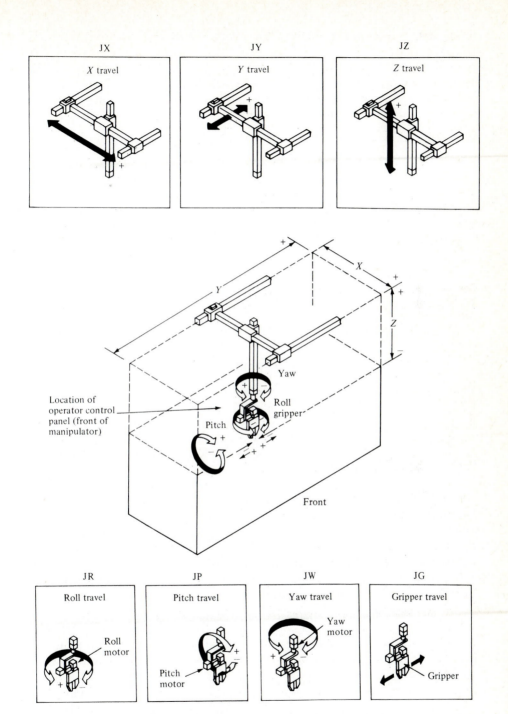

Figure 9D-1 The IBM 7565 robot, showing individual joints for maximum configuration. (Courtesy of International Business Machines Corporation)

Table 9D-2 Joint details for the IBM 7565 Robot system

Number	Joint name	Lower limit	Upper limit
1	JX	−8.9 in.	+8.9 in.
2	JY	−29.4 in.	29.4 in.
3	JZ	−8.75 in.	8.75 in.
4	JR	−135°	+135°
5	JP	−90°	90°
6	JW	−135°	135°
7	JG	0 in.	3.25 in.

the following statement

$$\text{MOVE}(\langle JX,JY,JZ \rangle, R(1));$$

where R(1) might refer to point number 1 in the array presented in Table 9D-1.

The DMOVE command is used to move relative to the current position. For example, the statement

$$\text{DMOVE}(JZ, 3.0);$$

would move the robot wrist 3 in. up in a vertical direction.

Two other move commands in AML are AMOVE and GUIDE. The AMOVE command provides the same basic motion and trajectory specifications as MOVE. The difference is that the MOVE statement completes the motion before proceeding with any subsequent statement(s) in the program. With AMOVE, the system can perform other work while the move is completing, such as data processing functions or calculations. Obviously, any subsequent moves must await completion of the current move. The format of the AMOVE statement is the same as the MOVE statement

$$\text{AMOVE}(\langle JX,JY \rangle, \langle 8.5,1.5 \rangle);$$

The GUIDE command allows the use of the teach pendant to move the specified set of joints under powered leadthrough control. The IBM 7565 teach pendant has control buttons corresponding to particular joints and directions. For each joint, there is a positive and a negative direction button that can be depressed to achieve the desired motion for the joint. The longer the button is depressed, the faster the joint moves. When the button is released, the joint motion stops. Once the robot has been moved to the desired position, the "End" button on the teach pendant is pressed and the set of axis positions is displayed on the monitor, and control is returned to the AML interpreter. For example, consider the following command

$$\text{GUIDE}(\langle JX,JY,JZ \rangle);$$

This would enable the x, y, and z axes for motion through the teach pendant.

The user could then move those axes with the teach pendant buttons. At the completion of the manipulation, the user depresses the "End" button, and the following type of information is displayed on the monitor

$$\langle 2.03700,27.6000,4.00300 \rangle$$

This information could then be used for defining a MOVE statement if desired.

The execution speed of a programmed move is set with the SPEED command. The statement

$$\text{SPEED(.5)}$$

sets the speed for any subsequent motions (until speed is respecified) to 0.5 of full speed. Valid speeds range from greater than 0.0 to 1.0 of full speed of the robot.

9D-6 SENSOR COMMANDS

The IBM 7565 robot system provides a sensor I/O interface to connect with various hardware in the work cell such as sensors, fixtures, tools, feeders, and other equipment. The system accommodates 1-bit digital input or output, as well as multiple-bit I/O (up to 16 bits). The IBM 7565 system supports up to 64 logical I/O definitions. If the robot is equipped with a sensored gripper, additional sensor commands are possible to interface with its sensors.

Sensor Input/Output

Three of the important commands in AML for sensor I/O are DEFIO, SENSIO, and MONITOR. The DEFIO command is used to define a logical sensor I/O device that can be accessed by SENSIO or MONITOR. The form of the DEFIO command is

$$\text{DEFIO(group,type,format,sbit,length,scale,offset)}$$

The first five parameters in parentheses are required and the last two parameters are optional. The group is an integer or aggregate specifying the I/O channel. The type refers to the input or output device (0 for input and nonzero for output). The format is an integer or aggregate specifying the format of the sensor I/O data. A 1 means that a closed contact yields a -1 and an open contact yields a 0. A 2 means that the closed contact yields a 1 and the open contact value is again zero. The sbit is an integer or aggregate specifying the bit number of the beginning of the I/O field. Bit numbers range from 0 to 15, with bit 0 being the left-most bit. The length is an integer or aggregate specifying the length of the bit field. Bit fields range from 1 to 16 bits.

The scale is an optional real number or aggregate specifying the scale factor that is to be used for scaling the I/O to make its size compatible with the system. The default value is one. Finally, the offset is an optional real number or aggregate specifying the constant to be used in offset the scaling. The default value is zero.

The DEFIO command returns the definition of a bit or a contiguous set of bits that can be accessed by the SENSIO command. The return is referred to as the ionum. The ionum is an integer or aggregate specifying the numerical identifier for each I/O defined. This value is to be used in all references to the particular logical input/output. The format of the SENSIO command is

SENSIO(ionum,template)

The ionum is the value (or values) that determines the input or output that is to be performed. Up to 10 ionums can be specified in a single SENSIO command, causing 10 I/Os to be executed from the same command. The template is an integer, real number, or aggregate that specifies the template for input operations or the data to be used for output operations.

An example of the use of the DEFIO and SENSIO statements would be

DEFIO(21,0,1,0,1);
1016
SENSIO(1016,INT)

The 21 in the DEFIO statement refers to channel 21; the first 0 indicates that the type is input; the first 1 indicates the format; the second 0 indicates the first bit of the field; and the second 1 indicates that the field is 1 bit long. The DEFIO statement returns the ionum to be referenced by the SENSIO statement (1016). The INT is used to reference the values under the control of the AML program.

The MONITOR command allows the use of one or more of the I/O channels for reading of sensors at regular time intervals. This can be used to design an interrupt system for safety monitoring or other purposes. The MONITOR command differs from SENSIO since it permits quasicontinuous monitoring of incoming sensor signals rather than at specific points in the AML program.

Gripper-Sensing Capabilities

Additional sensing capabilities are available with the optional sensored gripper for the IBM 7565 system. These capabilities include force sensing with strain gauges and optical sensing with a light emitting diode. The gripper is shown in Fig. 9D-2, indicating the tactile sensing and optical sensing features.

The sensors can be used with SENSIO and MONITOR commands. The strain gauges are used to detect positive and negative forces in three directions as indicated in Fig. 9D-2. The strain gauges corresponding to the three directions are

Tip sensors. These indicate whether there is a force pressing up from the bottom of the gripper. This might occur, for instance, if the gripper is

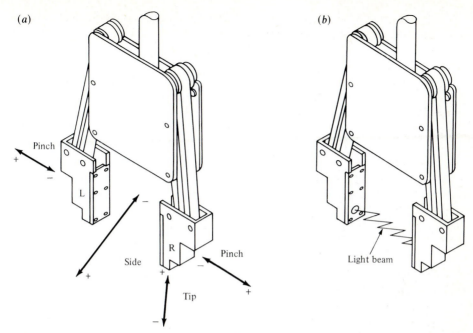

Figure 9D-2 Sensored gripper used with IBM 7565 robot system. (*a*) Tactile sensing features; (*b*) light emitting diode for optical sensing. (Courtesy of International Business Machines Corporation)

pressed against the worktable. It would also have a value when an object is being held in the gripper with the gripper in a vertical orientation as shown in the figure. The weight of the object would cause a force to be registered in these sensors. The AML words SLT and SRT are used as the logical I/O identifiers for the left and right sensors in SENSIO and MONITOR commands.

Pinch sensors. If the gripper is closed on an object, these pinch sensors will indicate the pressure with positive measurement values. If the gripper is used to grasp the inside of a cup by opening against the sides of the cup, negative values will be indicated. The AML words SLP and SRP are used as the logical I/O identifiers for SENSOR and MONITOR commands.

Side sensors. If the gripper is moved against an obstacle from the side, these sensors will indicate the presence of the obstacle. The AML words SLS and SRS are used as the logical I/O identifiers for SENSOR and MONI-TOR commands.

Two examples of the use of these gripper sensors will be used for illustration. The command

MONITOR(SLP,3,−1000.0,+1000.0);

would cause the monitor to trip when the force exerted on the left pinch strain gauge deviates by 1000 grams from its current value. The command

SENSIO(⟨SLP,SLT,SRP,SRT,SLS,SRS⟩,6 of REAL);

would return the following aggregate which contains the force on each of the six strain gauges in the gripper fingers.

$$\langle 236.432, 465.002, 21.5060, 17.8920, -15.0060, -23.7772 \rangle$$

These values are in grams of force.

The AML word LED can be used with the SENSIO and MONITOR commands to determine if an opaque object is blocking the light emitting diode. For example, the command

$$\text{SENSOR(LED,INT)}$$
$$-1$$

returns a 0 to indicate that no object is blocking the LED, whereas a value of -1 indicates that an object is blocking the light path. The word LED in the statement identifies the logical I/O number of the gripper light detection circuit.

9D-7 DATA PROCESSING

AML contains a number of data processing capabilities for interfacing with peripherals. The IBM 7565 robot includes the following data processing devices: display terminal (keyboard and CRT display), printer (optional), diskette, disk (optional), teach pendant (which incudes a 10-character display), and communications line. The communications line permits communication between different systems and workstations, including transfer of files between systems. There are 14 channels available for transferring data between the system and peripheral devices. Certain channels are reserved: channel 0 is the display terminal, channel 1 is the printer, channel 2 is the pendant, and channel 3 is the system message file. The remaining channels can be used to access user disk files.

Several commands can be used by the programmer to define communication from the system to the peripherals. These include the WRITE, PRINT, and DISPLAY commands. The WRITE command is used to write one logical record into a file through a defined channel. The command

$$\text{WRITE(0, 'Good Morning');}$$

would write the word hello to the display terminal through channel 0. The maximum length of the record that is transmitted depends on the output channel. If a WRITE to channel 0 is issued (display terminal), the maximum length is 79 characters. If a WRITE to channel 1 is issued (printer), the maximum length is 132 characters. If a WRITE to channel 2 is issued (teach pendant display), the maximum length is 10 characters. If the data are longer than the maximum record length, an error message is returned.

The PRINT command writes the values of a list of expressions on a currently open data channel. PRINT is different from WRITE in that the data in the expression are converted to character format before being displayed, and in the fact that more than one record can be transmitted to the device. With WRITE, only one record is written to the device. For example, the following sequence accomplishes the same as the above

X: NEW 'Good';
Y: NEW 'Morning';
PRINT(0,⟨X, Y, EOL⟩);

Another difference between PRINT and WRITE is that the output data is buffered until an EOL or EOP (end-of-print) character is detected, or the buffer is exceeded. Again, the channel number must be provided as one of the arguments in the command.

The DISPLAY command is the same as the PRINT directed to channel 0. It requires no identification of the channel. The following serves to illustrate

DISPLAY('Good Morning', EOL);

The READ command is used to receive data from a peripheral device. It reads one logical record from the specified channel. The following statement illustrates the format of the READ command

READ(0, DATA1);

This would read a file named DATA1.

Other system commands include ERASE (erase a file from a disk and free the space it occupies on the disk), COPYFILE (copy a file from one disk to another or the same disk), and RENAMEFILE (rename a file).

9D-8 SUMMARY

Additional statements are available in the AML language. The purpose of this appendix is to provide a sampling of the important features and statements that are contained in AML. The interested reader is referred to the user's manuals available from the IBM Corporation.[1,2]

REFERENCES

1. IBM Corporation, *A Manufacturing Language Concepts and User's Guide*, Base Publication 850912, Boca Raton, FL, December 1982.
2. IBM Corporation, *A Manufacturing Language Reference*, Base Publication 850915, Boca Raton, FL, October 1983.
3. IBM Corporation, *A Manufacturing Language Screen Editor*, Base Publication 850916, Boca Raton, FL, December 1982.
4. Taylor, R. H., Summers, P. D., and Meyer, J. M., "AML: A Manufacturing Language," *The International Journal of Robotics Research* **1**(3):19–41 (Fall 1982).

TEN

ARTIFICIAL INTELLIGENCE

Modern robot control concepts, as well as vision and tactile sensing have their roots in artificial intelligence (AI) research. In our discussion of future generation languages for programming robots in the previous chapter, we talked about world modeling and the need for a robot to receive an objective oriented command, such as ASSEMBLE TYPEWRITER, and translate it into a set of specific actions. The capacity to accomplish this translation will require a high level of intelligence by the robot. It is appropriate that we explore the subject of artificial intelligence and its opportunities in industrial robotics. We do so in this chapter.

AI is a vast topic and this chapter will only serve as an introduction. We will discuss some of the goals of AI research and some of the general techniques used for attaining these goals. We will then look at how these goals and techniques can be applied to robotics research. Finally we will introduce the programming language LISP which is uniquely suited for AI research.

10-1 INTRODUCTION

Artificial intelligence efforts aim at developing systems that appear to behave intelligently. Often this is described as developing machines that "think." Typically this work is carried out as a branch of computer science, although it also contains elements of psychology, linguistics, and mathematics.

In the introduction to *The Handbook of Artificial Intelligence*[1] the editors state:

> Artificial intelligence is the part of computer science concerned with the characteristics we associate with intelligence in human behavior—understanding language, learning, reasoning, solving problems, and so on.

Because of the difficulty in defining intelligence, AI research is often best described by discussing its goals, as we do in the next section.

10-2 GOALS OF AI RESEARCH

While the general objective of AI work is the development of machines which exhibit intelligent behavior, this statement is in itself too broad and ambiguous to be meaningful. The following paragraphs will describe some of the areas of AI which are presently being pursued as distinct areas of research.

1. *Problem solving*. Examples of problem-solving systems are the chess-playing programs. Given a set of rules and strategies these systems are capable of playing chess on a proficient level. Problem solving does not necessarily restrict itself to games. Consider, in the future, telling your household robot to fetch the paper. On the way to solving the problem of retrieving the paper the robot must identify and solve a number of smaller problems, hopefully in the most efficient sequence. First the robot must identify a possible path to reach the paper, then it must deal with obstacles along that path, such as opening the door. Finally it must deal with grasping the paper and returning it. Almost all tasks with which we are confronted daily involve problem solving. Design of a system requires the breaking down of the problem into successively smaller problems until the solutions begin to become evident. At this point we must "show" robots all of the motions required to perform an assembly task. A robot endowed with a form of "problem solver" may be able to develop the assembly strategy itself. We will discuss this example in greater detail in a later section.

The techniques of problem solving are the same whether developing a robotic assembly strategy solver or some other problem solver.

2. *Natural language*. In spite of the proliferation of computers into department stores, many people are still uneasy or uncomfortable dealing with computers. This is due in part to the need to talk to a computer in its language rather than the user's "natural" language. The problems facing natural language researchers are numerous. The computer must not only be able to understand the meanings of words, but how those meanings may differ in context with other words. The system must also be able to understand the syntax of the language so that the relationships of the words is understood. Imagine the possible meanings for "Time flew out the window" if taken purely from a grammatical standpoint.

3. *Expert systems*. This area of research is concerned with developing systems that appear to behave as human experts in specific fields. Through a dialog with a human operator an expert system can recommend tests to be performed and ask appropriate questions until it arrives at a conclusion. At present, expert systems are being used to configure computer systems, design circuits, and perform medical diagnosis. Some of the issues involved with the design of expert systems include the problems of dealing with vast amounts of data, explaining the system's "reasoning" on the way to reaching a conclusion,

representing the data collected from the human experts, and improving the "knowledge base" with experience.

4. *Learning*. One of the attributes of intelligence is the ability to learn from experience. If machines were capable of learning then the task of endowing them with knowledge, as in the case of expert systems, would be greatly simplified. Some systems have been developed which have shown the ability to learn from experience, but to this date limited progress has been made.

5. *Vision*. Most of the basic concepts employed in commercial vision systems are the result of AI research. One of the more interesting goals of AI vision research is to permit the systems to perform scene analysis. That is, present the vision system with a scene and allow the system to identify objects within the scene.

Some of the other areas of AI research include: automatic programming, hardware development, and deductive reasoning.

10-3 AI TECHNIQUES

AI is concerned with the use of data or knowledge. Therefore techniques must be developed for two basic tasks: data representation and data manipulation. In this section we will look at some of the approaches for representing data and using that data in some specific manner. We will not discuss actual programs for doing this, only the general techniques which might be employed by AI programs.

Knowledge Representation

When we discuss knowledge representation we are not concerned with the physical operation of the computer, that is, we are not discussing the storage of words as a series of 1s and 0s. Rather we are discussing the relationships of facts with respect to each other, for example, the statement, "Some birds have wings."

The material in this section is summarized to a large extent from Ref. 1. For a more detailed account the reader is directed to Ref. 1.

Before discussing the various representations of knowledge we must first describe the various types of knowledge which may require representation.

1. Objects. More specifically facts about objects, such as "robotics students drink heavily" or "birds have wings."
2. Events. Not only the event itself, such as "The robotics student broke his arm," but perhaps the time or cause-effect relation of the event. "The robotics student broke his arm yesterday and the nasty instructor made him pay for it."
3. Performance. If the AI system is one which is designed to control a robot

then it must have data on the performance of the arm, that is, its kinematics, dynamics, what bits to manipulate in the hardware, and so on.
4. Metaknowledge. This is the knowledge about our knowledge. This includes our knowledge of the origin of the information, its relative importance, its reliability, and so on. For example, one would give little weight to the following information "The study of robotics is painless" if that information came from a history student.

At this point let us look at some of the various techniques for representing knowledge.

1. *Logic*. Formal logic was developed by mathematicians and philosophers to provide a method for making inferences from facts. Formal logic allows facts to be represented in a specific syntax, and, by applying defined rules of inference to these facts, allows conclusions to be drawn. We are probably familiar with the following type of example:

Given the two statements

a. All roboticists play games.
b. Jack is a roboticist.

We can conclude by using a rule of inference that

c. Jack plays games.

Conceivably, if a computer were endowed with all of the possible facts about a subject, and also all of the applicable rules of inference, then it should be able to develop any new facts about the subject that may be inferred from the original set. We will discuss later in the chapter the concept of "combinatorial explosion" which may limit the usefulness of logic-based (or any other) representation schemes. As the number of facts become larger, the number of combinations of facts with the rules of inference which may apply also increases. This results in generating a problem too large for the computer to solve in a reasonable amount of time.

2. *Procedural representations*. In the previous discussion we examined a method for storing facts. It is also necessary to store information on how to use facts. In a logic-based system as above, every rule of inference must be applied to prove a new point. If it were possible to encode the information along with a way to use that information, it is possible that a more efficient system could be developed. For instance, if we wished to find out whether or not Jack plays games, the database of the computer could contain the following facts:

a. Jack is a roboticist.
b. All roboticists play games.

Additionally the system could have stored the following procedure about Jack

If need-prove plays (*x*)
show is-roboticist (*x*)

which states that if we need to prove some fact about *x*, we can do so by
showing that *x* is a roboticist. With this information the system would be able
to avoid any other information about Jack and jump instantly to the correct
solution. This technique of storing information as a procedure is commonly
used in computer programs in if–then statements. It offers the advantage of
directing the system to find the solution more directly. On the other hand, it
takes away from the generality of the system and makes changes difficult.

   3. *Semantic networks.* Semantic networks are representations of information which consist of nodes and arcs. Nodes typically represent objects,
concepts, or situations and the arcs represent the relationships between nodes.
The nodes and links are labeled with simple language descriptions. Figure
10-1 illustrates the following information as a semantic network:

Jack is a student.
Jack is a roboticist.
Roboticists play games.
Roboticists may be students.
Students may be roboticists.

Many inferences may be drawn about Jack, students, and roboticists by
investigating the network.

   Networks are used extensively in AI research as a means of knowledge
representation. Unfortunately they also have their drawbacks. They may be

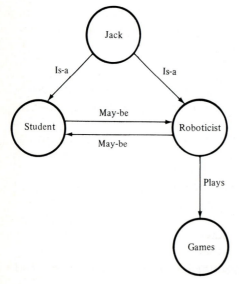

Figure 10-1 Semantic network describing jack.

too simplistic; how does a network deal with ideas or large amounts of knowledge? As with all representation techniques it cannot be pushed to an extreme.

4. *Production systems*. Production systems store information in the form of items called "productions." Productions all have the form of IF some expression is true THEN some action. For example, the information we have about Jack and roboticists would be presented as:

a. IF the person is a roboticist THEN he/she plays games.
b. IF the person is reading this book THEN he/she is a roboticist.

If we were able to catch Jack reading this book then we would know that he is, in fact, a game player.

Production systems are used to break the problem down into successively more manageable tasks. They are typical of the construction of "expert systems." They provide for uniformly designed systems using the IF-THEN construction as well as a modularity, in that each rule has no direct effect on other rules. Unfortunately, among other problems, these systems may become very inefficient as they become larger.

5. *Frames*. Another representation technique is the use of frames. Frames can be considered as predefined structures with empty slots which are filled to represent the specific facts. For example, the general frame for a student might look like the following:

> STUDENT Frame
>   Height: in inches
>   Weight: in pounds
>   Studies: Robotics or History

and for Jack the frame would look like

> JACK Frame
>   Height: 70 inches
>   Weight: 150 pounds
>   Studies: Robotics

The slots could be filled by references to other frames or by procedures defining how the slots are to be filled.

6. *Other representation techniques*. In many cases choosing the correct representation technique can greatly simplify the problem. As an example let us investigate a video view of an object. One way to represent the object would be as a semantic net with each pixel representing a node and each arc representing the "is-connected-to" property attached to each adjoining node. A simpler approach is to represent the picture as an array of values corresponding to the brightness of each individual pixel. The point is that the choice of a representation technique should be made only after giving consideration to the type of problem being solved. The above mentioned tech-

niques are applied to different circumstances and in some cases different representation schemes should be found.

Problem Representation and Problem Solving

Before discussing problem-solving techniques it is useful to explain what is meant by "problem solving." Problem solving is the task of reaching some specified goal. Examples of these goals may be:

Finding the proof to a mathematical theorem.
Solving a puzzle, such as Rubik's cube.
Determining a sequence of assembly steps.
Choosing the next move in a chess game.

In order to solve these problems it is necessary to represent the problem in some way that is amenable to discussion and solution. Two possible schemes for problem representation are:

1. The state-space representation
2. The problem-reduction representation

   1. *State-space representation.* In this method we can visualize the problem as all of the possible states found in developing the solution configured as a tree. The tree is made of nodes, which represent the states of the system after certain actions have been taken. The actions are represented by the arcs that connect the nodes. This can be illustrated by using the "traveling salesman" problem. A traveling salesman has to travel to four cities *A*, *B*, *C*, and *D*. The salesman wishes to travel to all four cities using the shortest possible path and by going to each city only once. He wishes to begin and end his trip at city *A*. The distances between the cities is given in Table 10-1. The state-space representation of the trip is shown in Fig. 10-2.
   By looking at the state-space representation, we can see that the problem becomes one of choosing the branch of the tree with the shortest sum of arc lengths. There are two paths which provide a solution of 15: *ABDCA* and *ACDBA*. Later in this section we will see how to search the state-space tree for this solution. This form of problem solving is called "forward reasoning,"

Table 10-1 Distance between cities

	A	B	C	D
A	0	3	4	5
B	3	0	7	6
C	4	7	0	2
D	5	6	2	0

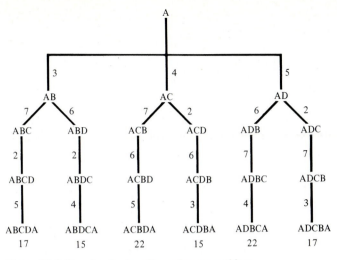

Figure 10-2 Tree for the traveling salesman problem.

because we worked our way forward through all of the states until we found the solution.

2. *Problem-reduction representation.* In this case we see an example of "backward reasoning." In the problem-reduction representation we present the goal as the primary data item and then reduce the problem until we have a set of primitive problems; that is, simpler problems for which we have the data available. This simplification may involve breaking the problem down into a set or sets of smaller problems which must all be solved or into alternative problems, any one of which may be solved. The scheme is graphically represented as an "and–or" graph. In an and–or graph arcs which are connected by a horizontal bar are "anded" and arcs which are not connected are "ored." Figure 10-3 illustrates a simple and–or graph which states:

a. *P* may be solved by solving *A* or *B* or *C*.
b. *A* may be solved by solving *D* and *E*.
c. *B* may be solved directly (*B* is a primitive).

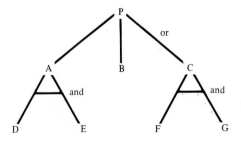

Figure 10-3 And-or graph.

d. *C* may be solved by solving *F* and *G*.

e. *D*, *E*, *F*, and *G* are primitives.

In this case, then, the simplest scheme for solving *A* is to solve *B*. In the next subsection we discuss the different techniques for searching for the solution.

Search Techniques in Problem Solving

Up to this point we have considered techniques for representing knowledge and problems. If we employ the proper representation technique, then it is often only a matter of searching for a solution to reach the desired goal. In many cases it is possible that the necessary search is too large to be successful in a reasonable amount of time. For example, consider the game of chess. Given that there is a known starting state and that the goal is to capture the opponent's king, it is possible to develop all of the possible move combinations in a state-space representation and then search for the best winning solution and follow that path. Unfortunately the number of move combinations in a game of chess may be on the order of 10^{120}. Because of this type of "combinatorial explosion," techniques that seek to minimize the search effort have been developed. This does not mean that search is the only available technique. Just as the method of knowledge and problem representation must be carefully considered for each task, so should the solution technique. In any case some of the developed search techniques follow:

1. *Depth-first search.* As the name implies, this search technique searches as deeply as possible in the tree network in an effort to find a solution. Figure 10-4 illustrates this principle. The objective is to find the node labeled "S" (solution). Using the depth-first search technique, the system searches through the network as deeply as possible until it finds a terminal node. If that node is not the desired solution, then it backtracks until it finds the first available branch point and goes down the next branch. In the example illustrated the system will eventually find the solution node. In some systems it is possible that the depth of the tree or certain branches of the tree are so deep that the solution is never found.

2. *Breadth-first search.* Figure 10-5 illustrates the concept of breadth-first search. The system evaluates all nodes at the same level in the tree before moving on to evaluate the next level. This system is more conservative than depth-first searches and is not as easily trapped. In cases where the tree is relatively deep on all branches it may be less efficient than a depth-first search.

3. *Hill climbing.* Hill-climbing techniques provide a variation on depth-first searches. Rather than moving in an arbitrary decision at each branch point, the hill-climbing algorithm attempts to make the best choice among the possible branches from the present node. This choice is based upon some selection technique. For example, in the traveling salesman problem, the next node chosen may be based on the total distance up to the next possible node. The risk here is that we are only looking at the next node, that is, at local

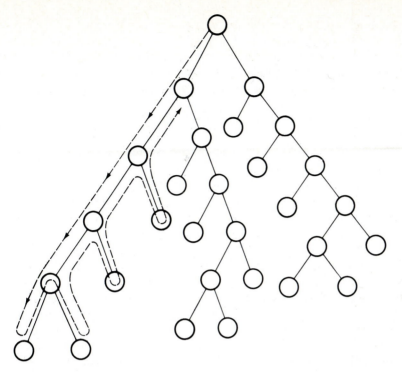

Figure 10-4 Depth-first search.

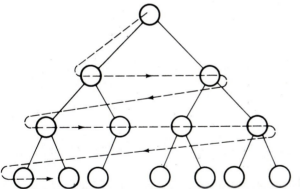

Figure 10-5 Breadth-first search.

information. While we may be making the best local choices, we may be missing a far better overall solution that had one bad arc.

4. *Best-first search*. This is a variation on hill climbing. In this case, rather than choosing the best next branch from a node, the system selects the best next node regardless of its position in the system. This generally provides the optimum solution, but does not guarantee it.

5. *Branch and bound*. This is similar to the best-first search. The system evaluates all of the partial paths to the first level and expands the most promising path to the next level(s). As the path ceases to be the optimum (it becomes too long) the system expands the new most promising path. In this way it is always investigating the optimum path to the deepest level necessary. In order to ensure that the optimum path is found all partial paths must be extended until they become longer than the solution path, or until a shorter solution is found.

6. *Constraints*. In some cases it is not necessary to consider all of the possible options in developing the tree. Very often in the real world, constraints are placed upon information by its context. We know, for instance, that mice and cats are not likely to live in the same cage. Applying these constraints as we are traveling through the tree may lead to a significant reduction in the possible number of nodes to be explored.

There are many other search techniques beyond those explained here but they are beyond the scope of this text. At this point it is worthwhile to review what has been covered so far.

Knowledge can be represented in a number of ways. Not only are the data important, but also the relation of data to other information and the rules for manipulating data are important. Problems may also be represented in a number of ways. Problems may be solved by forward or backward reasoning. By representing solutions as trees, optimal solutions may be found using search techniques.

10-4 LISP PROGRAMMING

The programming language traditionally used in AI research is called LISP, LISt Processing. LISP has many interesting features not commonly found in other programming languages. One of these is the fact that LISP data and LISP programs are both created out of lists." Because data and programs are identical, LISP programs may modify or generate other LISP programs. We will introduce a number of LISP commands in this section so that simple programs may be developed.

LISP programs are lists constructed from elements called atoms. The following are lists:

> (A, B, C, D)
> (Item1 Item2)
> ((Item1 A)(Item2 B))

The following are atoms:

> A
> B
> Item1
> NIL

LISP attempts to evaluate all expressions and return the value of the evaluation. It uses prefix notation when evaluating the expressions. The following expression

(PLUS 3 5)

returns

8.

LISP always tries to use the first element of a list as a command for evaluating the rest of the elements in the list which are the arguments. Since lists always take the form

(expression, expression),

where expression can itself be a list, we can write the following example:

(PLUS (PLUS 2 3)(PLUS 4 5))

LISP evaluates the two inner lists (expressions) first resulting in the equivalent of

(PLUS 5 9)

which returns

14.

The arithmetic functions of LISP are some of the easiest, so we will start with them. Some of the available functions can be demonstrated by examples. The value returned by LISP is to the right of the expression.

(PLUS 3 4)	7
(TIMES 5 7)	35
(DIFFERENCE 9 4)	5
(QUOTIENT 8 2)	4
(ADD1 8)	9
(SUB1 8)	7
(MAX 2 19 7)	19
(MIN 2 19 7)	2
(EXPT 2 4)	16
(SQRT 9)	3

As we saw with the first example, the arguments do not have to be actual numbers. They may be other expressions.

Earlier we stated that one of the interesting features of LISP was that it was designed to manipulate lists. Two functions which are used for this purpose are CAR and CDR (pronounced could-er). CAR returns the first element in a list and CDR returns everything else. Again let us consider some examples.

(CAR '(A B C))	A
(CAR '((A B)(C D)))	(A B)

```
(CDR '(A B C))          (B C)
(CDR '((A B)(C D)))     ((C D))
```

It should be noted that CAR may return an atom or a list and that CDR always returns a list. If CAR is asked to operate on an atom it will return an error. If CDR is asked to perform on a one-element list it will return NIL.

You may have noticed that we added some quote marks to the last set of examples. As stated earlier, LISP tries to evaluate expression by using the first element in each list as a function. The quote inhibits this evaluation so that LISP can tell the difference between the list and the function.

CAR and CDR can operate together in expressions. For example:

```
(CAR(CDR(CDR(CDR'(A B C D E)))))
```

returns

D

Of course, if we can take expressions apart there must be some way to put them back together. There are three functions that may be used for this purpose. They are APPEND, LIST, and CONS.

APPEND runs the elements of its lists together, for example,

```
(APPEND '(A B)'(C D))     (A B C D)
```

LIST makes a new list out of the arguments

```
(LIST '(A B)'(C D))     ((A B)(C D))
```

CONS inserts the first argument as the new first element in the list of the second argument

```
(CONS '(A B)'(C D))     ((A B)C D)
```

The following example shows that CONS can "redo" what CAR and CDR undo

```
(CONS(CAR'(A B C D))(CDR'(A B C D)))     (A B C D)
```

If we always had to explicitly state the list we wished to manipulate it would become very tedious. The LISP expression that saves us from calloused fingertips is SETQ. SETQ is used to "name" an expression. For example,

```
(SETQ VW '(A   TYPE OF CAR)),
```

assigns the value, "A TYPE OF CAR" to the argument VW. From now on, whenever LISP sees VW, it will return A TYPE OF CAR. For example,

```
(CAR VW)     A
(CDR VW)     TYPE OF CAR
```

In many instances it is useful to be able to define a new function. The command for this is DEFUN (DEFine FUNction). DEFUN uses the following format:

(DEFUN ⟨function name⟩
          (⟨parameter 1⟩,⟨parameter 2⟩...⟨parameter n⟩)
          ⟨function description⟩)

A function to convert height in the form of feet and inches to just inches might look like

(DEFUN FT-TO-IN (FT IN)
          (PLUS (TIMES FT 12) IN)).

Typing in

(FT-TO-IN (5 10))

would result in the response

70.

Earlier in the chapter we saw that decisions have to be made in order to evaluate the success of certain operations. For this LISP provides predicates. Predicates are functions that provide only two possible responses: T (for true) and NIL. Some examples of predicates are

(LESSP 2 3)	T
(LESSP 3 2)	NIL
(GREATERP 2 3)	NIL
(GREATERP 3 2)	T

Other predicates include: AND, which returns T if all the elements in a list are nonNIL; OR, which returns T if any element in a list is nonNIL; NOT, which returns T if its argument is NIL; ZEROP, which returns T if its argument is 0; EQUAL, which returns T if it has two arguments and they are equal.

Predicates are often used as tests in conjunction with the function COND. COND is used with the following syntax:

(COND (⟨test1⟩ ... ⟨result1⟩)
          (⟨test2⟩ ... ⟨result2⟩)
                              ⋮
                              ⋮
          (⟨testn⟩ ... ⟨resultn⟩))

A COND is made up of a series of lists each of which is called a clause. LISP evaluates the first element of each clause until it finds a nonNIL result (not necessarily a T). When it finds a nonNIL clause it evaluates that list until it reaches the last expression and returns that value.

Earlier in this chapter we discussed information representation. In LISP data are represented as lists and there are numerous commands available for manipulating these lists. For defining the structure of a property list there is the DEFSTRUCT command with the following syntax

> (DEFSTRUCT(⟨name⟩)
>   (⟨property name 1⟩⟨default property⟩)
>   (⟨property name 2⟩⟨default property⟩)
>   . . .
>   (⟨property name n⟩⟨default property⟩))

Evaluation of DEFSTRUCT not only sets up the property list structure but also generates a function called MAKE-name. MAKE-name is used to assign values to a property list. DEFSTRUCT also generates "selector" procedures called "property name n." The operation of DEFSTRUCT is probably best described using an example.

Example 10-1 Using **DEFSTRUCT** let us say that we want to set a property list that would be useful for describing dogs. We can assume that most dogs have four legs, so that would be a good default value. Let us also say that we wish to know the dog's color, but that it is variable so we will use NIL as the default. Finally we will define the breed as a sporting dog or a domestic dog. We can begin by setting up the structure

> (DEFSTRUCT (DOG)
>   (COLOR NIL)
>   (NUMB-OF-LEGS 4)
>   (TYPE NIL))

We can use MAKE-DOG to generate a property list with the structure we have developed called DOG using SETQ so that

> (SETQ BEAGLE (MAKE-DOG))

generates a property list named BEAGLE. We can use the selector procedures DOG-COLOR, DOG-NUMB-OF-FEET, and DOG-TYPE to examine the properties of BEAGLE. In this case the three procedures would return the three default values NIL, 4, and NIL. We can assign new values to the properties also by using SETQ and the field identifiers.

> (SETQ BEAGLE (MAKE-DOG :COLOR 'BROWN :TYPE 'SPORTING))

We have left the number of legs alone. We can now evaluate the selector procedures

> (DOG-COLOR BEAGLE)
> (DOG-TYPE BEAGLE)

which returns

> BROWN
> SPORTING.

A function used for setting specific fields within a property list is SETF with the following syntax

(SETF(⟨property⟩⟨name⟩)⟨new value⟩).

For example, we could change our Beagle to a purple dog by entering

(SEFT (DOG-COLOR BEAGLE) 'PURPLE).

We now have a way for associating properties with objects and for evaluating those properties. In the example at the end of the chapter we will see how this might be used.

At this point we are not going to introduce any more LISP functions. The references listed at the end of the chapter can provide additional detail for those interested. Example problems are given at the end of the chapter.

10-5 AI AND ROBOTICS

While the material covered in this chapter so far may appear to be far removed from the factory floor it is useful for providing insight into the operation of robotic systems in the future and in the function of sensory systems such as vision.

One of the problems involved with vision systems is the representation of data within the system. Rather than store pictures of the parts within the vision systems memory, certain features about the objects are stored, such as perimeter, area, number of holes, and so on. As systems become more complex and begin to be able to deal in crowded three-dimensional environments it will require greater intelligence on the part of the vision system. Techniques have been developed which allow vision systems to recognize specific types of objects (such as blocks) even when jumbled together. These systems rely heavily on search minimization methods to increase their processing speed.

Another promising area is research into task level programming. Rather than program the robot to make each required motion, it will be possible to make statements such as "Pick up the big red block." A program called SHRDLU, developed in 1971, allowed an operator to converse with a robot which lived in a world of blocks and pyramids. SHRDLU was able to respond and to plan out tasks. For example, if the red block was under a green block, SHRDLU would move the green block and set it down before getting the red block.

Factories are in fact, limited environments. This constrains the number of problems which must be solved to make robot intelligence practical.

10-6 LISP IN THE FACTORY

In this section we will develop an example LISP program which might be used to guide a robot in the performance of a simple task in the factory. The task which the robot must perform is the assembly of a gearbox consisting of three parts: a small gear, a large gear, and a plate with two shafts. The parts are presented to the robot by dumping the three parts out of a chute onto the worksurface in the view of a camera. Because of this feeding technique it is possible for the parts to be improperly oriented and one on top of the other. In order to make the problem simpler, we will make some assumptions: the camera is able to recognize the parts and their location, the camera can always find the three parts, and the robot can grasp all three parts in any orientation with its tool. Assume that the problem-solving procedure which we develop will not have to tell the robot how to perform simple tasks, only when to do them.

Let us begin by defining the task in more detail. The robot is to place the plate into an assembly fixture and then to place the large gear in place and then the small gear in place. If any part is in the way of another then the robot must clear that part before it can go on to complete the task. Let us also assume that the relationship among the parts is defined by the vision system; that is, the vision system can tell if one part is on top of another.

We may define the overall task as three smaller tasks: placing the plate in a fixture, placing the gear 1 on pin 1 of the plate and gear 2 on pin 2. We could define the task as

```
(PLACE 'PLATE FIXTURE)
(PLACE 'GEAR1 PIN1)
(PLACE 'GEAR2 PIN2)
```

We could define the locations by

```
(SETQ FIXTURE xyz
      PIN1 abc
      PIN2 lmn)
```

where xyz, abc, and lmn are the actual position values. Now all that must be done is to define the place function

```
(DEFUN PLACE (PART LOCATION)
       (GRAB (PART))
       (PUTON (PART LOCATION)))
```

Notice how we defined PLACE as two new procedures. We will continue to define new procedures until we can describe the task as primitive functions. For this example we will add some LISP primitives to make operation of our robot simpler. These are

OPEN	This opens the gripper
CLOSE	This closes the gripper
MOVE (LOCATION)	This moves the arm to "location"

In order to grab the part the system must first check the part to make sure that it is uncovered by another part, and the system must know its location. If we set up a property list for the part with these attributes then we can find these values.

```
(DEFSTRUCT (OBJECT)
    (POSITION NIL)
    (IS-UNDER NIL))
```

If the vision system is allowed to find the part locations and identify whether or not they are covered then we can use

```
(SETQ GEAR1 (MAKE-OBJECT :POSITION P1 :IS-UNDER U1)
      GEAR2 (MAKE-OBJECT :POSITION P2 :IS-UNDER U2)
      PLATE (MAKE-OBJECT :POSITION P3 :IS-UNDER U3))
```

where the vision system supplies values for P1, P2, P3, U1, U2, and U3.

We can now continue to define the necessary functions. GRAB checks for an obstruction, clears it off, and grasps the part.

```
(DEFUN GRAB (PART)
    (COND (OBJECT-IS-UNDER PART)(CLEAROFF PART))
    (GRASP PART))
```

If OBJECT-IS-UNDER returns a value, then CLEAROFF is performed, followed by GRASP; if not, GRASP is performed immediately.

```
(DEFUN GRASP (PART)
    (MOVE (OBJECT-POSITION PART))
    (CLOSE)
    (SETQ GRIPPER PART))
```

GRASP moves to the location of the part, closes the gripper, and assigns the part name to the gripper.

PUTON is used to position the part at the desired location

```
(DEFUN PUTON (PART LOCATION)
    (MOVE LOCATION)
    (UNGRASP PART))
```

UNGRASP releases the part and clears the gripper

```
(DEFUN UNGRASP (PART)
    (OPEN)
    (SETQ GRIPPER NIL))
```

Finally, the function CLEAROFF,

```
(DEFUN CLEAROFF (PART)
    (SETQ TROUBLE (OBJECT-IS-UNDER PART))
    (SETQ CLEARSPOT FREESPACE)
    (GRASP TROUBLE)
    (PUTON TROUBLE CLEARSPOT)
    (SETF (OBJECT-POSITION TROUBLE) CLEARSPOT)
    (SETF (OBJECT-IS-UNDER PART) NIL))
```

In this procedure, FREESPACE is a function which asks the vision system to return a value for an empty space where the system can drop off the obstruction which is called TROUBLE. The function also resets the location of the obstruction at its new location and clears off the part which is to be grabbed.

The example demonstrates the use of LISP to perform a simple problem-solving task as might be found in the factory. By considering the relationships of objects to each other we were able to perform the larger task by breaking it down into solvable primitive tasks.

REFERENCES

1. A. Barr, P. Cohen, and E. Feigenbaum, *The Handbook of Artificial Intelligence*, William Kaufmann, Los Altos, CA, 1981.
2. R. Brown and P. Winston, *Artificial Intelligence: An MIT Perspective*, MIT Press, Cambridge, MA, 1980.
3. B. Horn and P. Winston, *LISP*, Addison-Wesley, Reading, MA, 1984.
4. P. McCorduck, *Machines Who Think*, Freeman, San Francisco, CA, 1979.
5. K. Prendergast and P. Winston, *The AI Business*, MIT Press, Cambridge, MA, 1984.
6. P. Winston, *Artificial Intelligence*, Addison-Wesley, Reading, MA, 1984.

PROBLEMS

10-1 Describe three different search techniques.

10-2 Set up frames for the following:
   (*a*) Dogs.
   (*b*) Cars.
   (*c*) Professors.

10-3 Set up the and–or graph for the task performed in Sec. 10-6.

10-4 Evaluate the following expressions in the order given:

```
(SETQ STUFF '(ALL THINGS ARE SILLY))
STUFF
(CAR STUFF)
(CDR STUFF)
(CAR(CDR STUFF))
```

10-5 Write a LISP program which might be able to determine if one object is on top of another. Assume you are given the locations of two objects in the form of a list (x, y, z) for each object. Assume that if x and y are equal for both parts and z is greater for one than the other, that one is on top. Use CAR and CDR to get the individual x, y, and z values and then perform the comparisons.

10-6 The solution in Prob. 10-5 could have been used to determine the OBJECT-IS-UNDER property in Sec. 10-6. Implement this feature.

FOUR

APPLICATIONS ENGINEERING
FOR MANUFACTURING

This part and the one that follows are concerned with the applications of robotics in manufacturing. Applications engineering deals with problems such as the design of the physical layout of the workcell, the control of the various components in the cell, evaluating the anticipated performance of the cell, and the economic analysis required to justify the robot project. To discuss these topics, we divide this part of the book into two chapters.

Chapter Eleven addresses the physical design of the robot cell, and the coordination and control of the different pieces of equipment in the workcell. What are the various layout designs that can be used for the workcell? And what are the different control systems that can be used to coordinate the components of the cell? These systems are closely related to the robot programming methods discussed in Chaps. Eight and Nine. One of the important techniques in workcell control involves the use of interlocks (WAIT, SIGNAL, and REACT) to tie the different activities in the cell together. In this chapter, we also examine techniques for simulating the workcell.

Chapter Twelve deals with economic analyses used to justify robot projects. The analysis methods include the payback method, return on investment method, and equivalent uniform annual cost method. Robot and other programmable automation projects present certain unique problems in the economic justification of a project, and these issues are discussed in this chapter.

ELEVEN

ROBOT CELL DESIGN AND CONTROL

Industrial robots generally work with other pieces of equipment. These pieces of equipment include conveyors, production machines, fixtures, and tools. The robot and the associated equipment form a workcell. The term workstation can also be used, but this term is generally limited to mean either (1) a workcell with a single robot or (2) one work location along a production line consisting of several robot workstations. Sometimes, human workers are included within the robot workcell to perform tasks that are not easily automated. These tasks might consist of inspection operations or operations that require judgment or a sense of touch that robots do not possess.

Two of the problems in robot applications engineering are the physical design of the workcell and the design of the control system which will coordinate the activities among the various components of the cell. In this chapter, we consider these two problem areas. In an important way, these topics bring together many of the technology and programming topics in the previous chapters of the book in order to apply robotics for productive purposes.

11-1 ROBOT CELL LAYOUTS

Robot workcells can be organized into various arrangements or layouts. These layouts can be classified into three basic types:

1. Robot-centered cell
2. In-line robot cell
3. Mobile robot cell

The following subsections describe these workcell configurations.

Robot-Centered Workcell

In the robot-centered cell, illustrated in Fig. 11-1, the robot is located at the approximate center of the cell and the equipment is arranged in a partial circle around it. The most elementary case is where one robot performs a single operation, either servicing a single production machine, or performing a single production operation. Initial installations of industrial robots in the 1960s were illustrative of this case. Die casting, one of the very first applications for a robot, required the robot to unload the part from the die after each casting cycle and dip it into a quenching bath. Other production machine applications required the robot to both load and unload the workpart. For some of these applications, the cycle times of the machine were relatively long compared to the part-handling time of the robot. Metal-machining operations are examples of this imbalance condition. This required the robot to be idle for a high proportion of the cycle, causing low utilization of the robot. To increase the robot utilization, the workcell concept was developed in which one robot serviced several machines as pictured in Fig. 11-1.

An application of the robot-centered cell in which the robot performs the process is arc welding. In this case, the robot accomplishes the production operation itself, rather than servicing a production machine tool.

With these robot-centered cell arrangements, a method for delivering the workparts into and/or out of the cell must be provided. Conveyors, parts feeders with delivery chutes, and pallets are the means for accomplishing this function. Machining, die casting, plastic molding, and other similar production operations for discrete part production are examples of this case. These devices are used to present the parts to the robot in a known location and orientation for proper pickup. In arc welding, human operators are often used to accomplish the parts loading and unloading function for the robot. We discuss these kinds of cell layouts for arc welding in Chapter Fourteen.

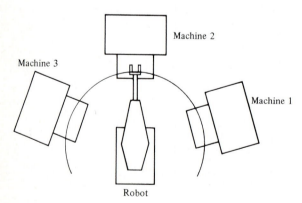

Figure 11-1 Robot-centered workcell layout.

In-Line Robot Cell

With the in-line cell arrangement, pictured in Fig. 11-2, the robot is located along a moving conveyor or other handling system and performs a task on the product as it travels past on the conveyor. Many of the in-line cell layouts involve more than a single robot placed along the moving line. A common example of this cell type is found in car body assembly plants in the automobile industry. Robots are positioned along the assembly line to spot weld the car body frames and panels. The three categories of transfer systems that can be used with the in-line cell configuration[3] are:

1. Intermittent transfer
2. Continuous transfer
3. Nonsynchronous transfer

An *intermittent transfer system* moves the parts with a start-and-stop motion from one workstation along the line to the next. It is sometimes called a synchronous transfer system because all of the parts are moved simultaneously and then registered at their next respective stations. In a robot layout using intermittent transfer, the robot is in a stationary location and constitutes one position along the line at which a part or product stops for processing. The advantage possessed by the intermittent transfer system in robot applications is that the part can be registered in a fixed location and orientation with respect to the robot during the robot's work cycle.

This registration of the part relative to the robot becomes a problem when the *continuous transfer system* is used to move parts in the cell. With this type of transfer system, the workparts are moved continuously along the line at constant speed. This means that the position and orientation of the part is

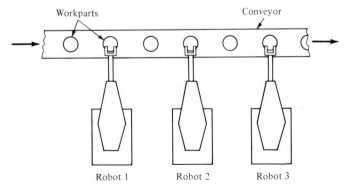

Figure 11-2 In-line robot workcell.

continuously changing with respect to any fixed location along the line. The problem can be solved by using either of two means[2]:

A moving baseline tracking system
A stationary baseline tracking system

The *moving baseline tracking system* involves the use of some sort of transport system to move the robot along a path parallel to the line of travel of the workpart while the operation is performed on the part. In this way, the relative position of the part and the robot remain constant during the work cycle. The problem with this arrangement is that an additional degree of freedom must be provided for the robot to move along the conveyor. This additional degree of freedom is usually accomplished by mounting the robot on a cart which can be moved along a track or rail parallel to the conveyor. This solution involves considerable capital expense to construct the system to maintain accurate registration between the robot and the part. One of the operational problems that must be taken into account in the design of a production line with several robots is the potential interference and collision problem between robots at adjacent stations along the line. The easiest way to solve the problem is to space the robots sufficiently to avoid the possibility of interference. However, this requires a significant amount of floor space. An alternative solution is to provide the workcell with enough intelligence that it knows where each robot is at any moment, and can control the sequence so as to avoid collisions.

In the *stationary baseline tracking system*, the robot is located in a stationary position along the line but its manipulator is capable of tracking the moving workpart. "Tracking" in this context means that the robot is able to maintain the positions of the programmed points, including the orientation of the end effector and the motion velocities, in relation to the workpart even though the part is moving along a conveyor. The engineering problems that must be solved to implement a stationary baseline tracking system are considerable although different from those encountered in the moving baseline system. First, the robot must have sufficient computational and control capability to accomplish tracking. This requires that the regular motion pattern of the manipulator be continuously translated in space in a direction parallel to the conveyor and at a speed equal to that of the conveyor. This allows the relative positions of the end effector and the part to be maintained during the cycle. A second problem, related to the first is concerned with the robot's "tracking window." The tracking window can be thought of as the intersection of the robot's work volume with the line of travel of the workpart along the conveyor. This concept is illustrated in Fig. 11-3. Allowances must be factored into this definition to account for manipulation of the wrist and end effector and for access to portions of the workpart that might be difficult to reach. For a robot with tracking capability, the total motion cycle in a particular application must be consistent with the tracking window for that application. A third problem involves the sensing of the part on the conveyor.

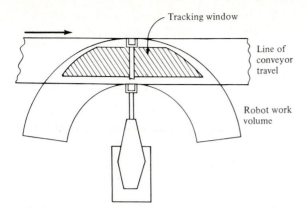

Figure 11-3 Concept of "tracking window".

For cells in which different product models will be processed, a sensor system must be used to identify which model is being delivered to the robot. A sensor is also required to determine that the part has entered the tracking window and that the robot can commence its work cycle. Other sensors are needed to track the position and velocity of the part during the cycle so as to coordinate with the robot tracking system. It is risky to presume that there will be no variations in the location and speed of the part as it is being processed.

The third type of transport system is *nonsynchronous transfer.* It is also referred to by the name "power-and-free" system. In this materials-handling system, each part moves independently along the conveyor in a stop-and-go fashion. When a particular workstation has completed its processing of a part, that part proceeds to move toward the next workstation in the line. Hence, at any given moment, some workparts are being processed while others are located between stations. The design and operation of this type of transfer system is more complicated than the other two because each part must be provided with its own independently operated, moving cart. However, the problem of designing and controlling the robot system used in conjunction with the power-and-free method is less complicated than for the continuous transfer method. For the irregular timing of arrivals on the nonsynchronous transfer system, sensors must be provided to indicate to the robot when to begin its work cycle. The more complex problems of registration between the robot and the part that must be solved in the continuously moving conveyor systems are not encountered on either the intermittent transfer or the nonsynchronous transfer.

Mobile Robot Cells

The third category of robot cell design is one in which the robot is capable of moving to the various pieces of equipment within the cell. This is typically accomplished by mounting the robot on a mobile base which can be transported on a rail system. The rail systems used in robot cells are either tracks fastened to the floor of the plant or overhead rail systems. Figure 11-4 illustrates the concept of the track-on-floor system, while the overhead rail

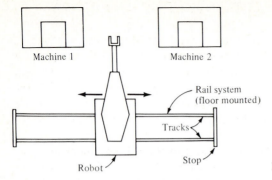

Machine 1

Machine 2

Rail system
(floor mounted)

Tracks

Stop

Robot

Figure 11-4 Mobile robot cell.

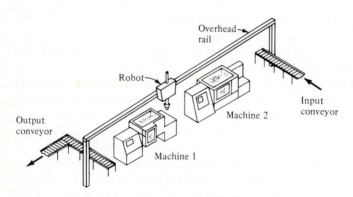

Overhead
rail

Robot

Input
conveyor

Output
conveyor

Machine 2

Machine 1

Figure 11-5 Overhead rail system for a mobile robot cell.

system is shown in Fig. 11-5. The advantage of the overhead rail system compared to the floor-mounted track system is that less floor space is required. The disadvantage is the increased cost of constructing the overhead system.

A mobile robot cell would be appropriate when the robot is servicing several machine tools with long processing cycles. In this situation, the robot would be able to share its time among the machines without significant idle time for either itself or the machines it is servicing. If a separate robot were to service each of the machines, the utilization of the robots would be low because most of their time would be spent waiting for the machine cycles to complete. Accordingly, one of the problems in the design of a mobile robot cell is to find the optimum number of machines for the robot to service. The objective in this problem is to maximize the number of machines in the cell without causing idle time on any of the machines.

11-2 MULTIPLE ROBOTS AND MACHINE INTERFERENCE

In some robot cells, there will be more than one robot required to perform the application. The in-line robot cell is a common example of this situation. In other cases, one robot will work with more than one machine in the cell. Either the robot-centered cell or the mobile robot cell are illustrative of this possibility. In either of these situations, care must be taken to ensure that the different pieces of equipment do not interfere with one another. There are two ways in which this inferference can occur.

The first case involves physical interference of the robots, where the work volumes of two robots in the cell overlap each other. In this situation, the danger of collision exists between the robot arms. This is most easily prevented by separating the robots by an adequate distance to avoid the problem. However, there are some applications in which it is desirable for two robots to share the same space. An example would be where one robot places a workpart at a certain location, and the second robot picks the part up. The location must be in the work envelopes of both robots. Accordingly, an alternative approach is to coordinate the programmed motion cycles of the two robots so that the arms are never close enough to risk a collision.

The second type of interference is when there are two or more machines being serviced by one robot, and the machine cycles are timed in such a way that idle time is experienced by one or more machines while another machine is being serviced by the robot. This is called *machine interference* and it is a common problem encountered when a human worker is assigned to service multiple machines. The difference between machine interference with a human worker and machine interference with a robot is that the amount of interference in the human cell is affected by variations in worker cycle times and by the worker's level of effort. With greater variation in the cycle time and a lower effort level, the machine interference will tend to be greater. In a robot cell, the robot's cycle time will not be affected by effort level and the amount of cycle time variation will be significantly less than for human work.

Machine interference can be measured as the total idle time of all the machines in the cell as compared to the operator (or robot) cycle time. The measure is most commonly expressed as a percent. To illustrate the problem, we will use an example of a three-machine cell in which a robot is used to load and unload the machines.

Example 11-1 Each of the three machines in the cell are identical and they have identical cycle times of 50 s. This cycle time is divided between run time (30 s) and service time (load/unload) by the robot (20 s). The organization of the cycle time is shown in the robot and machine process chart of Fig. 11-6. It can be seen that each machine has idle time during its cycle of 10 s while the robot is fully occupied throughout its work cycle. Total machine idle time of all three machines is $3 \times 10 = 30$ s and the cycle time of the robot is $3 \times 20 = 60$ s. Accordingly, the machine interference is $30 \text{ s}/60 \text{ s} = 50\%$.

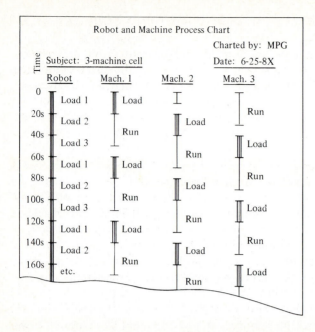

Figure 11-6 Robot and machine process chart for Example 11-1.

In this example the cycles of the three machines are the same. In this case, the question of whether or not machine interference will occur is determined by the relative values of machine cycle time and robot cycle time. The machine cycle time is the sum of service time and run time. The robot cycle time is equal to the number of machines multiplied by the service time. If the robot cycle time is greater than the machine cycle time, there will be resulting machine interference. If the machine cycle time is greater than the robot cycle time, there will be no machine interference, but the robot will be idle for part of the cycle.

In the case where the service and run times of the machines are different, the above relationships become complicated by the problem of determining the best sequence of servicing times for the machines into the robot cycle time. We will explore this problem in the exercises at the end of the chapter.

11-3 OTHER CONSIDERATIONS IN WORKCELL DESIGN

There are several other issues that must be considered in the design of the workcell. Among these considerations are the following:

1. *Changes to other equipment in the cell.* To implement the workcell and interface the robot to the other equipment in the cell, alterations will often have to be made to the equipment. Special fixtures and control devices must be devised to permit the cell to operate as a single, integrated mechanism. Examples of these fixtures and controls include work-holding nests and conveyor stops to position and orient the parts for the robot, changes in the

machines to allow the robot arm to gain access to the equipment, and limit switches and other devices to interface the various components in the cell.

2. *Part position and orientation.* For raw workparts being delivered into the cell, it is important that the robot have a precise pickup location to get the parts from the conveyor or other work-handling system. At this pickup point, the parts must be in a known orientation to enable the robot to grasp and hold it consistently and accurately. During subsequent processing within the cell, this part orientation should not be lost. A method for achieving these objectives of part positioning and orientation must be designed into the workcell.

3. *Part identification problem.* In cells where more than one type of part is processed or assembled, a method of identifying the particular part type must be determined. This can be done by any of a number of automated means, involving optical techniques or limit switches to sense differences in size or part geometry.

3. *Protection of the robot from its environment.* In certain types of applicaltions (e.g., spray painting, hot metal-working operations), a means of protecting the robot from the adverse effects of its environment must be provided.

5. *Utilities.* Providing the necessary utilities (e.g., electricity, air pressure, gas for furnaces, etc.) must be included among the factors considered in the design of the workcell layout.

6. *Control of the workcell.* The activities of the robot must be coordinated with those of the other equipment in the cell. This subject is referred to by the term workcell control, and we devote several sections to it and its related topics in this chapter.

7. *Safety.* A means of protecting human personnel from harm in and around the robot workcell must be provided. This is generally accomplished by means of fences or other barriers, and by designing a safety monitoring system to interrupt the cell operation if unsafe conditions are encountered. These kinds of mechanisms and other issues related to safety will be discussed in Chap. Seventeen.

11-4 WORKCELL CONTROL

In addition to the problem of designing the physical layout of the robot cell, another problem is concerned with coordinating the various activities that occur in the cell. Most of these activities occur sequentially, but simultaneous activities can also occur. There are other factors such as the safety of human personnel which must also be considered. Coordination of these various activities is accomplished by a device called the workcell controller or workstation controller. The functions performed by the workcell controller can be divided into three categories as suggested by Thomas[11]:

1. Sequence control
2. Operator interface
3. Safety monitoring
4. Breakdown Monitoring

These functions are accomplished either by the robot controller itself or by a higher-level control device, such as a programmable controller. The robot controller usually has a limited input/output capability to permit interfacing with other pieces of equipment in the cell. If the control requirements to operate the cell become at all complicated, then a higher-level controller is needed. We shall discuss the use of programmable controllers in a subsequent section of this chapter. Also, the implementation of an effective workcell controller is dependent largely on the programming capabilities of the robots used.

Sequence Control

Sequence control is the primary function of the workstation controller during regular automatic operation of the workcell. It includes the following kinds of control functions:

Control of the sequence of activities in the workcell
Control of simultaneous activities
Making decisions to proceed with the work cycle based on events that occur in the cell
Making decisions to stop or delay the work cycle based on events that occur in the cell

These functions occur during the normal operation of an average robot workcell. The following example illustrates these control requirements.

Example 11-2 This example is intended to demonstrate the importance of controlling the sequence of activities in the workcell during regular operation. The cell consists of a large robot, a numerically controlled machining center (which operates on an automatic cycle), and a belt conveyor for delivering raw workparts into the cell. Finished parts are placed on a pallet. The layout is shown in Fig. 11-7. The work cycle consists of the following sequence of activities:

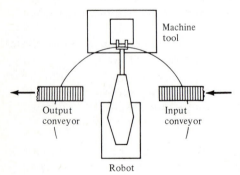

Figure 11-7 Workcell layout for Example 11-2.

1. Robot picks up raw workpart from conveyor which has delivered the part to a known pickup position (machine idle).
2. Robot loads part into fixture at machining center (machine idle).
3. Machining center begins automatic machining cycle (robot idle).
4. Machine completes automatic machining cycle. Robot unloads machine and places part on the pallet (machine idle).
5. Robot moves back to pickup point (machine idle).

In this cell operation, nearly all of the activities occur sequentially. The only significant exception is that the conveyor would probably operate continuously to deliver raw parts into the cell during the work cycle. The purpose of the workstation controller in this example would be to make sure that the activities occurred in the correct sequence and that each step in the sequence has been completed before the next one is initiated. For example, the start of the automatic machining cycle must not occur until the robot has loaded the raw workpart into the machining fixture.

Since all of the relevant activity in the preceding example is sequential, either the robot or the machine tool is idle for a significant portion of the cycle. The productivity of this work cell could be improved by using a double gripper. We have considered the different types of grippers in Chap. Five. A double gripper is a gripper with two grasping mechanisms for independent holding of two separate workparts. Use of a double gripper in our example would permit unloading and loading of the machine to be combined into a single (but more complicated) step. It would also allow the automatic machining cycle to occur at the same time that the robot was performing a portion of its motions. The following example illustrates how this would work.

Example 11-3 The same robot and equipment from Example 11-2 is used here except that a double gripper is employed for handling the workparts. This enables the robot to grasp the raw part and the finished part simultaneously. The revised work cycle would proceed as follows:

1. Robot picks up raw workpart using first gripper from conveyor which has delivered the part to a known pickup position. Robot moves its double gripper into ready position in front of machining center (machine cycle in progress).
2. At completion of machine cycle, robot unloads finished part from machine fixture with second gripper and loads raw part into fixture with first gripper (machine idle).
3. Machining center begins automatic machining cycle. Robot moves finished part to pallet and places it in programmed location on pallet.
4. Robot moves back to pickup point (machine cycle in progress).

The feature which distinguishes this work cycle from the previous example is that several activities occur simultaneously. Because a double gripper is used, the workstation can be organized so that the machine is processing the part at the same time the robot is performing much of its work. This reduces the production cycle time, but it has the apparent effect of increasing the complexity of the control function.

Simultaneous activities in the workcell usually do not pose as much a problem as they might seem. Although the activities in Example 11-3 occur simultaneously, they are initiated sequentially. Hence, the purpose of the controller is basically the same as in Example 11-2. The robot, the conveyor, and the machining center each work as separate units in the cell. The conveyor operates continuously to deliver parts into the cell. The machine tool operates under numerical control to perform its automatic cycle. And the robot is regulated by its controller. The function of workcell control is to make sure that the various control cycles begin at the required times. To perform this function, the workstation controller must be capable of communicating back and forth with the various pieces of equipment in the cell. Signals must be sent by the workcell controller to the various components of the cell, and other signals must be received from the components. These signals are called interlocks, and we shall discuss them in Sec. 11-5.

The workcell controller might be required to perform certain additional functions that are included within the general category of sequence control. These functions are usually associated only with more sophisticated workcells. They include:

Performing computations
Dealing with exceptional events such as equipment breakdowns
Perform irregular cycles, such as tool changing at periodic intervals

The two preceding examples serve to illustrate the possible need for logic computations in the work cycle. The application included a palletizing sequence after the part was unloaded from the machining center. This palletizing sequence requires that the finished parts be placed in a predetermined pattern of locations on the pallet. The most typical pattern of locations is a rectangular matrix (e.g., 5 parts in one direction by 6 parts in the perpendicular direction on the pallet—a total of 30 parts). To place the parts in this arrangement, the coordinate position of each part placed on the pallet must be determined by the workcell controller based on parameters provided by the programmer (e.g., the programmer would specify a 5 by 6 pallet with 3-in. spacing of part centers in each direction). The workcell controller must calculate the new position on each successive cycle of operation. In addition, the palletizing operation creates a slightly different motion pattern on each cycle.

Operator Interface

The purpose of the operator interface in workstation control is to provide a means for human operators to interact with the operation of the cell. There are several situations where this would be required. Among the most important cases are the following:

1. The human is an integral part of the workcell.
2. Emergency stop conditions.
3. Program editing or data input by operator.

The first situation is where the human worker plays an integral role in the operation of the cell. The human performs a portion of the work cycle and the robot performs a portion of the cycle. In this situation, there is a need to allow for variations in the time required by the operator to perform the manual part of the cycle. In a robotic cell, this can be easily accomplished by means of stop/start controls placed conveniently for the operator. The operator uses these controls to regulate the robot cycle as required by the situation. The following example illustrates the need for using operator controls to allow for variability in operating pacing.

Example 11-4 In this example, a human operator is used with a robot to perform a hot working operation on a metal billet. Hot forging is a common operation in this category. The operator's task in the cycle is to remove hot billets from random positions in a furnace and place them in a nest from which the robot can retrieve the parts for the hot working operation. There would probably be variations in the time required for the worker to perform the manual portion of the cycle because of differences in the locations of properly heated parts in the furnace. At a particular time during the day, it might turn out that none of the parts in the furnace are heated to a sufficient temperature for the operation. This would be a reason for delaying the next operation cycle. For these and various other reasons, there is a need for the operator to be able to signal the robot when to start its portion of the cycle.

The second situation involves an emergency in which a human worker needs to prevent continued operation of the robot cycle. The reason for the emergency may be a safety problem or an irregularity in the work cycle that is potentially destructive to the robot or other equipment in the cell. A safety problem might arise when some person walking through the plant has unwittingly intruded into the robot's space. An irregular event in the work cycle might be that the robot has grasped the workpart improperly for loading into a machine and this would cause damage to the machine or the associated tooling during processing. In either case, the human worker located in the cell would have reason to interrupt the operation of the robot cycle until the emergency situation was corrected. Thomas[11] stresses the need for the operator controls to respond in a consistent and predictable manner to emergency stop conditions.

The third requirement for operator control is to perform editing of the program or other similar input functions. Some robot controllers require that the robot be in a nonoperational mode when changes are made in the program. The more flexible controllers allow for editing to be accomplished while the robot is performing its regular cycle.

Safety Monitoring

In addition to the operator's ability to override the regular work cycle in the event of an observed safety hazard, the workcell controller should also be capable of monitoring its own operation for unsafe or potentially unsafe conditions in the cell. This function is called safety monitoring or hazard monitoring. We shall postpone our discussion of this topic until Chap. Seventeen when we provide a more comprehensive treatment of robot safety.

11-5 INTERLOCKS

An interlock in robotic workcell design is a method of preventing the work cycle sequence from continuing unless a certain condition or set of conditions are satisfied. It is a feature of workcell control which plays an important role in regulating the sequence in which the various elements of the cycle are carried out. Referring back to Examples 11-2 and 11-3, interlocks would be used for the following purposes:

To make sure that a raw workpart was at the pickup location on the conveyor before the robot tried to grasp the part
To determine when the machining cycle was completed before the robot attempts to load the part into the fixture
To indicate that the part has been successfully loaded so that the automatic machining cycle can begin

In each of these instances, it is critical that one element of the cycle has been completed before any attempt is made to begin the next element. The method of regulating the sequence of the elements would involve the use of interlocks.

Interlocks can be divided into two basic categories: output interlocks and input interlocks. An output interlock involves the use of a signal sent from the workstation controller to one of the machines or other devices in the workcell. It corresponds to the SIGNAL programming statement used in Chaps. Eight and Nine. For example, an output interlock would be used to signal the machining center in Examples 11-2 and 11-3 to commence the automatic cycle. The output signal originates from the workcell controller and is contingent upon certain conditions being satisfied. In our example, the conditions would be that the workpart has been properly loaded and the robot gripper has been removed to a safe distance. These conditions are usually determined by means of input interlocks.

An input interlock makes use of a signal sent from one of the components in the cell to the workstation controller. It corresponds to the WAIT command used in Chaps. Eight and Nine. It is employed to indicate that a certain condition or set of conditions have been met and that the programmed work cycle sequence can continue. As an illustration, an input interlock would be used in a machine-loading application to signal the workstation controller that the part has been properly loaded into the fixture on the machine tool table. This condition might readily be sensed by means of a simple limit switch mounted on the fixture which indicates that the part is in place.

Interlocks are essential in nearly all robotic workcells consisting of several operating pieces of equipment that must all work in a coordinated fashion. They serve to interface the various components of the workstation. Their use provides a synchronization and pacing of the activities in the cell which could not be accomplished through timing alone in the work cycle. Interlocks allow for variations in the times taken for certain elements in the cycle. They prevent work elements from starting before they should start. And they help to prevent damage of the various components of the cell.

In the design of the workcell, consideration must be given not only to the regular sequence of events that will occur during normal operation of the cell, but also to the possible irregularities and malfunctions that might happen. In the regular cycle, the various sequential and simultaneous activities must be identified, together with the conditions that must be satisfied in order for each activity to successfully take place. For the potential malfunctions, the applications engineer must determine a method of identifying that the malfunction has occurred and what action must be taken to respond to the malfunction. Then, for both the regular and irregular events in the cycle, interlocks must be provided to accomplish the required sequence control and hazard monitoring that must occur during the work cycle. In some cases, the interlock signals can be generated by the electronic controllers for the machines and other devices used in the workcell. For example, numerically controlled machine tools would be capable of being interfaced to the workcell controller to signal completion of the automatic machining cycle. In other cases, the applications engineer must design the interlocks using sensors to generate the required signals.

Interlocks are often implemented by means of limit switches and other simple devices that serve as sensors. In some cases, the robot cell must make use of more advanced sensors in order to successfully perform the work cycle. Examples of the latter cases would include seam tracking in robotic arc welding, identification of part position and orientation on a moving conveyor, and determining that a component had been properly assembled before proceeding with further assembly work on the product. Sensor systems for robotics were discussed in Chaps. Six and Seven.

11-6 ERROR DETECTION AND RECOVERY

Execution of the work cycle is expected to be repeated over and over for efficient operation of the robot cell. However, malfunctions and errors can occur during the cycle, for which some form of correction is needed to restore the workcell to regular automatic operation. In most robot cells, it is necessary to stop the workcell when errors occur, and to provide human assistance for the corrective action. This generally results in production delays before the maintenance crew arrives to diagnose the problem and make repairs. There is a trend in programmable automation technology to attempt to endow the robot (or other automated equipment) with the capability to sense errors and malfunctions when they occur, and to take the necessary compensating action to restore the system to normal operation. This capability is referred to as error detection and recovery.

By its name, error detection and recovery consists of two ingredients: error detection and error recovery. The detection problem is concerned with the use of the appropriate sensors to determine when an error has occurred. It also includes the associated intelligence to interpret the sensor signals so that errors can be properly recognized and classified. In general, errors in manufacturing can be classified as random errors, systematic errors, and illegitimate errors. Random errors are those that result from stochastic phenomena and are usually characterized by their statistical nature. For example, part size in a machining operation would be expected to vary randomly about some mean value. Depending on the amount of the variation, this could cause problems in a subsequent manufacturing process. Systematic errors are not determined by chance but by some bias that exists in the process. For example, an incorrect setting in the production machine or fixture would likely result in a systematic error in the product. The third class is the illegitimate error, typically resulting from an outright mistake, either by the equipment or by a human error. An error in the robot program would reflect this kind of mistake.

Although this general classification of errors may be helpful in determining potential sources of malfunctions in a process, it is usually not sufficient to design an error detection and recovery system for a specific application. An example will serve to illustrate how the errors might be classified into more specialized categories for a given process.

Example 11-5 In the context of an automated machining cell that is tended by a robot, the error categories would include: (1) tooling, (2) workpart, (3) process, (4) fixture, (5) machine tool, and (6) robot/end effector. This does not include the usual safety monitoring system that would probably be employed in the robot cell. In each of the categories, there are particular malfunctions and errors that could occur. The following list illustrates some of the possibilities:

Error source category	Particular malfunction or error
1. Tooling	Tool wear-out
	Tool breakage
	Vibration (chatter)
	Tool not present
	Wrong tool loaded

Error source category	Particular malfunction or error
2. Workpart	Workpart not present Wrong workpart Defective workpart Oversized or undersized part
3. Process	Wrong part program Wrong part Chip fouling No coolant when there should be Vibration (chatter) Excessive force Cutting temperature too high
4. Fixture	Part in fixture (yes or no) Part located properly (yes or no) Clamps actuated Part dislodged during processing Part deflection during processing Part breakage Chips causing location errors Hydraulic or pneumatic failure
5. Machine tool	Vibration Loss of power Power overload Thermal deflection Mechanical failure Hydraulic or electrical failure
6. Robot/end effector	Improper grasping of workpart No part present at pickup Hydraulic or electrical failure Loss of positioning accuracy Robot drops part during handling

Given that an error has occurred and that the error detection system has correctly sensed and classified the error, then certain corrective procedures can be initiated. The error recovery problem is concerned with defining and implementing the strategies that can be employed by the robot to correct or compensate for the malfunction that has occurred. The classification of errors is required during error detection because a specific recovery strategy must usually be developed to deal with each specific type of error. Recovery strategies can be grouped into the following general categories:

1. **Adjustments at the end of the current cycle.** This recovery strategy would represent a relatively low level of urgency. At the end of the current cycle, the robot program would branch to a subroutine to make the required corrections, then branch back to the main program.

2. **Adjustments during current cycle.** The error is sufficiently serious that corrective action must be taken during the current cycle of operation. However, it is not so urgent that the process must be stopped. The corrective action is typically accomplished by calling a special subroutine that has been designed to deal with the particular error.
3. **Stop process and invoke corrective algorithm.** The error in this case requires that the process be stopped, and that a subroutine be called to correct the error. At the end of the correction algorithm, the process can be resumed or restarted.
4. **Stop process—call for help.** This action is usually taken either because the malfunction is one that cannot be corrected by the robot or because an unclassified error is identified for which no corrective algorithm has been designed. In either case, human assistance is required to restore the system.

The following will continue the previous example and show some specific recovery strategies that might be used for certain errors.

Example 11-6 Among the errors that can occur during the operation of the workcell, one of the possibilities is that the robot will drop the part. In terms of error recovery strategy, this would normally be classified as a category 1 situation, in which corrective action would be taken during the cycle. To search for the part on the factory floor would probably be a hopeless and time-consuming task for the robot. Instead, the logical recovery strategy would be to reach for the next part.

An example of a category 2 error recovery situation would be when a part is detected to be dimensionally oversized—too large for the normal machining sequence. The logical recovery strategy might be to invoke a subroutine to provide an additional machining pass to remove the extra material.

A tool failure during the machining operation would be a category 3 situation. The process would have to be stopped in order to replace the broken tool with a sharp tool before resuming the cut. The fact that the tool broke during the cut may result in damage to the work surface. If the likelihood of this is high, then it might be necessary to change the part.

An example of a category 4 error recovery situation might be a hydraulic failure of the robot drive system. An automatic recovery from this malfunction may not be possible, and this would necessitate calling for human assistance.

The error detection and recovery system is implemented by means of the sensors used in the workcell together with the robot programming system. In terms of the programming requirements, the textual languages have a distinct advantage in developing the sometimes complex logic that is often needed. For highly sophisticated robot cells, a majority of the programming may be required for the error detection and recovery system.

11-7 THE WORKCELL CONTROLLER

The control systems and components discussed in Chaps. Three and Four were concerned with controlling the motions of the robot's manipulator. The workcell control system is concerned with the coordination of the robot's activities with those of the other equipment in the cell.

A number of options are available to satisfy the requirements of the workcell controller. These options include the use of the robot controller itself, relays, programmable controllers, and small stand-alone computers (minicomputers or microcomputers). The decision of which option to select depends on the complexity of the cell (e.g., the number of separate pieces of equipment, the number of separate control actions that must be controlled, and the number of robots in the cell), and whether the robot controller alone is capable of handling all of the activities. In the subsections below, we compare the various alternatives.

The Robot Controller

There are various types of control technology used for robot controllers. These include the simpler limited sequence controllers, electronic controllers, and computer controls. The more sophisticated types usually have a limited input/output capability to interface with other equipment. This input/output interface is provided specifically for the incorporation of interlocks in the workcell. The robot controller has the capability to tie the incoming signals to the work cycle program, so that the proper sequencing of output signals and robot motions can take place. The number of input/output ports might range between 10 and 20 for playback robots. A typical arrangement for the input/output module of the robot controller would be as follows:

Input ports (perhaps 10 to 25 input lines)—These would be used for incoming signals from external pieces of equipment. The signals would be binary (voltage on or off) and could be referenced as logical conditions in the robot program for purposes of interlocking. On newer controllers, the input ports would include the capacity to read in analog signals.

Output ports (perhaps 10 to 25 output lines)—These ports would be used for output interlock signals to the external equipment. The signals would be initiated or terminated according to logical conditions in the robot program, thus resulting in some response by the external equipment. Again, some newer robots would have the capacity for analog outputs as well as binary signal outputs.

Input port (perhaps five input lines)—These would be reserved for safety interlocks. Upon receipt of a signal from the external safety sensor on one of these lines, the controller would immediately interrupt the program, thus stopping the robot. In some cases, these input ports might be used to simply turn the power off to the manipulator.

This represents a limited input/output capacity for a workcell with any degree of complexity. Also, today's robot controllers are generally limited to sequence control and, as the above list indicates, often do not possess the capability to incorporate any significant safety monitoring or operator interfacing into the workcell control system. With the growing use of computer controls and the need by competing robot manufacturers to increase the control capabilities of their products, it is expected that future robot controllers will be equipped with enhanced input/output capacity and the capability to control intelligent robots.

Electromechanical Relays

An electromechanical relay is a control device used to actuate electrical circuits in response to changes in incoming signals. They are commonly used in industrial applications to provide sequence control of electrically operated equipment although they are gradually being displaced by more modern devices such as programmable controllers.

Relays can be used to augment the capabilities of a robot controller in the design of a workcell control system. Their use would typically be reserved for simple robot cells, such as pick-and-place applications, and where the robot has very limited input/output capacity. With relays, it would be relatively easy to include a simple safety monitoring scheme in the workcell. Such a scheme might consist of a fence surrounding the work place with a safety gate to gain access to the cell. Using the appropriate sensors (e.g., a limit switch to indicate closure of the safety gate) the relays could be set up to stop the robot, perhaps by interrupting its power source, as soon as a hazardous condition was sensed.

The limitations of relay control include the difficulty in interfacing with plant computer systems, their hard-wired configuration which makes it difficult to change over to a new workstation control task, and the fact that they are susceptible to mechanical wear and are less reliable than computer-type controls. The functions of a relay panel can be accomplished by a programmable controller, which avoids the above problems.

Programmable Controllers

Programmable controllers were introduced in the late 1960s as a replacement for systems of electromechanical relays. Up until that time, relay panels constituted the standard technique for accomplishing sequence control in industrial operations. The programmable controller was smaller in size, more reliable, more flexible, and its use could be readily learned by shop personnel who were familiar with the logic diagrams used for relay control panels.

A programmable controller (PC) can be defined as a digitally operating device with programmable memory that is capable of generating output signals according to logic operations and other functions performed on input signals. The program for a PC determines the sequence of operations and the generation of input and output signals. A PC is programmed by specifying the same kinds of logic diagrams, called ladder diagrams, used for years to set up relay control panels. Other programming methods are also possible on many programmable controllers, including the use of symbolic notation similar to computer programming. The functions that can be accomplished on a programmable controller typically include:

Control relay functions—The generation of an output signal based on logic rules applied to one or more input signals.

Timing functions—For example, the generation of an output signal for a specified length of time.

Counting functions—An internal counter in the PC is used to sum the number of contact closures and generate an output signal when the sum reaches a certain level.

Arithmetic functions—Some PCs can perform the basic arithmetic operations such as addition, subtraction, multiplication, and division.

Analog control functions—Another feature which is available on some PCs is the capability to simulate analog functions, such as proportional, integral, and derivative control.

These functions permit the PC to perform as a powerful robot workcell controller. Some PCs have the capacity to accept several hundred input/output connections, significantly more than a typical robot controller. This means that the PC can control a more complex workcell with more activities taking place in the cell. An automobile body spot-welding line, in which many robots perform various welding operations, would use a programmable controller as the overall cell control device. In addition to handling a greater number of input and output signals, the programmable controller also possesses other features that are beyond the capability of most robot controllers. These features include:

Maintenance and diagnostic functions—The CRT terminal used to program the PC can also be used in some systems to monitor the operation of the workcell. Some PCs have sophisticated diagnostic capabilities to quickly determine the origin of a problem when it occurs.

Operator interface—The use of the PC as the robot cell controller allows greater capacity and flexibility to implement the operator interface. Display terminals can be included in complex cells to provide operating performance information about production rates, tool usage, equipment breakdowns, and other data. Printers can be included at the control station to provide hard copy reports about the cell performance.

Safety monitoring—More sophisticated hazard monitoring systems can be implemented with programmable controllers. A greater number of safety conditions can be observed while the cell is operating than is possible with the robot controller alone.

A Computer as the Workcell Controller

Some robot applications have requirements for which a digital computer is the most appropriate method of workcell control. We are referring to the use of a stand-alone computer (generally a minicomputer or microcomputer) rather than the computer which is used as the robot control unit. In cases where a computer is the workcell controller, it would be used either in series with a programmable controller or as a substitute for the PC. The computer might perform other functions in the plant, and so it would be implemented to control the robot cell in a time-sharing mode of operation. Also, the computer would probably form a component in a hierarchical computer network in the factory, connected down to the programmable controller(s) and/or robot controller(s) in the cell, and connected up to the next hierarchical level in the plant.

Programmable controllers are specialized devices that are designed to be interfaced with industrial processes. They are provided with input/output ports that can be directly wired to the plant equipment. This is an advantage over the digital computer, and special arrangements must be made to interface the computer to the industrial equipment in the cell. However, the PC has certain limitations in data processing and programming languages which give the computer an advantage in applications requiring these capabilities. Some examples of the kinds of robot application features that might tend to favor the use of computers for workcell control would include the following:

Cases in which there are several cells whose operations must be coordinated, and significant amounts of data must be communicated between the cells.

Cells in which the error detection and recovery problem constitutes a significant portion of the coding that must be programmed into the workcell operation.

Where several different products are made on the same robot-automated production line, the operations at the different stations have to be coordinated and sequenced properly. Computers would be well suited to the data processing chores required in this type of application. In cases where the production lines are used for assembly operations, the various sizes and styles of the component parts must be sorted and matched to the particular model being assembled at each respective workstation along the line.

Situations in which a high level of production scheduling and inventory control are required in the operation of the cell. Again, this type of data processing function might require the use of a computer in addition to or as a substitute for a programmable controller.

The differences between digital computers and programmable controllers are principally differences in applications rather than differences in basic technology. The PC can, in fact, be considered to be a specialized form of digital computer with dedicated features for input/output control of industrial equipment. The technologies of the two types of control devices are quite similar.

11-8 ROBOT CYCLE TIME ANALYSIS

The amount of time required for the work cycle is an important consideration in the planning of the workcell. The cycle time determines the production rate for the job, which is a significant factor in the economic success of the robot installation. In the case of work performed by a human operator, the time required to accomplish the cycle would be determined by one of several work measurement techniques. One of these work measurement techniques is called MTM (for Methods Time Measurement). With MTM, the work cycle is divided into its basic motion elements and standard time values are assigned to each of the elements to construct the time for the total cycle. The standard time values have been previously compiled by studying similar elements and analyzing the factors that determine the time required to perform the elements. For example, the time required for a human operator to transport an object from one place to another depends on such factors as the weight of the object, the distance the object is moved, and the precision with which the object is located at the end of the move.

An approach similar to MTM has been developed by Nof and Lechtman[8] at Purdue University for analyzing the cycle times of robot work. The method, called RTM (for Robot Time and Motion), is useful for estimating the amount of time required to accomplish a certain work cycle before setting up the workstation and programming the robot. This would allow an applications engineer to compare alternative methods of performing a particular robot task. It could even be utilized as an aid in selecting the best robot for a given application by comparing the performance of the different candidates on the given work cycle.

The methodology of RTM is similar to MTM. There are 10 general categories of robot work cycle elements as presented in Table 11-1. The 10 categories can be collected into four major groups:

1. *Motion elements*. These are the manipulator movements, performed either with or without load.
2. *Sensing elements*. These are sensory activities performed by robots equipped with sensing capabilities. Examples include vision sensing, force sensing, and position sensing.
3. *End effector elements*. These elements relate to the action of the gripper or tool attached to the robot wrist as its end effector.
4. *Delay elements*. These are delay times resulting from waiting and processing conditions in the work cycle.

To use Table 11-1, the robot work cycle must be divided into its corresponding elements, and each element is specified with its associated parameters such as distance, velocity, and so forth. Different models of robots will be capable of performing the various elements at different times. According to Nof and Lechtman, element time values must be determined for each available robot in order to use RTM. There are four possible approaches that can be used to determine the element times and analyze a robot cycle with RTM.[8] The first involves tables of elements, in which time values are determined for the different elements listed in Table 11-1. This is the basic approach used to analyze human work with MTM. The second approach is to

Table 11-1 The 10 elements (and corresponding symbols) in RTM†

Element	Symbol	Definition of element	Element parameters
1	Rn	*n*-segment reach: Move unloaded manipulator along a path comprised of *n* segments	Displacement and velocity (or path geometry and velocity)
2	Mn	*n*-segment move: Move object along path comprised of *n* segments	Displacement and velocity (or path geometry and velocity)
3	ORn	*n*-segment orientation: Move manipulator mainly to reorient	Displacement and velocity (or path geometry and velocity)
4	SEi	Stop on position error	Error bound
4.1	SE1	Bring the manipulator to rest immediately without waiting to null out joint errors	
4.2	SE2	Bring the manipulator to rest within a specified position error tolerance	
5	SFi	Stop on force or moment	Force, torque, and touch
5.1	SF1	Stop the manipulator when the force conditions are met	
5.2	SF2	Stop the manipulator when the torque conditions are met	
5.3	SF3	Stop the manipulator when either the force or torque conditions are met	
5.4	SF4	Stop the manipulator when the touch conditions are met	
6	VI	Vision operation	Time function
7	GRi	Grasp an object	Distance to open/close
7.1	GR1	Simple grasp of object by closing fingers	
7.2	GR2	Grasp object while centering hand over it	
7.3	GR3	Grasp object by closing one finger at a time	
8	RE	Release object by opening fingers	
9	T	Process time delay when robot is part of the process	Time function
10	D	Time delay when robot is waiting for a process completion	Time function

† *Source*: (Reprinted with permission from *Industrial Engineering* Magazine, April, 1982. Copyright © Institute of Industrial Engineers, 25 Technology Park/Atlanta, Norcross, GA 30092.)

develop regression equations for the more complicated elements whose values are functionally related to several factors. Once the equation is developed for a given element, the user simply plugs the factor values into the equation to calculate the element time. Both of the preceding approaches have been applied in work measurement of traditional human performance.

The third approach is called "motion control," and it can be applied to the group 1 elements involving robot motions. Motion control is concerned with the kinematic and dynamic analysis of the manipulator movement. It determines the element time values by considering the distances moved and the velocities to make the moves. It also considers acceleration and deceleration at the beginning and end of the moves. For example, if acceleration and deceleration are ignored for the moment, the time required to move the manipulator will be the distance S divided by the velocity V. For some robots, the acceleration and deceleration times can be approximated closely by a constant value.

Table 11-2 Hypothetical values for selected elements in RTM for a hypothetical robot model.† (Refer to Table 11-1 for element definitions.)

Element	Symbol	Element time, s		Parameters
1	R1	$S/V + 0.40$	for $S > V/2.5$	S = distance moved (ft)
				V = velocity (ft/sec)
		0.40	for $S < V/2.5$	This is used for short moves
2	M1		For payloads of less than 1.0 lb	
		$S/V + 0.40$	for $S > V/2.5$	S = distance moved (ft)
				V = velocity (ft/sec)
		0.40	for $S < V/2.5$	This is used for short moves
			For payloads between 1 and 5 lb	
		$S/V + 0.60$	for $S > V/2.5$	S = distance moved (ft)
				V = velocity (ft/sec)
		0.60	for $S < V/2.5$	This is used for short moves
			For payloads between 5 and 15 lb	
		$S/V + 0.90$	for $S > V/2.5$	S = distance moved (ft)
				V = velocity (ft/sec)
		0.90	for $S < V/2.5$	This is used for short moves
4.1	SE1	$0.1V$		V = previous velocity (ft/sec)
7.1	GR1	0.1		Assumed to be independent of any parameters
8	RE	0.1		Assumed to be independent of any parameters
9	T	T		T = robot delay time
10	D	D		D = process delay time

† The values and equations listed in this table are fictitious and are not intended to indicate performance values for any robot model. Also, although the listings are based on the RTM research at Purdue, the values derived from this table should not be interpreted to represent the results of the Purdue research.

The fourth modeling approach in RTM is called "path geometry" by Nof and Lechtman. This approach is similar to motion control and requires the specification of the motion path to be followed by the manipulator together with the robot joint and arm velocities. It turns out that most robot motions involve the simultaneous actuation of several joints, but one of the joints usually predominates because its relative move is the largest. This can be analyzed by one of the computer programs developed at Purdue to determine the time for the move.

Table 11-2 presents a listing of element time values and equations for calculating the times for selected elements given in Table 11-1. These values represent the element times for a hypothetical robot, and do not reflect actual values developed through the research of Lechtman, Nof, and others. We are using the listings in Table 11-2 to demonstrate the RTM method in a simplified form for the purposes of an example and exercises at the end of the chapter. Potential users of RTM should consult the original research reports[6,7,8] for guidance on applying the method in actual projects.

Example 11-7 This example will illustrate the use of the RTM method. The work cycle consists of a simple task in which the robot must move parts weighing 3 lb from one conveyor to another conveyor. The sequence of the work cycle proceeds as follows:

1. Robot picks up part from first conveyor which has delivered the part to a known pickup position.
2. Robot transfers part to second conveyor and releases part.
3. Robot moves back to ready position at first conveyor.

A sketch of the workstation showing distances that the robot must move from one position to the next is shown in Fig. 11-8. The detailed sequence of elements to accomplish the work cycle is presented in Table 11-3. The conveyor delivers one part every 15 s, so the work cycle is limited by the conveyor feed rate. The RTM analysis in this problem would be useful for determining whether the time for the robot motion cycle is compatible

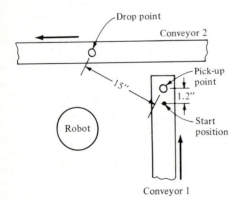

Figure 11-8 Workcell layout for Example 11-7.

Table 11-3 Detailed listing of work cycle elements for Example 11-7

Sequence	Element description
1	Conveyor delivers a part to a fixed position every 15 s. Robot in ready position above conveyor must await part delivery before executing its motion cycle.
2	Robot approaches part with gripper in open position. Speed setting is 0.1 ft/sec.
3	Robot gripper closes on part.
4	Robot lifts part 0.1 ft above conveyor. Speed is 0.1 ft/sec.
5	Robot moves part to a position 0.1 ft above second conveyor. Robot arm speed is 0.5 ft/sec. Distance traveled is 15 in.
6	Robot moves part to conveyor surface. Gripper would have to be oriented so that conveyor motion does not cause immediate tipping of the part when released. Speed is 0.1 ft/sec.
7	Robot gripper opens to release part on conveyor surface.
8	Robot moves empty gripper away from conveyor surface by 0.1 ft. Speed is 0.1 ft/sec.
9	Robot arm returns to ready position 0.1 ft above first conveyor surface. Distance traveled between conveyor is 15 in. Robot arm speed is 0.5 ft/sec.

Table 11-4 Elements from Table 11-3 in RTM notation with resulting times

Sequence	RTM symbol	Distance, ft	Velocity, ft/sec	Delay time, s	Element time, s	Description
1	D	—	—	15.0	15.0	Await delivery
2	R1	0.1	0.1	—	1.4	Approach part
3	GR1				0.1	Grasp part
4	M1	0.1	0.1	—	1.6	Lift part
5	M1	1.25	0.5	—	3.1	Move part
6	M1	0.1	0.1	—	1.6	Approach release point
7	RE	—	—	—	0.1	Release part
8	R1	0.1	0.1	—	1.4	Depart release point
9	R1	1.25	0.5	—	2.9	Reposition
Total					12.2	

with the conveyor feed rate. If the robot motion takes longer than 15 s, then the conveyor might have to be slowed down, or an alternative robot cycle developed. For example, the distances moved could be reduced to shorten the move element times. If the robot motion cycle takes less than 15 s, then the possibility of speeding up the parts delivery to the conveyor could be investigated.

Table 11-4 reduces the data contained in Table 11-3 to the RTM symbol notation, and presents the calculated element times as determined from the hypothetical robot values in Table 11-2. The total cycle time is 15 s. It turns out that the conveyor is the limiting factor in the cycle, requiring 2.8 s more time than the robot motion cycle. It might be possible to reduce the feed rate on the conveyor down to one part every 12.2 s. This would provide a perfect match between the feed rate and the robot cycle.

11-9 GRAPHICAL SIMULATION OF ROBOTIC WORKCELLS

RTM can be considered a method of simulating, in terms of time, the activities in the robot workcell. Another method of simulation involves graphical modeling on a CAD/CAM system. Simulation based on computer graphics can be used not only to analyze cycle times, but to design the cell itself. It turns out that a substantial amount of time is spent in designing and laying out the cell, designing or selecting the equipment, and similar activities. One industry estimate[10] is that 60 to 80 percent of the total cell implementation time is spent on these design-related problems and cell fabrication. (The remaining 20 to 40 percent of the time is spent in programming and refining the cell.) With so much effort expended on the design of the robot cell, it is reasonable to utilize labor saving tools to make the process as efficient as possible.

This section will discuss the use of computer graphics to simulate the design and operation of the robot and the workcell. We will provide an example of university research in this area and an example of a commercial package for designing and simulating the robot cell.

Research in Graphics Modeling for Robotics

Research in Lehigh University's Computer-Aided Design Laboratory in conjunction with our Institute for Robotics has led to the development of a graphics simulator of the PUMA 600 robot and the VAL language used to program the PUMA.[1] The simulator makes use of a FORTRAN callable graphics language to display the kinematic behavior of the PUMA in response to VAL motion statements. Algorithms for computing the positioning of the manipulator along segmented paths are used to simulate joint coordinate and straight line motions. Sequences of VAL commands can be entered interactively and their resulting motions shown on the graphics monitor.

The PUMA model was constructed by means of a series of extruded polyhedrons. To simulate motion, the vertices of each polyhedron are transformed and the model is redrawn. Since the PUMA consists of revolute joints, rotation transformations are used predominantly. The model can be scaled up or down, and other capabilities of the CAD/CAM system were exploited to facilitate viewing of the model. The research has explored both wire-frame and solid models, and the alternatives are illustrated in Figs. 11-9 and 11-10.

The motivation for the graphics simulation research derives from our interest in several engineering issues related to robotics applications. These

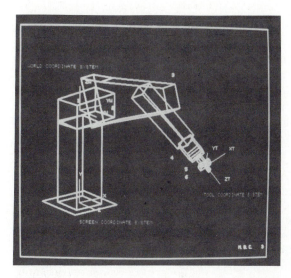

Figure 11-9 Computer graphics simulation of the PUMA robot with wire-frame model. (Photo courtesy Computer-Aided Design Laboratory, Lehigh University)

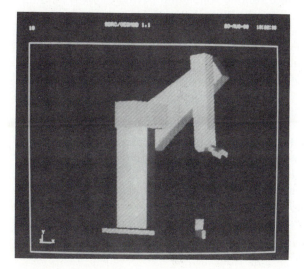

Figure 11-10 Computer graphics simulation of the PUMA robot with solids model. (Photo courtesy Computer-Aided Design Laboratory, Lehigh University)

issues include:

- Collision detection between the robot and other objects in the workcell. This problem is difficult to check visually with a wire-frame model. An algorithm was developed to accomplish a coarse and fine check to determine interference.
- Effects of acceleration, deceleration, arm member mass, payload mass, and other related factors on the dynamic performance of the robot.
- Problems of off-line programming on the CAD/CAM system and then downloading the program directly to the PUMA. This problem was discussed in our programming chapters.
- Input/output to CAD/CAM systems for assembly simulation.

The simulator has also been found useful in teaching the principles of both computer graphics modeling and robotics to students at Lehigh. Even though the PUMA is not a large robot, there is a substantial safety problem involved in exposing students in significant numbers to the actual machine for "hands-on" training. Simulation on the CAD/CAM system provides a safe trial run of the program, thus reducing hazards to the student and the robot.

The PLACE System

Several commercial packages are available for graphical simulation, and it is anticipated that these systems will grow in availability and use. At the time of this writing, the commercial simulation products include PLACE (McDonnell Douglas Manufacturing Industry Systems Company), Robographix (Computervision Corp.), and Robot-SIM (General Electric's Calma Co.). It seems appropriate to conclude our discussion of workcell design and control by describing the operation of these robotic simulation packages and the opportunities offered by them. We will use the PLACE system[5] as our example of these systems. Although the other commercial systems may not be organized in exactly the same way as the PLACE system, many of its features that we will describe are similar to those of other available systems.

The PLACE graphic simulation package consists of four modules. The first module to be released was PLACE, and the others were made available subsequently. The four modules are:

1. PLACE—This stands for Positioner Layout and Cell Evaluation. It is used to construct a three-dimensional model of the robot workcell in the CAD/CAM data base and to evaluate the operation of the cell.
2. BUILD—This module is used to construct models of the individual robots that might be used in a cell.
3. COMMAND—This is used to create and debug programs off-line that would be downloaded to the robot to save on-line programming time.
4. ADJUST—It must be expected that there will be a difference between the

computer graphics model of the workcell and the actual workcell. The ADJUST module is used to calibrate the cell.

The configuration of the four software modules in relation to the design and programming of the workcell is presented in Fig. 11-11.

PLACE is used to develop a computer graphics model of the robot and other workcell components in three dimensions. Figures 11-12 and 11-13 illustrate two models displayed on the McDonnell Douglas Unigraphics system. The system also permits the user to test out the motion sequence of the robot by means of a controlled animated simulation on the graphics monitor. Some of the specific capabilities of the system include:

Model the cell components. The user can enter the geometric data into the CAD data base to construct models of the fixtures, conveyors, machine tools, and other components. The robot models can be entered through the BUILD module and are available for use by PLACE.

Model the workcell. The various components can be called from the CAD system and assembled into a workcell in various ways to examine alternative design configurations. By permitting the user-designer to conveniently examine the alternatives in three dimensions, a better workcell design can be developed in less time compared to manual planning methods.

Define the robot motions. The user specifies a "working point" on the end effector and commands the working point to move to individual target points in the cell. These target points represent positions in the cell which the end effector must visit in the execution of the work cycle. The user can verify the physical capacity of the robot to reach all of the desired points in the cell.

Build motion sequences. This is accomplished using the animation capability. When the various target points of the cell have been defined the user is

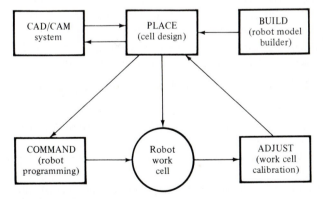

Figure 11-11 Configuration of the PLACE system for simulation and off-line programming of the workcell.

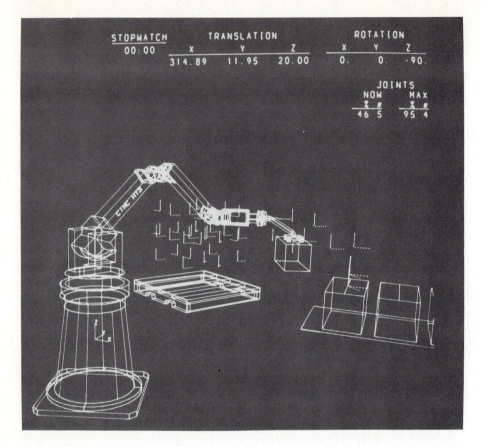

Figure 11-12 PLACE graphics simulation of robot cell for unloading cartons from conveyor onto pallet. (Photo courtesy Computer-Aided Design Laboratory, Lehigh University)

able to combine these individual moves into motion sequences. The sequences can then be played back to permit viewing of the robot work cycle. The animation speed can be controlled and the user can utilize the zooming capability of the computer graphics system so as to visually verify that the robot arm clears the various obstacles that might be in the way during the motion sequences.

Analyze cycle times. Cycle time analysis can be performed in PLACE to determine the time required to accomplish the work cycle. The basic computations for element times are similar to the RTM calculations from the previous section in this chapter.

The BUILD module can be used to construct three-dimensional computer graphics models of various robots that might be components of a workcell. The geometric characteristics of the robot (e.g., size, shape, etc.) are entered into the data base along with the design specifications of the robot. These

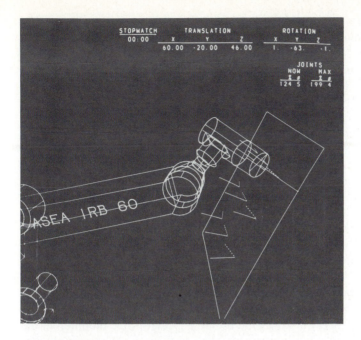

Figure 11-13 Computer graphics simulation of robot with tool as end effector showing close-up capability of PLACE. (Photo courtesy Computer-Aided Design Laboratory, Lehigh University)

specifications must include the number of degrees of freedom, the joint motions, joint travel limits, and other similar constraints on the robot's mechanical operation. Based on these parameters, the BUILD program creates the kinematic motion control equations that are used in PLACE. This relieves the workcell designer from the need to perform a detailed kinematic analysis for the robot motion cycle.

The COMMAND module is designed to permit off-line programming of robots. The limitations of current programming methods were discussed in Chaps. Eight and Nine. Certainly one of the biggest limitations from a production viewpoint is that the robot itself must be employed during the programming procedure to teach it the locations of the various points in space, usually with a teach pendant. Even the textual languages require this definition of point locations using the teach pendant. It is a generally held belief among robotics engineers that off-line programming of robotics will probably require some form of three-dimensional computer graphics simulation in order to be practicable. COMMAND has been developed as step in the direction of off-line robot programming.

The COMMAND programming module permits certain portions of the robot program to be written in the textual language for the particular robot model. Opening and closing the gripper and other process-oriented commands are examples of the kinds of statements permitted. To program the motion cycle, the COMMAND module interacts with the PLACE module by allowing the user to call the various motion sequences that have been developed during cell design and evaluation. When the program has been developed, it is then translated into the language code for the particular robot controller to be used in the application. This step is similar to postprocessing in numerical control part programming. Its disadvantage is that a unique translator (postprocessor) must be available for each robot controller.

The final module in the PLACE package is ADJUST. This module is designed to address a problem area that relates to off-line programming. The problem is the likely existence of discrepancies between the actual robot workcell and the computer model of the cell that resides in the CAD/CAM data base. These discrepancies were mentioned previously in our discussion of world modeling in Chap. Nine. There will undoubtedly be certain positional errors that exist between the actual physical objects in the workcell and the corresponding objects in the model. ADJUST provides the mechanism to calibrate the computer model for these errors by permitting positional data for the actual cell to be entered to update the PLACE model. This, of course, must be done after the robot cell has been installed in the factory. The original PLACE model was created during the design of the cell before its installation.

The calibration process involves the use of a special probe mounted in the position of the "working point" (either attached to the wrist mounting plate or held by the gripper). A calibration program is written using the COMMAND module to move the robot through a series of specific test points in the cell. Corrections to the geometric model in PLACE are made according to the positional errors at each of the test points. These corrections are in turn made to the COMMAND program, thereby updating the robot application program.

REFERENCES

1. M. B. Clifton and J. B. Ochs, "An Interactive Computer Graphics Simulator of VAL Programming Language of the Unimation PUMA Robot," *Proceedings*, IEEE Computer Society of the COMPCON Fall 1983 Conference, Arlington, VA, September 1983.
2. J. F. Engelberger, *Robotics in Practice*, AMACOM (American Management Association), New York, 1980, chaps. 4 and 5.
3. M. P. Groover, *Automation, Production Systems, and Computer-Aided Manufacturing*, Prentice-Hall, Englewood Cliffs, NJ, 1980, chaps. 4, 6, and 11.
4. M. P. Groover and E. W. Zimmers, Jr., *CAD/CAM: Computer-Aided Design and Manufacturing*, Prentice-Hall, Englewood Cliffs, NJ, 1984, chap. 10.
5. P. Howie, "Graphic Simulation for Off-line Robot Programming," *Robotics Today*, February 1984, pp. 63–66.
6. H. Lechtman and S. Y. Nof, "A User's Guide to the RTM Analyzer," *Technical Report*, Research Program on Advanced Industrial Robot Control, School of Industrial Engineering, Purdue University, 1981.
7. H. Lechtman and S. Y. Nof, "Robot Performance Models Based on the RTM Method," *Technical Report*, Research Program on Advanced Industrial Robot Control, School of Industrial Engineering, Purdue University, 1981.
8. S. Y. Nof and H. Lechtman, "The RTM Method of Analyzing Robot Work," *Industrial Engineering*, April 1982, pp. 38–48.

9. N. G. Odrey, "Operational Strategies for Error Recovery Within a Manufacturing Work-station," *Proposal for Research*, Lehigh University, May 1984.
10. R. N. Stauffer, "Robot System Simulation," *Robotics Today*, June 1984, pp. 81–90.
11. R. Thomas, "Designing Controls for Robotics Work Cells," *Industrial Engineering*, May 1983, pp. 34–39.
12. L. L. Toepperwein, M. T. Blackman, et al., "ICAM Robotics Application Guide," *Technical Report AFWAL-TR-80-4042*, Vol. II, Materials Laboratory, Air Force Wright Aeronautical Laboratories, Ohio, April 1980.
13. Unimation, Inc., *Programming Manual—User's Guide to VAL II* (398T1), Version 1.1, Danbury, CT, August 1984.

PROBLEMS

11-1 For the machine cycle times from Example 11-1, determine the amount of machine interference and the amount of robot idle time (expressed as a percent) in a robot cell composed of two machines. Sketch the robot and machine process time chart similar to Fig. 11-6 to analyze the problem.

11-2 For the machine cycle times from Example 11-1, determine the amount of machine interference and the amount of robot idle time (expressed as a percent) in a robot cell composed of four machines. Sketch the robot and machine process time chart similar to Fig. 11-6 to analyze the problem.

11-3 Three machines will be organized in a machine cell using a robot to load and unload the machines. The cycle times of the three machines are given as follows:

Machine 1: Run time = 30 s, service time = 20 s
Machine 2: Run time = 15 s, service time = 10 s
Machine 3: Run time = 20 s, service time = 10 s

Determine the best sequencing of these activities using a robot and machine process time chart similar to Fig. 11-6 to analyze the problem. Determine the amount of machine interference and the amount of robot idle time (expressed as a percent) in the cell.

11-4 Make a list of the interlocks required for the workcell of Example 11-3 in the text. For each interlock indicate whether it is an input interlock or an output interlock. For each of the input interlocks, define what the interlock should sense before the signal is sent to the workcell controller. For each output interlock, define what conditions must be satisfied before the signal is sent from the workcell controller.

11-5 For each of the input interlocks identified in Prob. 11-4, determine an appropriate means of implementing the interlock. Use the list of sensors in Table 6-2 of Chap. Six for reference.

11-6 Make a list of the interlocks required for the workcell of Example 11-4 in the text. For each interlock indicate whether it is an input interlock or an output interlock. For each of the input interlocks, define what the interlock should sense before the signal is sent to the workcell controller. For each output interlock, define what conditions must be satisfied before the signal is sent from the workcell controller.

11-7 For each of the input interlocks identified in Prob. 11-6, determine an appropriate means of implementing the interlock. Use the list of sensors in Table 6-2 of Chap. Six for reference.

11-8 A robot workcell is to be installed for a plastic molding operation. The cell will consist of a large robot, the molding machine (which operates on an automatic cycle), and a belt conveyor for delivering the molded parts out of the cell. The robot has a jointed-arm configuration with a fully extended reach of 80 in., and the arm swivels a full 300'. The robot base occupies a floor space of 30 in.². The molding machine has overall dimensions of 50 in. in width by 120 in. in length. When the mold opens, the opening is 18 in. wide. The center of the mold is located at the center of the 50-in. machine width. The belt conveyor is 12 in. wide. *Note*: For this problem the reader may want to refer ahead to Chap. Thirteen (Sec. 13-3) in which the plastic molding operation is discussed.

(*a*) Determine the likely sequence of activities in the work cycle. Use a list of steps similar to the format in Examples 11-2 and 11-3.

(*b*) What type of workcell layout is the most logical for this case?

(*c*) Make a sketch of the workcell, showing relative positions of the different pieces of equipment in the cell.

(*d*) Make a list of the interlocks required for the workcell. For each interlock indicate whether it is an input interlock or an output interlock. For each of the input interlocks, define what type of sensor might be used to implement the interlock. For each output interlock, define what conditions must be satisfied before the signal is sent from the workcell controller.

(*e*) For each of the input interlocks identified in part (*d*), determine an appropriate means of implementing the interlock. Use the list of sensors in Table 6-2 of Chap. Six as a reference.

11-9 The sketch in Fig. P11-9 shows a proposed robot cell designed to process workparts through a certain industrial operation. The cell consists of the robot, a conveyor, and a processing machine. These components all operate under the robot cell controller. Parts (each part weighing 3 lb) arrive on the conveyor in a pallet (four parts to a pallet), are picked out of the pallet by the robot, loaded into the processing machine, processed, and unloaded by the robot into the same position on the pallet. When all four parts are processed in this manner, an interlock signal from the robot cell controller activates the conveyor to deliver the current pallet out of the cell and to deliver a new pallet. The sequence is as follows:

Pallet in position with four raw workparts for start of cycle.
Robot moves from point P0 as 0.5 ft/sec to grasp first workpart.
Robot grasps first part, lifts 4 in. to clear pallet at 0.2 ft/sec, moves to machine at 0.6 ft/sec, and loads into machine. The machine loading adds 2.0 s to the move. Robot arm moves to safe position 6 in. from part at a speed of 0.3 ft/sec.
Machine processes part for 20 s.
Robot arm moves from safe position at 0.3 ft/sec, unloads part, and moves it back to same pallet position at a speed of 0.6 ft/sec. Unloading requires 2.0 s. Part must be inserted back into pallet from 4 in. above. Speed of insertion is 0.2 ft/sec. Depart to 4 in. above pallet at 0.2 ft/sec.
Move to second pallet position (speed = 0.5 ft/sec) and repeat cycle.
Move to third pallet position (speed = 0.5 ft/sec) and repeat cycle.

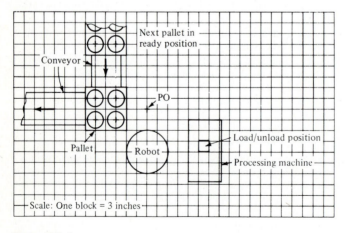

Figure P11-9

Move to fourth pallet position (speed = 0.5 ft/sec) and repeat cycle. When finished, move to position P0 (speed = 0.5 ft/sec).

Activate conveyor to remove current pallet and present new pallet. This movement of pallets requires 15 s.

For the cell operation as described, write the list of RTM symbols and corresponding times that describe the sequence using Tables 11-1 and 11-2. What is the total cycle time for the cell to process four parts in a pallet?

11-10 In Prob. 11-9, consider the possibility of improvements in the cell layout that will reduce the cycle time and increase production rate. You need not draw a new layout but estimate the revised distances that result from your improvements and determine the corresponding reduction in estimated cycle time.

11-11 In Prob. 11-9, consider possible improvements other than cell layout changes, and estimate their effect on the overall cycle time.

11-12 A simple industrial robot whose element times are determined by Table 11-2 in the text is being applied to perform a certain pick-and-place operation which requires that parts be picked from a fixed location on a moving conveyor and dropped into a parts magazine at a fixed location. Each magazine holds 12 parts. The parts are stacked on top of each other in the magazine so the robot must deliver each part that goes into the magazine to a fixed location. The robot uses an "approach" and "depart" command for each pickup which is 2 in. away from the conveyor surface. The parts weigh 1.5 lb each. The distance between the pickup (assume 2 in. above the pickup point) and the magazine loading point is 14 in. After the 12 parts are loaded into the magazine, an automatic transfer element is activated to replace the full magazine with an empty one. This element takes 10 s. A constant speed setting V is to be used for the robot throughout its work cycle. Determine an equation with speed V as an independent variable that can be used to compute the total cycle time for this operation. Assume $S > V/2.5$.

TWELVE

ECONOMIC ANALYSIS FOR ROBOTICS

In addition to the technological considerations involved in applications engineering for a robotics project, there is also the economic issue. Will the robot justify itself economically? The economic analysis for any proposed engineering project is of considerable importance in most companies because management usually decides whether to install the project on the basis of this analysis. In the present chapter, we consider the economic analysis of a robot project. We discuss the various costs and potential benefits associated with the robot installation, and we describe several methods for analyzing these factors to determine the economic merits of the project.

12-1 ECONOMIC ANALYSIS: BASIC DATA REQUIRED

To perform the economic analysis of a proposed robot project, certain basic information is needed about the project. This information includes the type of project being considered, the cost of the robot installation, the production cycle time, and the savings and benefits resulting from the project.

Type of Robot Installation

There are two basic categories of robot installations that are commonly encountered. The first involves a new application. This is where there is no existing facility. Instead, there is a need for a new facility, and a robot installation represents one of the possible approaches that might be used to satisfy that need. In this case, the various alternatives are compared and the best alternative is selected, assuming it meets the company's investment criteria. The second situation is the robot installation to replace a current method of operation. The present method typically involves a production operation that is performed manually, and the robot would be used somehow to substitute for the human labor. In this situation, the economic justification of the robot installation often depends on how inefficient and costly the manual method is, rather than the absolute merits of the robot method.

In either of these situations, certain basic cost information is needed in order to perform the economic analysis. The following subsection discusses the kinds of cost and operating data that are used to analyze the alternative investment projects. The methods by which the analysis is accomplished are explained later in the chapter.

Cost Data Required for the Analysis

The cost data required to perform the economic analysis of a robot project divide into two types: investment costs and operating costs. The investment

Table 12-1 Direct costs associated with robot project

A. Investment costs

1. *Robot purchase cost*—The basic price of the robot equipped from the manufacturer with the proper options (excluding end effector) to perform the application.

2. *Engineering costs*—The costs of planning and design by the user company's engineering staff to install the robot.

3. *Installation costs*—This includes the labor and materials needed to prepare the installation site (provision for utilities, floor preparation, etc.).

4. *Special tooling*—This includes the cost of the end effector, parts positioners, and other fixtures and tools required to operate the work cell.

5. *Miscellaneous costs*—This covers the additional investment costs not included by any of the above categories (e.g., other equipment needed for the cell).

B. Operating costs and savings

6. *Direct labor cost*—The direct labor cost associated with the operation of the robot cell. Fringe benefits are usually included in the calculation of direct labor rate, but other overhead costs are excluded.

7. *Indirect labor cost*—The indirect labor costs that can be directly allocated to the operation of the robot cell. These costs include supervision, setup, programming, and other personnel costs not included in category 6 above.

8. *Maintenance*—This covers the anticipated costs of maintenance and repair for the robot cell. These costs are included under this separate heading rather than in category 7 because the maintenance costs involve not only indirect labor (the maintenance crew) but also materials (replacement parts) and service calls by the robot manufacturer. A reasonable "rule of thumb" in the absence of better data is that the annual maintenance cost for the robot will be approximately 10 percent of the purchase price (category 1).

9. *Utilities*—This includes the cost of utilities to operate the robot cell (e.g., electricity, air pressure, gas). These are usually minor costs compared to the above items.

10. *Training*—Training might be considered to be an investment cost because much of the training required for the installation will occur as a first cost of the installation. However, training should be a continuing activity, and so it is included as an operating cost.

costs include the purchase cost of the robot and the engineering costs associated with its installation in the workcell. In many robot application projects, the engineering costs can equal or exceed the purchase cost of the robot. Table 12-1 presents a list of the investment costs typically encountered in robot projects. The operating costs include the cost of any labor needed to operate the cell, maintenance costs, and other expenses associated with the robot cell operation. The table lists most of the major operating costs for a robot application project. In the case of the operating costs, it is often convenient to identify the cost savings that will result from the use of a robot as compared to an existing method, rather than to separately identify the operating costs of the alternative methods. Material savings, scrap reductions, and advantages resulting from more consistent quality are examples of these savings. Items 6 through 10 in Table 12-1 should be interpreted to allow for this possible method of declaring cost savings between the alternatives.

The manner in which these investment costs and operating costs play out over the life of the robot installation can be conceptualized as illustrated in Fig. 12-1. At the beginning of the project, the investment costs are being paid into the project with no immediate return. When the installation is completed and the project begins operation, the operating costs begin. However, there is also a compensating cash flow representing revenues to the company which should exceed the amount of the operating cost. The difference between the revenues and the operating costs is the net cash flow. At the beginning of operations, there are usually startup problems to be solved and "bugs" to be worked out of the system. These difficulties often prevent the net cash flow from immediately reaching the steady-state value anticipated for the project.

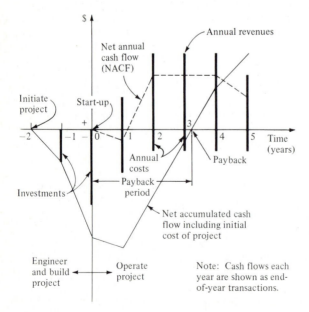

Note: Cash flows each year are shown as end-of-year transactions.

Figure 12-1 Life cycle of cash flows in a capital investment project.

If the robot project is a good investment, the net cash flow will allow the company to recover its investment costs in the project in a relatively short period of time. The point at which the investment is recovered is displayed in Fig. 12-1 as the payback period, and this payback period represents one of several methods for evaluating investment alternatives. The payback period method, as well as several other methods for analyzing the economics of robot projects are discussed in the following section.

12-2 METHODS OF ECONOMIC ANALYSIS

We shall describe three methods for analyzing investments and comparing investment alternatives that are in common use in industry. The three methods are:

1. Payback (or payback period) method
2. Equivalent uniform annual cost (EUAC) method
3. Return on investment (ROI) method

Each of the methods accomplishes the analysis using a slightly different twist. Ideally, the same decision should be reached no matter which method is used; however, this is not always the case. We assume that the reader has some familiarity with the principles of engineering economy.

Payback Method

The payback method uses the concept of the payback period as shown in Fig. 12-1. The payback period is the length of time required for the net accumulated cash flow to equal the initial investment in the project. Under the assumption that the net annual cash flows are equal from year to year, this notion can be reduced to the following simple formula

$$n = \frac{IC}{NACF} \qquad (12\text{-}1)$$

where $\quad n$ = the payback period
$\quad IC$ = the investment cost
$\quad NACF$ = the net annual cash flow

In most investment projects it would be unlikely that the cash flows would be exactly equal each year. When there are year-to-year differences in the cash flows, Eq. (12-1) must be altered slightly to account for the differences. The subscript i is used to identify the year in the following.

$$0 = -(IC) + \sum_{i=1}^{n} (NACF) \qquad (12\text{-}2)$$

In this equation, the value of n is determined so that the sum of the annual cash flows is equal to the initial investment cost. In the special case when the net annual cash flows are equal, Eq. (12-2) can be recast as

$$0 = -(IC) + n(NACF)$$

This is equivalent to Eq. (12-1).

The reader should note that we have adopted the logical convention that costs are treated as negative values and revenues or savings (as well as profits) are treated as positive values in these equations. The NACF is assumed to be a positive cash flow since revenues derived from the robot project would be greater than the operating costs (we hope). We have also assumed that all cash flows occur either at the beginning of the year or at the end of the year. Any investments are assumed to be transactions that occur at the beginning of the year, while the net annual cash flows are assumed to be end-of-year transactions.

Most companies today require paybacks of no more than two or three years. An investment whose cash flow pays back the investment in less than one year is considered excellent. Let us illustrate the payback method by means of the following example.

Example 12-1 Suppose that the total investment cost is estimated to be $100,000 for a particular robot project. The total operating costs (labor, maintenance, and other annual expenses) are expected to be $20,000 per year, and the anticipated revenues from the robot installation are $65,000 annually. It is expected that the robot project will have a service life of 5 years. Determine the payback period that is expected of the investment.

The net annual cash flow for the robot project is $65,000 − $20,000 = $45,000. Using Eq. (12-1),

$$n = \frac{100,000}{45,000} = 2.22 \text{ years}$$

One of the disadvantages of the payback period method is that it ignores the time value of money. It does not consider the objective of the company to derive a certain minimum rate of return from its investments. The other two methods to be discussed do include this consideration.

Equivalent Uniform Annual Cost Method

The equivalent uniform annual cost (EUAC) method converts all of the present and future investments and cash flows into their equivalent uniform cash flows over the anticipated life of the project. It does this by making use of the various interest factors associated with engineering economy calculations. We present a tabulation of these interest factors in an appendix to this chapter, and we save a considerable amount of explanation by assuming that the reader is familiar with their use.

To begin with, the company must select a minimum attractive rate-of-return (MARR) which is used as a criterion to decide whether a potential investment project should be funded. Today, MARR values of 20 to 50 percent are not unusual for robot projects. Using the interest factors for the MARR to make the conversions, the uniform annual cost method then sums up the EUAC values for each of the various investments and cash flows associated with the project. If the sum of the EUACs is greater than zero, this is interpreted to mean that the actual rate of return associated with the investment is greater than the MARR used by the company as the criterion. If the EUAC sum is less than zero, then the project is considered unattractive.

Example 12-2 We will illustrate the EUAC method using the same data from Example 12-1. The company uses a 30 percent MARR as a criterion for selecting its investment projects. As mentioned in Example 12-1, the robot project is expected to have a 5-year service life, and that is what we shall use in determining the values for any interest factors required in our calculations.

The annual operating cost ($20,000) and the annual revenues ($65,000) are already expressed as uniform annual cash flows. The initial investment cost ($100,000) must be converted to its equivalent uniform annual cash value using the capital recovery factor from the appendix. The sum of the annual cash flows would be figured as follows.

$$EUAC = 100,000(A/P, 30\%, 5) + 65,000 - 20,000$$
$$EUAC = -100,000(0.41058) + 45,000$$
$$EUAC = +\$3942$$

Since the resulting uniform annual cost value is positive, this robot project would be a good investment.

Return on Investment (ROI) Method

The return on investment (ROI) method determines the rate of return for the proposed project based on the estimated costs and revenues. This rate of return is then compared with the company's minimum attractive rate of return to decide whether the investment is justified. The determination of the rate of return involves setting up an equivalent uniform annual cost equation similar to the one used in Example 12-2. The difference is that the EUAC sum on the left-hand side of the equation is made equal to zero. Then the values of the interest factors (and correspondingly, the interest rates) are found that make the right-hand side of the equation sum to zero. The following example will illustrate this procedure.

Example 12-3 Again the same data are used from our previous two examples. The company's MARR is 30 percent as before. The EUAC equation would be set up as follows:

$$EUAC = -100,000(A/P, i, 5) + 65,000 - 20,000 = 0$$

$$(A/P, i, 5) = 45,000/100,000 = 0.45$$

Looking through the interest factor tables for a match of the A/P factor for $n = 5$ years, we find the following values:

$$\text{For } i = 30\%, \quad (A/P, 30\%, 5) = 0.41058$$

$$\text{For } i = 35\%, \quad (A/P, 35\%, 5) = 0.45046$$

By interpolation, the rate of return for our problem turns out to be $i = 34.94$ percent. It can be seen that our calculated value of $(A/P, i, 5) = 0.45$ is very close to $(A/P, 35\%, 5) = 0.45046$, so it stands to reason that the rate of return for our problem should be close to 35 percent.

In Example 12-3, the determination of the rate of return was a straight-forward computation. However, many problems involve a trial-and-error calculation procedure because there is more than one interest factor that must be used in the EUAC equation. We will demonstrate this procedure in Example 12-4. Indeed, there are several complications that are encountered in the economic analysis of robot applications problems. These complications are not necessarily unique to robotics problems, but we will discuss them in the context of robotics.

12-3 SUBSEQUENT USE OF THE ROBOT

Many automation projects involve pieces of equipment that have service lives corresponding to the life cycle of the product that will be made on the equipment. The automated equipment is very specialized to manufacture the particular product as efficiently as possible. However, after the product is no longer produced, the equipment often becomes obsolete and is of no further use or value to the company.

By contrast, robots represent programmable automation that can be used again after the current product life cycle is over. This is an attractive feature of an industrial robot because it means that the service life of the robot can be extended beyond the current production use. On the other hand, this feature tends to promote the use of robots in applications with shorter life cycles, thus making it more difficult to justify the investment cost of the robot for the application. One way of dealing with this problem is to recognize the oppor-tunity for subsequent use of the robot by assigning it a salvage value at the end of the current project. The current project may have a relatively short service life (perhaps 2 or 3 years), whereas the robot itself might be expected to last for 8 or 10 years. Accordingly, it might be reusable for three or four projects before either wearing out or becoming technologically obsolete.

The question that arises is this: How is the salvage value at the end of the current project determined? A reasonable procedure for assessing the salvage value at the end of the current application project is to use the straight line method of depreciation for the robot. Using this method we must estimate the actual number of years that the robot can be used and reused. In other words, we must determine the robot's service life. This is different from the service life of the current project on which the new robot will be used. In our previous examples, the service life for the project is 5 years. The service life for the robot would be longer than that.

Suppose the anticipated service life for the robot is 8 years. At the end of that time, the company expects the robot to be worn out, unreliable, unmaintainable, technologically obsolete, and of no further value. Using the straight line method, we simply divide the initial cost of the robot ($100,000) by the number of years in the robot's expected service life to obtain an annual depreciation on the robot.

$$\text{Annual depreciation} = 100,000/8 = 12,500$$

To get the salvage value at the end of the project life, we multiply this annual depreciation by the project life and subtract the amount from the initial cost.

$$\text{Salvage value} = 100,000 - 5(12,500) = 37,500$$

The company may choose to raise or lower this value to reflect factors such as the risk of not finding another project on which the robot can be used, or the possibility that the robot may become technologically obsolete before the end of its expected service life. Also, the company may elect to use a different depreciation method to determine the salvage value for the robot. Using the $37,500 determined above, the following example will show the effect of taking the salvage value into account.

Example 12-4 The service life given in our previous examples was 5 years. Let us make the reasonable assumption that the robot will be reusable for another project when the life of the current project is over. The robot will therefore possess a salvage value at the end of the current project. Using the salvage value of $37,500 determined above, compute the rate of return expected from the current project.

The $37,500 can be interpreted to represent a positive cash flow at the end of 5 years. The EUAC equation for determining the rate of return would be set up as follows:

$$\text{EUAC} = -100,000(A/P, i, 5) + 65,000 - 20,000 + 37,500(A/F, i, 5) = 0$$

We know that the interest rate (rate of return) is greater than the 35 percent from Example 12-3 because there is a greater positive cash flow due to the salvage value. The EUAC equation must be solved by trial-and-error because there is more than one unknown term.

Try $i = 40$ percent.

$$\text{EUAC} = -100,000(0.49136) + 45,000 + 37,500(0.09136)$$

$$\text{EUAC} = -49,136 + 45,000 + 3426 = -710$$

Try $i = 35$ percent.

$$\text{EUAC} = -100,000(0.45046) + 45,000 + 37,500(0.10046)$$

$$\text{EUAC} = -45,046 + 45,000 + 3767 = +3721$$

Interpolating between these two values, the rate of return is 39.20 percent. The presence of the salvage value, representing an evaluation of the robot's capacity to be reused on a subsequent application project, has increased the rate of return on the project from 34.94 to 39.20 percent.

12-4 DIFFERENCES IN PRODUCTION RATES

In comparing automated production methods with their manual counterparts, an issue which often arises is the difference in production rates between the alternative methods. The automated method usually outproduces the manual method, and this advantage must be taken into account in the analysis. The same issue is relevant when a robot is used to automate an operation. The robot may be capable of working faster and with fewer rest breaks than the human operator. On the other hand, what sometimes happens in a robot application is that the robot cannot work quite as fast as a human operator, but the company may decide to use the robot for two or three shift operations, whereas it felt limited to one shift when the task was performed manually.

These changes create differences in daily production rates which would presumably affect revenues assignable to the alternatives. The easiest case to analyze is where the value added by the operation is known per product. This value added can then be used to determine revenues for the alternatives.

Example 12-5 Suppose two alternative methods are to be compared, one a manual method and the other a robot cell. The robot cell will be capable of producing 300 units per day while the manual method will be capable of producing 200 units per day. The value added for each unit made is known to be $1.00. The investment cost of the robot cell is $100,000, and the operating cost is expected to be $20,000 annually. It will require an investment of $29,000 to set up the manual operation and the annual operating costs will be $36,000. A 25 percent MARR is used by the company and the anticipated service life of either method is 3 years. At the end of the 3 years the robot is expected to have a salvage value of $50,000. Determine the payback period and the ROI for each alternative. We will assume 250 days of operation per year.

Considering the manual method first, the annual revenues from the operation will be (200 units/day)(250 days/year)($1/unit) = $50,000/year. Accordingly, the payback period n can be determined from

$$-29{,}000 + n(50{,}000 - 36{,}000) = 0$$

$$n = \frac{29{,}000}{14{,}000} = 2.07 \text{ years}$$

The rate of return attributable to the manual operation would be determined by setting up the EUAC equation as follows:

$$EUAC = -29,000(A/P, i, 3) + (50,000 - 36,000) = 0$$

$$(A/P, i, 3) = 14,000/29,000 = 0.48276$$

The corresponding rate of return is 21.07 percent. This does not meet the minimum attractive rate of return criterion for the company.

For the automated robot method, the annual revenues from the operation will be (300 units/day)(250 days/year)($1/unit) = $75,000/year. The payback period would therefore be

$$n = \frac{100,000}{75,000 - 20,000} = 1.82 \text{ years}$$

The rate of return equation would be set up as follows:

$$EUAC = -100,000(A/P, i, 3) + (74,000 - 20,000) + (50,000)(A/F, i, 3)$$
$$= 0$$

By trial and error, the rate of return is determined to be 43.91 percent.

The robot method seems preferable on the basis of both payback period and rate of return. Note that the salvage value is a significant factor in the rate of return calculation, whereas its effect is ignored in the payback method. One could conceive of a situation in which the payback method favored one alternative while the rate of return method favored the other alternative. In the application of the payback method, the salvage value does not enter the analysis unless the calculated payback period turns out to be equal to (or greater than) the life of the project. In that case, the payback period becomes equal to the project life, so long as the total project costs (investment costs plus annual costs) are less than or equal to the total revenues (including salvage valuation) during the project life. If the total project costs are greater than the total revenues during the project life, then the project is clearly not worthwhile and the concept of payback period becomes blurred.

The feature of the preceding example which makes the calculations straightforward is that the value added per unit of production is known to be $1.00 per product made. In practice, it is often quite difficult to determine a specific figure for the value added by a given operation because the operation is one of a sequence of processing steps required to make the product. The final value or price of the product may be known, but the contribution of each individual production operation is not. In this case, one approach is to determine the unit cost per product made in the operation for the two (or more) alternatives. The alternative with the lowest unit cost would be selected.

Example 12-6 Let us revisit the previous example and compute the unit costs for the manual method and the robot method. We will ignore the information given in Example 12-5 that the value added is $1.00 per unit of production.

The rate of return criterion is 25 percent which we will use in determining the values of equivalent uniform annual cost for the two production methods. Writing the EUAC equation for the manual method excluding any revenue derived from the units sold, we have

$$EUAC = -29,000(A/P, 25\%, 3) - 36,000 = 0$$

$$EUAC = -14,857 - 36,000 = -50,857/\text{year}$$

The number of units produced annually in the operation would be

$$(200 \text{ units/day})(250 \text{ days/year}) = 50,000 \text{ units per year}$$

$$\text{Unit cost} = 50,857/50,000 = \$1.017 \text{ per unit}$$

Note that the value added is slightly higher than the \$1.00 value used in Example 12-5. The reason is that rate of return of 25 percent used here is greater than the rate of return found for the manual method in the previous example. A value added in this operation of \$1.017 per unit would produce revenues sufficient to meet the 25 percent minimum attractive rate of return.

For the robot method, the equivalent uniform annual cost (again, ignoring any consideration of revenue derived from units produced, but including salvage value) would be

$$EUAC = -100,000(A/P, 25\%, 3) - 20,000 + (50,000)(A/F, 25\%, 3)$$

$$EUAC = -51,230 - 20,000 + 13,115 = -58,115/\text{year}$$

The number of units produced annually in the operation would be

$$(300 \text{ units/day})(250 \text{ days/year}) = 75,000 \text{ units per year}$$

$$\text{Unit cost} = 58,115/75,000 = \$0.775 \text{ per unit}$$

The robot method would be favored.

There is an implicit assumption in the case illustrated by Example 12-6 where revenues are not known. The assumption is that the operation must be performed by one method or another, and our problem is simply to determine the method whose cost is the lowest.

12-5 OTHER FACTORS MORE DIFFICULT TO QUANTIFY

In addition to labor, equipment, and other sources of costs that are readily quantifiable in a robotics application project, there are other possible sources of costs and savings that are more difficult to evaluate. These factors are sometimes referred to as indirect costs and savings, and they include such factors as inventory savings, scrap savings, and reduced downtime. A comprehensive listing of these potential factors is presented in Table 12-2. These factors should not be ignored simply because their assessment is more difficult

Table 12-2 Indirect costs and savings in a robot application project

1. *In-process inventories*—The savings in in-process inventory result from a reduced manufacturing lead time with a robot installation. Shorter operation cycle times, use of the second and third shifts, and the possibility for combining separate operations into one robot cell are reasons why the manufacturing lead time is reduced.

2. *Finished inventories*—The technical feasibility of using robots in flexible, adaptable manufacturing cells and assembly systems provides the opportunity for reducing the production batch size. Smaller lot sizes translate into lower final inventories.

3. *Materials savings*—In some applications, robots use the raw materials more efficiently in the production process. This leads to a lower usage rate of these materials. Robotic spray painting operations are an example of these savings; the consistency with which the paint is applied by a robot allows a reduction in the total amount of paint consumed as compared to a manual spray paint operation.

4. *Less scrap and rework*—The avoidance of human error in the operation, the consistency of the robot cycle (both in terms of timing and positional repeatability) are some of the factors that contribute to a more uniform product and a reduction in the scrap and rework rates when robots are used.

5. *Equipment utilization*—When robots are used to automate an operation, the utilization of the existing equipment generally increases. The reasons for the increase include the opportunity to convert to multishift operation of the equipment when robots are integrated into the operation, fewer breaks in the shift as compared to the requirements in a manual operation.

6. *Material handling*—When several operations are combined into a single robot cell, the amount of material handling in the plant is reduced.

7. *Floor space*—A well-designed robot cell typically reduces the amount of floor space required for the operation. This is especially true when several operations, previously accomplished at separate workstations, are combined into a single robot workcell.

and less exact than the cost factors listed in Table 12-1. Their value might be significant enough to turn a marginal project into one whose return on investment is much larger than the rate of return criterion used by the company. Meyer[4] presents an approach for estimating some of these indirect factors, and we recommend this paper as a reference to the interested reader.

Finally, there are considerations in deciding about a robot installation that are virtually impossible to quantify. These considerations include improvements in safety by removing human operators from the immediate dangers of the production operation, reduced dependence on direct labor and the associated personnel problems, better customer relations from improved delivery schedules and better quality, and the ability to use the automated factory as a showplace to impress customers and potential customers. Another possible consideration is production flexibility. Some robot cells are designed to be highly adaptable to changes in the product mix that is processed in the cell. These robot applications allow the company a significant amount of scheduling flexibility, both in terms of product mix and production volume that can be accommodated on the system. This flexibility, although its value is difficult to quantify, is a definite advantage to the company. While these factors cannot be

Robot Economic Analysis Form

Project Number _____ Prepared by: _____ Date _____

A – Investment costs

    1. Robot purchase cost   _____
    2. Engineering costs   _____
    3. Installation costs   _____
    4. Special tooling (e.g., and effector)   _____
    5. Miscellaneous costs   _____
    6. Gross investment cost (sum of items 1 through 5)   _____

    7. Disposal value of any equipment retired   _____
    8. Capital required in absence of proposed project   _____
    9. Total investment released or avoided (6 + 7)   _____

  10. Total net investment (6 − 9)   _____

B – Operating Costs and Savings

	Costs	Savings
11. Direct labor cost	_____	_____
12. Indirect labor costs	_____	_____
13. Maintenance	_____	_____
14. Utilities	_____	_____
15. Training	_____	_____
16. In-process inventories	_____	_____
17. Finished inventories	_____	_____
18. Materials and supplies	_____	_____
19. Less scrap and rework	_____	_____
20. Equipment utilization	_____	_____
21. Material handling	_____	_____
22. Floor space	_____	_____
23. Safety (estimate a value)	_____	_____
24. Flexibility (estimate a value)	_____	_____
25. Other (specify)	_____	_____
26. Totals (sum 11 through 25)	_____	_____

C – Other Data

  27. Estimated service life of installation_____years
  28. Estimated life of robot used in installation_____years
  29. Estimated salvage value of robot at end of service life_____
  30. Estimated salvage value of other equipment at end of project_____
  30. Minimum attractive rate of return criterion_____%

D – Calculated Results of Analysis

  31. Payback period_____years
  32. Return on investment over the project life_____%
  33. Equivalent uniform annual cost_____

Figure 12-2 A possible robot economic analysis form.

quantified into costs and savings, they should nevertheless be considered in a company's overall strategy for implementing robotics and automation. They are generally considerations which tend to favor the application of robotics.

12-6 ROBOT PROJECT ANALYSIS FORM

Many companies have developed a standard investment analysis form that they use to show the results of their economic evaluation of a proposed project. Several of the references at the end of the chapter give examples of these kinds of forms designed specifically for projects devoted to robotics and related automation areas (Ref. 3, p. 184; Ref. 4; Ref. 5, p. 126; and Ref 8, p. 130). In Fig. 12-2, we present an economic analysis form that can be used to display the costs and savings for a proposed robot project according to the procedures we have developed in this chapter.

In closing, it should be noted that our development of robot investment analysis has omitted certain considerations, such as depreciation methods, income taxes, and capital gains and losses. We leave these considerations to the many texts available on engineering economy, several of which are listed among the Refs. 2, 6, and 7. Our purpose in this chapter has been to present an introductory treatment of investment analysis and to examine the particular issues that arise in robot projects.

REFERENCES

1. J. F. Engelberger, *Robotics in Practice*, AMACOM, 1980, chaps. 5, 6, 7.
2. E. L. Grant, W. G. Ireson, and R. S. Leavenworth, *Principles of Engineering Economy*, 6th ed., Wiley, New York, 1976.
3. M. P. Groover, *Automation, Production Systems, and Computer-Aided Manufacturing*, Prentice-Hall, Englewood Cliffs, NJ, 1980, chaps. 3 and 7.
4. R. J. Meyer, "A Cookbook Approach to Robotics and Automation Justification," Soc. Mfg. Engrs, *Technical Paper MS82-192*, 1982.
5. K. Susnjara, *A Manager's Guide to Industrial Robots*, Corinthian Press, Shaker Heights, OH, 1982, chap. 9.
6. A. J. Tarquin and L. T. Blank, *Engineering Economy*, McGraw-Hill, New York, 1976.
7. G. J. Thuesen, W. Fabrycky, and H. G. Thuesen, *Engineering Economy*, 5th ed., Prentice-Hall, Englewood Cliffs, NJ, 1977.
8. L. L. Toepperwein, M. T. Blackman, et al., *ICAM Robotics Application Guide*, Technical Report AFWAL-TR-80-4042, Vol. II, Air Force Wright Aeronautical Laboratories, Wright-Patterson Air Force Base, OH, April 1980.
9. W. Vogel, "Economic Analysis for Robotics," *Special Report*, IE 398: Industrial Robotics, Lehigh University, May 1984.

PROBLEMS

12-1 A robot installation has an initial (investment) cost of $85,000 and the expected annual costs for operation and maintenance are estimated to be $24,000. The project is expected to generate revenues of $60,000 per year for 4 years, at which time the project will be retired. The estimated salvage value of the robot at the end of the 4 years is $40,000.

(*a*) Determine the payback period for this project.

(*b*) Determine the equivalent uniform annual cost for the project, using a 25 percent rate of return in the interest calculations.

(*c*) Determine the expected return on investment to be derived from the investment.

12-2 Solve Prob. 12-1 except that in the first year of operation for the project, the revenues will be $20,000 rather than $60,000. This is due to the anticipation that there will be "startup" problems in the installation during the first year which will prevent the full revenues from being realized.

12-3 Two production methods are to be compared, one a robot cell and the other a manual operation. The robot cell has a first cost of $120,000, annual operating and maintenance costs of $25,000 and revenues of $75,000. The manual method has a first cost of $30,000, annual operating and maintenance costs of $60,000, and the same revenues of $75,000. In both cases the anticipated service life for the project is 3 years.

(*a*) Determine which alternative should be selected using the payback method.

(*b*) Determine which alternative should be selected using the return on investment method.

(*c*) Explain why each method selects opposite alternatives. Which method would you place more reliance on and why?

12-4 A robot cell has been proposed for a certain industrial operation. The robot cell will cost $130,000 installed. Operating costs are estimated to be $14,000 per year for the 4-year expected life of the project. Revenues will be generated from the installation of $60,000 per year. For purposes of the analysis, the robot will have a zero salvage value at the end of the project. Determine the payback period and the rate of return from the investment.

12-5 The robot in Prob. 12-4 has a purchase price of $60,000. The engineering and installation costs for the project will be $70,000. The robot has an expected operating life of 6 years. If the appropriate salvage value of the robot at the end of the 4-year project is figured into the analysis, determine the payback period and the rate of return from the investment.

12-6 Give the information in Prob. 12-5, and given that the company uses a minimum attractive rate of return of 25 percent, determine the equivalent uniform annual cost for the proposed project, if the salvage value is figured into the problem.

12-7 Two manually operated workstations are currently used to produce a certain part. The two operators are each paid $10.00 per hour, including fringe benefits, for 8 hours per day. Considering work breaks and miscellaneous lost time, there are only seven actual hours of production during the day. Each operator can produce at the rate of 25 units per hour during each hour of production. A robot has been proposed to replace the two human operators. The robot is capable of working an average of 22 hours per day at a production rate of 20 units per hour. The price of the robot is $75,000 and its anticipated service life is 8 years while this project has an expected life of 3 years. The engineering and installation costs will be $50,000. The annual operating costs of the robot will be $10,000. The minimum attractive rate of return used by the company for projects of this type is 30 percent. Assume 250 days of operation per year. Determine the payback period and the return on investment (rate of return) for the proposed robot. Should the company install this robot?

12-8 Three years ago, the initial cost of a robot was $90,000, and it was anticipated that the robot would have a service life of 10 years. The project for which the robot was originally installed is now at the end of its life cycle, and the robot is being proposed for a new project which will last 4 years. The engineering costs to install the new project will be $45,000. The robot will work an average of 15 hours per day for 5 days per week (250 days per year) and the annual maintenance and operating costs are $12,000. The robot's production rate will be 30 pieces per hour.

(*a*) Determine the required value added per piece in order for the project to earn a rate of return (return on investment) of 20 percent.

(*b*) With your answer determined in part (*a*), what is the payback period for the robot?

(*c*) The robot is being proposed to replace a manual operation currently performed at $12.00 per hour for 7.5 hours per day (250 days per year) at a production rate of 35 pieces per hour. Will the robot be an improvement over the manual operation? Support your answer.

APPENDIX TABLE OF INTEREST FACTORS

Tables of interest factors for several interest rates are presented on the following pages. The tables are reprinted by permission from L. T. Blank and A. J. Tarquin, *Engineering Economy*, Second Edition, McGraw-Hill, Inc., New York, 1983.

Table 12A-1 10.00 percent compound interest factors

	Single payments		Uniform series payments				
n	Compound amount F/P	Present worth P/F	Sinking fund A/F	Compound amount F/A	Capital recovery A/P	Present worth P/A	n
1	1.1000	0.9091	1.00000	1.000	1.10000	0.9091	1
2	1.2100	0.8264	0.47619	2.100	0.57619	1.7355	2
3	1.3310	0.7513	0.30211	3.310	0.40211	2.4869	3
4	1.4641	0.6830	0.21547	4.641	0.31547	3.1699	4
5	1.6105	0.6209	0.16380	6.105	0.26380	3.7908	5
6	1.7716	0.5645	0.12961	7.716	0.22961	4.3553	6
7	1.9487	0.5132	0.10541	9.487	0.20541	4.8684	7
8	2.1436	0.4665	0.08744	11.436	0.18744	5.3349	8
9	2.3579	0.4241	0.07364	13.579	0.17364	5.7590	9
10	2.5937	0.3855	0.06275	15.937	0.16275	6.1446	10
11	2.8531	0.3505	0.05396	18.531	0.15396	6.4951	11
12	3.1384	0.3186	0.04676	21.384	0.14676	6.8137	12
13	3.4523	0.2897	0.04078	24.523	0.14078	7.1034	13
14	3.7975	0.2633	0.03575	27.975	0.13575	7.3667	14
15	4.1772	0.2394	0.03147	31.772	0.13147	7.6061	15
16	4.5950	0.2176	0.02782	35.950	0.12782	7.8237	16
17	5.0545	0.1978	0.02466	40.545	0.12466	8.0216	17
18	5.5599	0.1799	0.02193	45.599	0.12193	8.2014	18
19	6.1159	0.1635	0.01955	51.159	0.11955	8.3649	19
20	6.7275	0.1486	0.01746	57.275	0.11746	8.5136	20
22	8.1403	0.1228	0.01401	71.403	0.11401	8.7715	22
24	9.8497	0.1015	0.01130	88.497	0.11130	8.9847	24
25	10.8347	0.0923	0.01017	98.347	0.11017	9.0770	25
26	11.9182	0.0839	0.00916	109.182	0.10916	9.1609	26
28	14.4210	0.0693	0.00745	134.210	0.10745	9.3066	28
30	17.4494	0.0573	0.00608	164.494	0.10608	9.4269	30
32	21.1138	0.0474	0.00497	201.138	0.10497	9.5264	32
34	25.5477	0.0391	0.00407	245.477	0.10407	9.6086	34
35	28.1024	0.0356	0.00369	271.024	0.10369	9.6442	35
36	30.9127	0.0323	0.00334	299.127	0.10334	9.6765	36
38	37.4043	0.0267	0.00275	364.043	0.10275	9.7327	38
40	45.2593	0.0221	0.00226	442.593	0.10226	9.7791	40
45	72.8905	0.0137	0.00139	718.905	0.10139	9.8628	45
50	117.391	0.0085	0.00086	1163.909	0.10086	9.9148	50
55	189.059	0.0053	0.00053	1880.591	0.10053	9.9471	55

Table 12A-1 (*continued*)

	Single payments		Uniform series payments				
n	Compound amount *F/P*	Present worth *P/F*	Sinking fund *A/F*	Compound amount *F/A*	Capital recovery *A/P*	Present worth *P/A*	*n*
60	304.482	0.0033	0.00033	3034.816	0.10033	9.9672	60
65	490.371	0.0020	0.00020	4893.707	0.10020	9.9796	65
70	789.747	0.0013	0.00013	7887.470	0.10013	9.9873	70
75	1271.895	0.0008	0.00008	12708.954	0.10008	9.9921	75
80	2048.400	0.0005	0.00005	23474.002	0.10005	9.9951	80
85	3298.969	0.0003	0.00003	32979.690	0.10003	9.9970	85
90	5313.023	0.0002	0.00002	53120.226	0.10002	9.9981	90
95	8556.676	0.0001	0.00001	85556.760	0.10001	9.9988	95

Table 12A-2 15 percent compound interest factors

	Single payments		Uniform series payments				
n	Compound amount *F/P*	Present worth *P/F*	Sinking fund *A/F*	Compound amount *F/A*	Capital recovery *A/P*	Present worth *P/A*	*n*
1	1.1500	0.8696	1.00000	1.000	1.15000	0.8696	1
2	1.3225	0.7561	0.46512	2.150	0.61512	1.6257	2
3	1.5209	0.6575	0.28798	3.472	0.43798	2.2832	3
4	1.7490	0.5718	0.20027	4.993	0.35027	2.8550	4
5	2.0114	0.4972	0.14832	6.742	0.29832	3.3522	5
6	2.3131	0.4323	0.11424	8.754	0.26424	3.7845	6
7	2.6600	0.3759	0.09036	11.067	0.24036	4.1604	7
8	3.0590	0.3269	0.07285	13.727	0.22285	4.4873	8
9	3.5179	0.2843	0.05957	16.786	0.20957	4.7716	9
10	4.0456	0.2472	0.04925	20.304	0.19925	5.0188	10
11	4.6524	0.2149	0.04107	24.349	0.19107	5.2337	11
12	5.3503	0.1869	0.03448	29.002	0.18448	5.4206	12
13	6.1528	0.1625	0.02911	34.352	0.17911	5.5831	13
14	7.0757	0.1413	0.02469	40.505	0.17469	5.7245	14
15	8.1371	0.1229	0.02102	47.580	0.17102	5.8474	15
16	9.3576	0.1069	0.01795	55.717	0.16795	5.9542	16
17	10.7613	0.0929	0.01537	65.075	0.16537	6.0472	17
18	12.3755	0.0808	0.01319	75.836	0.16319	6.1280	18
19	14.2318	0.0703	0.01134	88.212	0.16134	6.1982	19
20	16.3665	0.0611	0.00976	102.444	0.15976	6.2593	20
22	21.6447	0.0462	0.00727	137.632	0.15727	6.3587	22
24	28.6252	0.0349	0.00543	184.168	0.15543	6.4338	24
25	32.9190	0.0304	0.00470	212.793	0.15470	6.4641	25
26	37.8568	0.0264	0.00407	245.712	0.15407	6.4906	26
28	50.0656	0.0200	0.00306	327.104	0.15306	6.5335	28

Table 12A-2 (*continued*)

| n | Single payments | | | Uniform series payments | | | | n |
	Compound amount F/P	Present worth P/F	Sinking fund A/F	Compound amount F/A	Capital recovery A/P	Present worth P/A	
30	66.2118	0.0151	0.00230	434.745	0.15230	6.5660	30
32	87.5651	0.0114	0.00173	577.100	0.15173	6.5905	32
34	115.805	0.0086	0.00131	765.365	0.15131	6.6091	34
35	133.176	0.0075	0.00113	881.170	0.15113	6.6166	35
36	153.152	0.0065	0.00099	1014.346	0.15099	6.6231	36
38	202.543	0.0049	0.00074	1343.622	0.15074	6.6338	38
40	267.864	0.0037	0.00056	1779.090	0.15056	6.6418	50
45	538.769	0.0019	0.00028	3585.128	0.15028	6.6543	45
50	1083.657	0.0009	0.00014	7217.716	0.15014	6.6605	50

Table 12A-3 20.00 percent compound interest factors

| n | Single payments | | | Uniform series payments | | | | n |
	Compound amount F/P	Present worth P/F	Sinking fund A/F	Compound amount F/A	Capital recovery A/P	Present worth P/A	
1	1.2000	0.8333	1.00000	1.000	1.20000	0.8333	1
2	1.4400	0.6944	0.45455	2.200	0.65455	1.5278	2
3	1.7280	0.5787	0.27473	3.640	0.47473	2.1065	3
4	2.0736	0.4823	0.18629	5.368	0.38629	2.5887	4
5	2.4883	0.4019	0.13438	7.442	0.33438	2.9906	5
6	2.9860	0.3349	0.10071	9.930	0.30071	3.3255	6
7	3.5832	0.2791	0.07742	12.916	0.27742	3.6046	7
8	4.2998	0.2326	0.06061	16.499	0.26061	3.8372	8
9	5.1598	0.1938	0.04808	20.799	0.24808	4.0310	9
10	6.1917	0.1615	0.03852	25.959	0.23852	4.1925	10
11	7.4301	0.1346	0.03110	32.150	0.23110	4.3271	11
12	8.9161	0.1122	0.02526	39.581	0.22526	4.4392	12
13	10.6993	0.0935	0.02062	48.497	0.22062	4.5327	13
14	12.8392	0.0779	0.01689	59.196	0.21689	4.6106	14
15	15.4070	0.0649	0.01388	72.035	0.21388	4.6755	15
16	18.4884	0.0541	0.01144	87.442	0.21144	4.7296	16
17	22.1861	0.0451	0.00944	105.931	0.20944	4.7746	17
18	26.6233	0.0376	0.00781	128.117	0.20781	5.8122	18
19	31.9480	0.0313	0.00646	154.740	0.20646	4.8435	19
20	38.3376	0.0261	0.00536	186.688	0.20536	4.8696	20
22	55.2061	0.0181	0.00369	271.031	0.20369	4.9094	22
24	79.4968	0.0126	0.00255	392.484	0.20255	4.9371	24
25	95.3962	0.0105	0.00212	471.981	0.20212	4.9476	25
26	114.4755	0.0087	0.00176	567.377	0.20176	4.9563	26
28	164.8447	0.0061	0.00122	819.223	0.20122	4.9697	28

Table 12A-3 (*continued*)

	Single payments		Uniform series payments				
	Compound amount	Present worth	Sinking fund	Compound amount	Capital recovery	Present worth	
n	F/P	P/F	A/F	F/A	A/P	P/A	n
30	237.3763	0.0042	0.00085	1181.882	0.20085	4.9789	30
32	341.8219	0.0029	0.00059	1704.109	0.20059	4.9854	32
34	492.2235	0.0020	0.00041	2456.118	0.20041	4.9898	34
35	590.6682	0.0017	0.00034	2948.341	0.20034	4.9915	35
36	708.8019	0.0014	0.00028	3539.009	0.20028	4.9929	36
38	1020.675	0.0010	0.20020	5098.373	0.20020	4.9951	38
40	1469.772	0.0007	0.20014	7343.858	0.20014	4.9966	40
45	3657.262	0.0003	0.20005	18281.310	0.20005	4.9986	45
50	9100.438	0.0001	0.20002	45497.191	0.20002	4.9995	50

Table 12A-4 25.00 percent compound interest factors

	Single payments		Uniform series payments				
	Compound amount	Present worth	Sinking fund	Compound amount	Capital recovery	Present worth	
n	F/P	P/F	A/F	F/A	A/P	P/A	n
1	1.2500	0.8000	1.00000	1.000	1.25000	0.8000	1
2	1.5625	0.6400	0.44444	2.250	0.69444	1.4400	2
3	1.9531	0.5120	0.26230	3.813	0.51230	1.9520	3
4	2.4414	0.4096	0.17344	5.766	0.42344	2.3616	4
5	3.0518	0.3277	0.12185	8.207	0.37185	2.6893	5
6	3.8147	0.2621	0.08882	11.259	0.33882	2.9514	6
7	4.7684	0.2097	0.06634	15.073	0.31634	3.1611	7
8	5.9605	0.1678	0.05040	19.842	0.30040	3.3289	8
9	7.4506	0.1342	0.03876	25.802	0.28876	3.4631	9
10	9.3132	0.1074	0.03007	33.253	0.28007	3.5705	10
11	11.6415	0.0859	0.02349	42.566	0.27349	3.6564	11
12	14.5519	0.0687	0.01845	54.208	0.26845	3.7251	12
13	18.1899	0.0550	0.01454	68.760	0.26454	3.7801	13
14	22.7374	0.0440	0.01150	86.949	0.26150	3.8241	14
15	28.4217	0.0352	0.00912	109.687	0.25912	3.8593	15
16	35.5271	0.0281	0.00724	138.109	0.25724	3.8874	16
17	44.4089	0.0225	0.00576	173.636	0.25576	3.9099	17
18	55.5112	0.0180	0.00459	218.045	0.25459	3.9279	18
19	69.3889	0.0144	0.00366	273.556	0.25366	3.9424	19
20	86.7362	0.0115	0.00292	342.945	0.25292	3.9539	20
22	135.5253	0.0074	0.00186	538.101	0.25186	3.9705	22
24	211.7582	0.0047	0.00119	843.033	0.25119	3.9811	24
25	264.6978	0.0038	0.00095	1054.791	0.25095	3.9849	25
26	330.8722	0.0030	0.00076	1319.489	0.25076	3.9879	26
28	516.9879	0.0019	0.00048	2063.952	0.25048	3.9923	28

Table 12A-4 (*continued*)

	Single payments			Uniform series payments			
	Compound amount	Present worth	Sinking fund	Compound amount	Capital recovery	Present worth	
n	F/P	P/F	A/F	F/A	A/P	P/A	n
30	807.7936	0.0012	0.00031	3227.174	0.25031	3.9950	30
32	1262.177	0.0008	0.00020	5044.710	0.25020	3.9968	32
34	1972.152	0.0005	0.00013	7884.609	0.25013	3.9980	34
35	2465.190	0.0004	0.00010	9856.761	0.25010	3.9984	35
36	3081.488	0.0003	0.00008	12321.952	0.25008	3.9987	36
38	4814.825	0.0002	0.00005	19255.299	0.25005	3.9992	38
40	7523.164	0.0001	0.00003	30088.655	0.25003	3.9995	40
45	22958.87	0.0000	0.00001	91831.496	0.25001	3.9998	45

Table 12A-5 30.00 percent compound interest factors

	Single payments			Uniform series payments			
	Compound amount	Present worth	Sinking fund	Compound amount	Capital recovery	Present worth	
n	F/P	P/F	A/F	F/A	A/P	P/A	n
1	1.3000	0.7692	1.00000	1.000	1.30000	0.7692	1
2	1.6900	0.5917	0.43478	2.300	0.73478	1.3609	2
3	2.1970	0.4552	0.25063	3.990	0.55063	1.8161	3
4	2.8561	0.3501	0.16163	6.187	0.46163	2.1662	4
5	3.7129	0.2693	0.11058	9.043	0.41058	2.4356	5
6	4.8268	0.2072	0.07839	12.756	0.37839	2.6427	6
7	6.2749	0.1594	0.05687	17.583	0.35687	2.8021	7
8	8.1573	0.1226	0.04192	23.858	0.34192	2.9247	8
9	10.6045	0.0943	0.03124	32.015	0.33124	3.0190	9
10	13.7858	0.0725	0.02346	42.619	0.32346	3.0915	10
11	17.9216	0.0558	0.01773	56.405	0.31773	3.1473	11
12	23.2981	0.0429	0.01345	74.327	0.31345	3.1903	12
13	30.2875	0.0330	0.01024	97.625	0.31024	3.2233	13
14	39.3738	0.0254	0.00782	127.913	0.30782	3.2487	14
15	51.1859	0.0195	0.00598	167.286	0.30598	3.2682	15
16	66.5417	0.0150	0.00458	218.472	0.30458	3.2832	16
17	86.5042	0.0116	0.00351	285.014	0.30351	3.2948	17
18	112.4554	0.0089	0.00269	371.518	0.30269	3.3037	18
19	146.1920	0.0068	0.00207	483.973	0.30207	3.3105	19
20	190.0496	0.0053	0.00159	630.165	0.30159	3.3158	20
22	321.1839	0.0031	0.00094	1067.280	0.30094	3.3230	22
24	542.8008	0.0018	0.00055	1806.003	0.30055	3.3272	24
25	705.6410	0.0014	0.00043	2348.803	0.30043	3.3286	25
26	917.3333	0.0011	0.00033	3054.444	0.30033	3.3297	26
28	1550.293	0.0006	0.00019	5164.311	0.30019	3.3312	28

Table 12A-5 (*continued*)

	Single payments		Uniform series payments				
	Compound amount	Present worth	Sinking fund	Compound amount	Capital recovery	Present worth	
n	F/P	P/F	A/F	F/A	A/P	P/A	n
30	2619.996	0.0004	0.00011	8729.985	0.30011	3.3321	30
32	4427.793	0.0002	0.00007	14755.975	0.30007	3.3326	32
34	7482.970	0.0001	0.00004	24939.899	0.30004	3.3329	34
35	9727.860	0.0001	0.00003	32422.868	0.30003	3.3330	35

Table 12A-6 35.00 percent compound interest factors

	Single payments		Uniform series payments				
	Compound amount	Present worth	Sinking fund	Compound amount	Capital recovery	Present worth	
n	F/P	P/F	A/F	F/A	A/P	P/A	n
1	1.3500	0.7407	1.00000	1.000	1.35000	0.7407	1
2	1.8225	0.5487	0.42553	2.350	0.77553	1.2894	2
3	2.4604	0.4064	0.23966	4.172	0.58966	1.6959	3
4	3.3215	0.3011	0.15076	6.633	0.50076	1.9969	4
5	4.4840	0.2230	0.10046	9.954	0.45046	2.2200	5
6	6.0534	0.1652	0.06926	14.438	0.41926	2.3852	6
7	8.1722	0.1224	0.04880	20.492	0.39880	2.5075	7
8	11.0324	0.0906	0.03489	28.664	0.38489	2.5982	8
9	14.8937	0.0671	0.02519	39.696	0.37519	2.6653	9
10	20.1066	0.0497	0.01832	54.590	0.36832	2.7150	10
11	27.1439	0.0368	0.01339	74.697	0.36339	2.7519	11
12	36.6442	0.0273	0.00982	101.841	0.35982	2.7792	12
13	49.4697	0.0202	0.00722	138.485	0.35722	2.7994	13
14	66.7841	0.0150	0.00532	187.954	0.35532	2.8144	14
15	90.1585	0.0111	0.00393	254.738	0.35393	2.8255	15
16	121.7139	0.0082	0.00290	344.897	0.35290	2.8337	16
17	164.3138	0.0061	0.00214	466.611	0.35214	2.8398	17
18	221.8236	0.0045	0.00158	630.925	0.35158	2.8443	18
19	299.4619	0.0033	0.00117	852.748	0.35117	2.8476	19
20	404.2736	0.0025	0.00087	1152.210	0.35087	2.8501	20
22	736.7886	0.0014	0.00048	2102.253	0.35048	2.8533	22
24	1342.797	0.0007	0.00026	3833.706	0.35026	2.8550	24
25	1812.776	0.0006	0.00019	5176.504	0.35019	2.8556	25
26	2447.248	0.0004	0.00014	6989.280	0.35014	2.8560	26
28	4460.109	0.0002	0.00008	12740.313	0.35008	2.8565	28
30	8128.550	0.0001	0.00004	23221.570	0.35004	2.8568	30
32	14814.28	0.0001	0.00002	42323.661	0.35002	2.8569	32
34	26999.03	0.0000	0.00001	77137.223	0.35001	2.8570	34
35	36448.69	0.0000	0.00001	104136.25	0.35001	2.8571	35

Table 12A-7 40.00 percent compound interest factors

	Single payments		Uniform series payments				
n	Compound amount F/P	Present worth P/F	Sinking fund A/F	Compound amount F/A	Capital recovery A/P	Present worth P/A	n
1	1.4000	0.7143	1.00000	1.000	1.40000	0.7143	1
2	1.9600	0.5102	0.41667	2.400	0.81667	1.2245	2
3	2.7440	0.3644	0.22936	4.360	0.62936	1.5889	3
4	3.8416	0.2603	0.14077	7.104	0.54077	1.8492	4
5	5.3782	0.1859	0.09136	10.946	0.49136	2.0352	
6	7.5295	0.1328	0.06126	16.324	0.46126	2.1680	6
7	10.5414	0.0949	0.04192	23.853	0.44192	2.2628	7
8	14.7579	0.0678	0.02907	34.395	0.42907	2.3306	8
9	20.6610	0.0484	0.02034	49.153	0.42034	2.3790	9
10	28.9255	0.0346	0.01432	69.814	0.41432	2.4136	10
11	40.4957	0.0247	0.01013	98.739	0.41013	2.4383	11
12	56.6939	0.0176	0.00718	139.235	0.40718	2.4559	12
13	79.3715	0.0126	0.00510	195.929	0.40510	2.4685	13
14	111.1201	0.0090	0.00363	275.300	0.40363	2.4775	14
15	155.5681	0.0064	0.00259	386.420	0.40259	2.4839	15
16	217.7953	0.0046	0.00185	541.988	0.40185	2.4885	16
17	304.9135	0.0033	0.00132	759.784	0.40132	2.4918	17
18	426.8789	0.0023	0.00094	1064.697	0.40094	2.4941	18
19	597.6304	0.0017	0.00067	1491.576	0.40067	2.4958	19
20	836.6826	0.0012	0.00048	2089.206	0.40048	2.4970	20
22	1639.898	0.0006	0.00024	4097.245	0.40024	2.4985	22
24	3214.200	0.0003	0.00012	8032.999	0.40012	2.4992	24
25	4499.880	0.0002	0.00009	11247.199	0.40009	2.4994	25
26	6299.831	0.0002	0.00006	15747.079	0.40006	2.4996	26
28	12347.67	0.0001	0.00003	30866.674	0.40003	2.4998	28
30	24201.43	0.0000	0.00002	60501.081	0.40002	2.4999	30
32	47434.81	0.0000	0.00001	118584.52	0.40001	2.4999	32
34	92972.22	0.0000	0.00000	232428.06	0.40000	2.5000	34
35	130161.1	0.0000	0.00000	325400.28	0.40000	2.5000	35

Table 12A-8 45.00 percent compound interest factors

	Single payments		Uniform series payments				
n	Compound amount F/P	Present worth P/F	Sinking fund A/F	Compound amount F/A	Capital recovery A/P	Present worth P/A	n
1	1.4500	0.6897	1.00000	1.000	1.45000	0.6897	1
2	2.1025	0.4756	0.40816	2.450	0.85816	1.1653	2
3	3.0486	0.3280	0.21966	4.552	0.66966	1.4933	3
4	4.4205	0.2262	0.13156	7.601	0.58156	1.7195	4
5	6.4097	0.1560	0.08318	12.022	0.53318	1.8755	5

Table 12A-8 (*continued*)

	Single payments		Uniform series payments				
n	Compound amount F/P	Present worth P/F	Sinking fund A/F	Compound amount F/A	Capital recovery A/P	Present worth P/A	*n*
6	9.2941	0.1076	0.05426	18.431	0.50426	1.9831	6
7	13.4765	0.0742	0.03607	27.725	0.48607	2.0573	7
8	19.5409	0.0512	0.02427	41.202	0.47427	2.1085	8
9	28.3343	0.0353	0.01646	60.743	0.46646	2.1438	9
10	41.0847	0.0243	0.01123	89.077	0.46123	2.1681	10
11	59.5728	0.0168	0.00768	130.162	0.45768	2.1849	11
12	86.3806	0.0116	0.00527	189.735	0.45527	2.1965	12
13	125.2518	0.0080	0.00362	276.115	0.45362	2.2045	13
14	181.6151	0.0055	0.00249	401.367	0.45249	2.2100	14
15	263.3419	0.0038	0.00172	582.982	0.45172	2.2138	15
16	381.8458	0.0026	0.00118	846.324	0.45118	2.2164	16
17	553.6764	0.0018	0.00081	1228.170	0.45081	2.2182	17
18	892.8308	0.0012	0.00056	1781.846	0.45056	2.2195	18
19	1164.105	0.0009	0.00039	2584.677	0.45039	2.2203	19
20	1687.952	0.0006	0.00027	3748.782	0.45027	2.2209	20
22	3548.919	0.0003	0.00013	7884.264	0.45013	2.2216	22
24	7461.602	0.0001	0.00006	16579.115	0.45006	2.2219	24
25	10819.32	0.0001	0.00004	24040.716	0.45004	2.2220	25
26	15688.02	0.0001	0.00003	34860.038	0.45003	2.2221	26
28	32984.06	0.0000	0.00001	73295.681	0.45001	2.2222	28
30	69348.98	0.0000	0.00001	154106.62	0.45001	2.2222	30
32	145806.2	0.0000	0.00000	324011.62	0.45000	2.2222	32
34	306557.6	0.0000	0.00000	681236.87	0.45000	2.2222	34
35	444508.5	0.0000	0.00000	987794.46	0.45000	2.2222	35

Table 12A-9 50.00 percent compound interest factors

	Single payments		Uniform series payments				
n	Compound amount F/F	Present worth P/F	Sinking fund A/F	Compound amount F/A	Capital recovery A/P	Present worth P/A	*n*
1	1.5000	0.6667	1.00000	1.000	1.50000	0.6667	1
2	2.2500	0.4444	0.40000	2.500	0.90000	1.1111	2
3	3.3750	0.2968	0.21053	4.750	0.71053	1.4074	3
4	5.0625	0.1975	0.12308	8.125	0.62308	1.6049	4
5	7.5938	0.1317	0.07583	13.188	0.57583	1.7366	5
6	11.3906	0.0878	0.04812	20.781	0.54812	1.8244	6
7	17.0859	0.0585	0.03108	32.172	0.53108	1.8829	7
8	25.6289	0.0390	0.02030	49.258	0.52030	1.9220	8
9	38.4434	0.0260	0.01335	74.887	0.51335	1.9480	9
10	57.6650	0.0173	0.00882	113.330	0.50882	1.9653	10

Table 12A-9 (*continued*)

	Single payments		Uniform series payments				
	Compound amount	Present worth	Sinking fund	Compound amount	Capital recovery	Present worth	
n	*F/P*	*P/F*	*A/F*	*F/A*	*A/P*	*P/A*	*n*
11	86.4976	0.0116	0.00585	170.995	0.50585	1.9769	11
12	129.7463	0.0077	0.00388	257.493	0.50388	1.9846	12
13	194.6195	0.0051	0.00258	387.239	0.50258	1.9897	13
14	291.9293	0.0034	0.00172	581.859	0.50172	1.9931	14
15	437.8939	0.0023	0.00114	873.788	0.50114	1.9954	15
16	656.8408	0.0015	0.00076	1311.682	0.50076	1.9970	16
17	985.2613	0.0010	0.00051	1968.523	0.50051	1.9980	17
18	1477.892	0.0007	0.00034	2953.784	0.50034	1.9986	18
19	2216.838	0.0005	0.00023	4431.676	0.50023	1.9991	19
20	3325.257	0.0003	0.00015	6648.513	0.50015	1.9994	20
22	7481.828	0.0001	0.00007	14961.655	0.50007	1.9997	22
24	16834.11	0.0001	0.00003	33666.224	0.50003	1.9999	24
25	25251.17	0.0000	0.00002	50500.337	0.50002	1.9999	25
26	37876.75	0.0000	0.00001	75751.505	0.50001	1.9999	26
28	85222.69	0.0000	0.00001	170443.39	0.50001	2.0000	28
30	191751.1	0.0000	0.00000	383500.12	0.50000	2.0000	30
32	431439.9	0.0000	0.00000	862877.77	0.50000	2.0000	32
34	970739.7	0.0000	0.00000	1941477.5	0.50000	2.0000	34

FIVE

ROBOT APPLICATIONS IN MANUFACTURING

Current-day robot applications include a wide variety of production operations. For purposes of organization in this book, we will classify the operations into the following categories:

1. Material transfer and machine loading/unloading. These are applications in which the robot grasps and moves a workpart from one location to another. This category includes applications in which the robot transfers parts into and out of a production machine. Examples of these load/unload operations include metal-machining operations, die casting, plastic molding, and certain forging operations. Material transfer and machine loading and unloading applications are covered in Chap. Thirteen.

2. Processing operations. These are operations in which the robot uses a tool as an end effector to accomplish some processing operation on a workpart that is positioned for the robot during the work cycle. Spot welding, arc welding, spray painting, and certain machining operations fall into this application category. Chapter Fourteen deals with these robot applications.

3. Assembly and inspection. Assembly and inspection are relatively new applications for robots. (We are excluding welding operations from this group though they are assembly operations.) The robot is used to put components together into an assembly, or the robot is used to perform some form of automated inspection operation. More and more robots in the future will be equipped with vision capability to facilitate the performance of these operations. We discuss robot assembly and inspection in Chap. Fifteen.

THIRTEEN

MATERIAL TRANSFER AND MACHINE
LOADING/UNLOADING

There are many robot applications in which the robot is required to move a workpart or other material from one location to another. The most basic of these applications is where the robot picks the part up from one position and transfers it to another position. In other applications, the robot is used to load and/or unload a production machine of some type. In this book we divide material-handling applications into two specific categories:

1. Material transfer applications
2. Machine loading/unloading applications

There are other robot applications which involve parts handling. These include assembly operations and holding parts during inspection. Assembly and inspection applications are treated in Chap. Fifteen. Before discussing material transfer and machine loading/unloading applications, let us review some of the considerations that should be examined when robots are used for material handling.

13-1 GENERAL CONSIDERATIONS IN ROBOT MATERIAL HANDLING

In planning an application in which the robot will be used to transfer parts, load a machine, or other similar operation, there are several considerations that must be reviewed. Most of these considerations have been discussed in previous chapters of the book, and we itemize them below as a reference checklist.

1. *Part positioning and orientation.* In most parts-handling applications the parts must be presented to the robot in a known position and orientation. Robots used in these applications do not generally possess highly sophisticated sensors (e.g., machine vision) that would enable them to seek out a part and identify its orientation before picking it up.

2. *Gripper design.* Special end effectors must be designed for the robot to grasp and hold the workpart during the handling operation. Design considerations for these grippers were discussed in Chap. Five.

3. *Minimum distances moved.* The material-handling application should be planned so as to minimize the distances that the parts must be moved. This can be accomplished by proper design of the workcell layout (e.g., keeping the equipment in the cell close together), by proper gripper design (e.g., using a double gripper in a machine loading/unloading operation), and by careful study of the robot motion cycle.

4. *Robot work volume.* The cell layout must be designed with proper consideration given to the robot's capability to reach the required extreme locations in the cell and still allow room to maneuver the gripper.

5. *Robot weight capacity.* There is an obvious limitation on the material-handling operation that the load capacity of the robot must not be exceeded. A robot with sufficient weight-carrying capacity must be specified for the application.

6. *Accuracy and repeatability.* Some applications require the materials to be handled with very high precision. Other applications are less demanding in this respect. The robot must be specified accordingly.

7. *Robot configuration, degrees of freedom, and control.* Many parts transfer operations are simple enough that they can be accomplished by a robot with two to four joints of motion. Machine-loading applications often require more degrees of freedom. Robot control requirements are unsophisticated for most material-handling operations. Palletizing operations, and picking parts from a moving conveyor are examples where the control requirements are more demanding.

8. *Machine utilization problems.* It is important for the application to effectively utilize all pieces of equipment in the cell. In a machine loading/unloading operation, it is common for the robot to be idle while the machine is working, and the machine to be idle while the robot is working. In cases where a long machine cycle is involved, the robot is idle a high proportion of the time. To increase the utilization of the robot, consideration should be given to the possibility for the robot to service more than a single machine. One of the problems arising in the multimachine cell is machine interference, discussed in Chap. Eleven.

We now proceed to deal with the specific cases of material transfer and machine loading/unloading applications in the following two sections.

13-2 MATERIAL TRANSFER APPLICATIONS

Material transfer applications are defined as operations in which the primary objective is to move a part from one location to another location. They are usually considered to be among the most straightforward of robot applications to implement. The applications usually require a relatively unsophisticated robot, and the interlocking requirements with other equipment are typically uncomplicated. These applications are sometimes called pick-and-place operations because the robot simply picks the part from one location and places it in another location. Some material transfer applications have motion patterns that change from cycle to cycle, thus requiring a more sophisticated robot. Palletizing and depalletizing operations are examples of this more complicated case. In this type of application, the robot must place each part in a different location on the pallet, thus forcing the robot to remember or compute a separate motion cycle until the pallet is fully loaded.

Pick-and-Place Operations

As defined above, pick-and-place operations involve tasks in which the robot picks up the part at one location and moves it to another location. In the simplest case, the part is presented to the robot by some mechanical feeding device or conveyor in a known location and orientation. The known location is a stationary location, achieved either by stopping the conveyor at the appropriate position, or by using a mechanical stop to hold the part at the stationary location. An input interlock (commonly based on using a simple limit switch) would be designed to indicate that the part is in position and ready for pickup. The robot would grasp the part, pick it up, move it, and position it at a desired location. The orientation of the part remains unchanged during the move. The desired location is usually at a position where there is the capability to move the part out of the way for the next delivery by the robot. This basic case is illustrated in Fig. 13-1. In this simple case, the robot

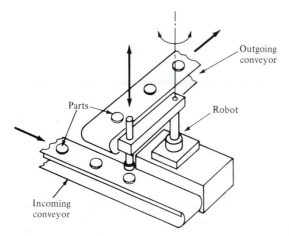

Figure 13-1 Simple pick-and-place operation.

needs only 2 degrees of freedom. As shown in the figure, 1 degree of freedom is needed to lift the part from the pickup point and put it down at the dropoff point. The second degree of freedom is required to move the part between these two positions. In some pick-and-place operations, a reorientation of the workpart is accomplished during the move. This part reorientation requires a robot with one or more additional degrees of freedom.

One complication encountered in material transfer operations is when the robot is required to track a moving pickup point. The general problems related to robot tracking systems were described in Chap. Eleven. In robotic materials handling, tracking arises when parts are carried along a continuously moving conveyor, typically an overhead hook conveyor, and the robot is required to pick parts from the conveyor. The opposite case is when the robot must put parts onto the moving conveyor. In either case a more sophisticated sensor-interlock system is required to determine the presence and location of the parts in the robot's tracking window.

Another complication arises in material transfer operations (as well as palletizing and other operations) when different objects are being handled by the same robot. In material transfer, a single conveyor might be used to move more than one type of part. The robot must be interfaced to some type of sensor system capable of distinguishing between the different parts so that the robot can execute the right program subroutine for the particular part. For example, there may be differences in the way the part must be retrieved from the conveyor due to part configuration, and the placement of the part by the robot may vary for different parts. In other cases where multiple items are handled, the information system which supports the workcell can be used to keep track of where each item is located in the workcell. The information support system would be used either in lieu of or in addition to sensors. The following example illustrates a material-handling situation which uses both sensors and a hierarchical information system to support the control of the robot cell.

Example 13-1 This is a good example of a complex material-handling robot application located at the Poughkeepsie plant of IBM Corporation. The application is described in an article by Giacobbe.[3] The purpose of the operation is to write and package 8-in. diskettes for the large computer systems produced at the Poughkeepsie plant. Prior to the robot cell, the operation was performed manually and was prone to error and damage to the diskettes. A robot cell was designed and installed to overcome these problems.

The robot cell includes: a Unimation PUMA 560 robot, two diskette writers, a box storage carousel, two envelope printers, and two label printers. The workcell controller is an IBM Series/1 computer. Figure 13-2 shows an overall view of the workcell. The individual tasks performed by the robot include getting the diskettes, labeling and packing them, setting up the boxes, closing them, labeling them, and storing them. The end effector is designed with mechanical fingers to handle the boxes

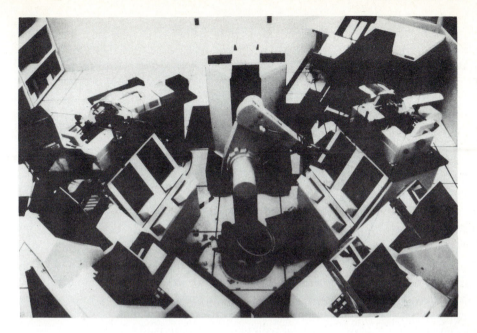

Figure 13-2 Robot workcell for writing diskettes. PUMA 560 robot is used for handling diskettes, envelopes, and boxes in the cell. (Photo courtesy of IBM Corporation)

and box lids, and vacuum pads to handle the flexible diskettes and diskette envelopes.

The Series/1 computer provides the overall coordination of the activities in the workcell. It is linked to a host computer in the plant which supplies diskette data and order information to the diskette writers. The host computer also receives data requests and status information from the robot cell controller. Giacobbe reports that about 90 percent of the coding for the system is for automatic error recovery. For example, if one of the diskettes is dropped during handling, sensors detect this error and the system requests that a duplicate diskette be written.

Palletizing and Related Operations

The use of pallets for materials handling and storage in industry is widespread. Instead of handling individual cartons or other containers, a large number of these containers are placed on a pallet, and the pallet is then handled. The pallets can be moved mechanically within the plant or warehouse by fork lift trucks or conveyors. Shipments of palletized product to the customer are very common because of the convenience in handling, both at the manufacturer's warehouse and the customer's receiving department.

The only handling of the individual cartons arises when the product is placed onto the pallet (palletizing) or when it is removed from the pallet

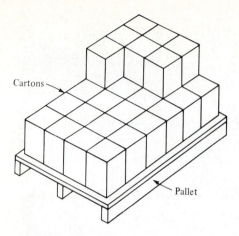

Cartons

Pallet

Figure 13-3 A typical pallet configuration.

(depalletizing). We will discuss palletizing as the generic operation. The loading of cartons onto pallets is typically heavy work, performed manually by unskilled labor. It is also repetitive work (picking cartons up at one location and putting them down at another location) except that the locations change from carton to carton. A typical pallet configuration is illustrated in Fig. 13-3, showing how each container must be placed at a different location on the pallet. The variation in carton location is in three dimensions, not simply two dimensions, since the pallets are usually stacked on top of each other in layers.

Robots can be programmed to perform this type of work. Because the motion pattern varies in the palletizing operation, a computer-controlled robot using a high-level programming language is convenient. This feature facilitates the mathematical computation of the different pallet locations required during the loading of a given pallet. The kinds of programming capabilities required in palletizing were discussed in our previous chapters on robot programming. A less sophisticated robot limited to leadthrough programming can also be used, but the programming becomes laborious because each individual carton location on the pallet must be individually taught. Another technical problem that must be addressed is that when humans perform palletizing, the cartons are often randomly located prior to loading. Unless some sensor scheme is used to identify these carton locations, they must be delivered to a known pickup point for the robot.

There are a number of variations on the palletizing operation, all of which use a similar work cycle and a robot with the same general features needed for the generic case. These other operations include:

Depalletizing operations, the reverse of palletizing, in which the robot removes cartons from a pallet and places them onto a conveyor or other location.

Inserting parts into cartons from a conveyor. This is similar to palletizing. Figure 13-4 shows an example of this operation.

Removing parts from cartons. This is the reverse of the preceding situation.

Figure 13-4 MAKER 100 Robot used to unload plastic bottles from a collector and to insert them into a cardboard carton. (Photo courtesy of United States Robots, subsidiary of Square D Company)

Stacking and unstacking operations, in which objects (usually flat objects such as metal sheets or plates) are stacked on top of each other.

In palletizing and related operations, the robot may be called on to load different pallets differently. Reasons for these differences would include the following: The pallets may vary in size; different products may be loaded onto the pallets; and there may be differences in the numbers and combinations of cartons going to different customers. To deal with these variations, methods of identifying the cartons and/or pallets and the way in which they are to be loaded or unloaded must be devised. Bar codes and other optical schemes are sometimes used to solve the identification problem. Differences in the loading or unloading of the pallets must be accomplished by means of program subroutines which can be called by the workcell controller. For depalletizing operations, the optical reader system would identify the pallet and the appropriate unloading subroutine would then be applied for that pallet. For palletizing operations, the systems problems can become more complicated because there may be an infinite number of different situations that could arise. For example, if the robot were used to palletize cartons for different customer orders, it is conceivable that each customer order would be different. A method would have to be devised for delivering the correct combination of cartons to the palletizing workstation, and integrating that process with the robot loading procedure. In the future, robots may be equipped with sufficient intelligence to figure out how to load the different cartons onto the pallet. At the time of this writing, it is a systems problem of significant proportions.

13-3 MACHINE LOADING AND UNLOADING

These applications are material-handling operations in which the robot is used to service a production machine by transferring parts to and/or from the machine. There are three cases that fit into this application category:

Machine load/unload. The robot loads a raw workpart into the process and unloads a finished part. A machining operation is an example of this case.

Machine loading. The robot must load the raw workpart or materials into the machine but the part is ejected from the machine by some other means. In a pressworking operation, the robot may be programmed to load sheet metal blanks into the press, but the finished parts are allowed to drop out of the press by gravity.

Machine unloading. The machine produces finished parts from raw materials that are loaded directly into the machine without robot assistance. The robot unloads the part from the machine. Examples in this category include die casting and plastic modeling applications.

The application is best typified by a robot-centered workcell which consists of the production machine, the robot, and some form of parts delivery system. To increase the productivity of the cell and the utilization of the robot, the cell may include more than a single production machine. This is desirable when the automatic machine cycle is relatively long, hence causing the robot to be idle a high proportion of the time. Some cells are designed so that each machine performs the same identical operation. Other cells are designed as flexible automated systems in which different parts follow a different sequence of operations at different machines in the cell. In either case, the robot is used to perform the parts handling function for the machines in the cell.

Robots have been successfully applied to accomplish the loading and/or unloading function in the following production operations:

Die casting
Plastic molding
Forging and related operations
Machining operations
Stamping press operations

We will discuss these applications in the following subsections. For each application, a brief description of the manufacturing process will be given. More detailed descriptions of the processes are to be found in other references.[1,2,7,8]

Die Casting

Die casting is a manufacturing process in which molten metal is forced into the cavity of a mold under high pressure. The mold is called a die (hence the name, die casting). The process is used to cast metal parts with sufficient accuracy so that subsequent finishing operations are usually not required. Common metals used for die-casted parts include alloys of zinc, tin, lead, aluminum, magnesium, and copper.

The die consists of two halves that are opened and closed by a die casting machine. During operation the die is closed and molten metal is injected into the cavity by a pump. To ensure that the cavity is filled, enough molten metal is forced into the die that it overflows the cavity and creates "flash" in the space between the die halves. When the metal has solidified, the die is opened and the cast part is ejected, usually by pins which push the part away from the mold cavity. When the part is removed from the machine, it is often quenched (to cool the part) in a water bath. The flash that is created during the casting process must be removed subsequently by a trimming operation which cuts around the periphery of the part. Thus, the typical die-casting production cycle consists of casting, removing the part from the machine, quenching, and trimming.

The production rates in the die-casting process range from about 100 up to 700 openings of the die per hour, depending on type of machine, the metal being cast, and the design of the part. For small parts, the die can be designed with more than one cavity, thus multiplying the number of parts made for each casting cycle. The die-casting machines have traditionally been tended by human operators. The work tends to be hot, repetitive, dirty, and generally unpleasant for humans.

Perhaps because of these conditions, die casting was one of the very first processes to which robots were applied. The first use of a robot in die casting was around 1961. Engelberger cites one instance in which a Unimate robot had been used in a die casting application for over 90,000 hours.[2]

The die-casting process represents a relatively straightforward application for industrial robots. The alterations required of the die-casting machine are minimal, and the interlocking of the robot cycle with the machine cycle can be accomplished by simple limit switches. Few problems are encountered in either the programming of the robot or the design of the gripper to remove the part from the machine when the die is opened. The process requires only that the robot unload the die-casting machine, since the metal is in the molten state before the part is formed. On some die-casting machines (called cold-chamber die-casting machines), the molten metal must be ladled from the melting container into the injection system. This part of the process is more difficult for robots to accomplish.

Plastic Molding

Plastic molding is a batch-volume or high-volume manufacturing process used to make plastic parts to final shape and size. The term plastic molding covers a number of processes, including compression molding, injection molding, thermoforming, blow molding, and extrusion. Injection molding is the most important commercially, and is the process in this group for which robots are most often used. The injection-molding operation is quite similar to die casting except for the differences in materials being processed. A thermoplastic material is introduced into the process in the form of small pellets or granules from a storage hopper. It is heated in a heating chamber to 200 to 300°C to transform it into semifluid (plastic) state and injected into the mold cavity under high pressure. The plastic travels from the heating chamber into the part cavity through a sprue-and-runner network that is designed into the mold. If too much plastic is injected into the mold, flash is created where the two halves of the mold come together. If too little material is injected into the cavity, sink holes and other defects are created in the part, rendering it unacceptable. When the plastic material has hardened sufficiently, the mold opens and the part(s) are removed from the mold.

Injection molding is accomplished using an injection-molding machine, a highly sophisticated production machine capable of maintaining close control over the important process parameters such as temperature, pressure, and the amount of material injected into the mold cavity. Traditionally, injection-molding machines have been operated on a semiautomatic cycle, with human operators used to remove the parts from the mold. Many injection-molding operations can be fully automated so long as a method can be developed for removing the parts from the mold at the end of the molding cycle. If a part sticks in the mold, considerable damage to the mold can occur when it closes at the beginning of the next cycle. Methods of removing the parts from the mold include: gravity to cause the parts to drop out of the mold, directing an air stream to force the parts out of the mold, and the use of robots to reach into the mold and remove the parts. The selection of the method depends largely on the characteristics of the molding job (part size, weight, how many parts to be molded per shot).

Industrial robots are sometimes employed to unload injection-molding machines when other less expensive automatic methods are deemed to be insufficiently reliable. One of the robot application problems in injection molding is that the production times are considerably longer than in die casting, hence causing the robot to be idle for a significant portion of the cycle. When humans tend the molding machines, this time can be utilized to perform such tasks as cutting the parts from the sprue-and-runner system, inspecting the parts, and removing the flash from the parts if that is necessary. However, some of these tasks are difficult for a robot to perform, and methods must be devised to accomplish these activities that do not rely on a human operator performing the unloading function. Cutting the parts from the sprue-and-runner system can be readily accomplished by the robot using a trimming apparatus similar to the setup used in die casting for trimming the flash from the casting. Part inspection and flash removal are not as easily accomplished by the robot.

Another issue arising when long molding cycle times are involved is whether the robot should be used to tend one machine or two. If two molding machines are tended by the robot, there is a significant likelihood that the two molding cycles will be different. This creates machine interference problems, in which one machine must wait for the robot because it is presently engaged in unloading the other machine. This waiting can lead to problems in overheating of the plastic and upsetting of the delicate balance between the various process parameters in injection molding.

Forging and Related Operations

Forging is a metalworking process in which metal is pressed or hammered into the desired shape. It is one of the oldest processes and derived from the kinds of metalworking operations performed by blacksmiths in ancient times. It is most commonly performed as a hot working process in which the metal is heated to a high temperature prior to forging. It can also be done as a cold working process. Cold forging adds considerable strength to the metal and is used for high-quality products requiring this property such as hand tools (e.g., hammers and wrenches). Even in hot forging, the metal flow induced by the hammering process adds strength to the formed part.

The term forging includes a variety of metalworking operations, some of which are candidates for automation using robots. These operations include die forging and upset forging. Other processes in the forging category include press forging and roll die forging. Generally these processes do not lend themselves to the use of robots for parts loading and unloading of the machines.

Die forging is a process accomplished on a machine tool called a drop hammer in which the raw billet is hit one or more times between the upper and lower portions of a forging die. The die often has several cavities of different shapes which allow the billet to be gradually transformed from its elementary form into the desired final shape. The drop hammer supplies mechanical energy to the operation by means of a heavy ram to which the upper portion of the forging die is attached. The ram is dropped onto the part, sometimes being accelerated by steam or air pressure. Die forging can be carried out either hot or cold.

Upset forging, also called upsetting, is a process in which the size of a portion of the workpart (usually a cylindrical part) is increased by squeezing the material into the shape of a die. The formation of the head on a bolt is usually made by means of an upsetting operation. The process is performed by an upsetting machine, also called a header. The blank (unformed raw workpart) is clamped by the two halves of a die possessing the desired shape of the product. The die is open on one end, and a plunger is forced by the upsetting machine into the blank causing it to take the shape of the die. Upsetting is often used in high-volume production of hardware items such as bolts, nails, and similar items. In these cases, the economics permit the use of fixed automation to produce the parts. In other cases, where the production of parts is in medium-sized batches, automation can sometimes be accomplished using industrial robots.

Forging, especially hot forge operations, is one of the worst industrial jobs for humans. The environment is noisy and hot, with temperatures at the workplace well above 100°F for hot forging. The air in the forge shop is generally filled with dirt, furnace fumes, and lubricant mist. The operation itself is repetitive, often requiring considerable physical strength to move and manipulate the heavy parts during the operation. The human operator experiences the blows from the drop hammer directly through the grasping tongs used to hold the part and in the form of vibration through the floor.

Unfortunately, the process is not easily adapted for robots. Some of the technical and economic problems include:

The forging hammers and upsetting machines used for low- and medium-production runs are typically older machines, designed for manual operation, and do not lend themselves to the interfacing required for robotics automation.

Short production runs are typical in many forge shops, thus making it difficult to justify the robot setup and programming effort for any single part.

The parts occasionally stick in the dies. This can be readily detected by a human operator but poses problems for the robot. To minimize the frequency of sticking, the human operator periodically sprays lubricant into the die openings. The robot would have to be equipped and programmed to do this also.

The design of a gripper for forging is a significant engineering problem for several reasons. First, the parts are hot, perhaps 2000°F, and the gripper must be protected against these temperatures. Second, the gripper must be designed to withstand the shock from the hammer blows because the parts must typically be held in position by the robot during the process. Third, the gripper must be designed to accommodate substantial changes in the shape of the parts during successive hits in the forging cycle. Some aspects of the forging process require operator judgment. The part must be heated to a sufficient temperature in order to successfully perform the hot working operation and the human operator often makes this judgment. A cold part would probably ruin the die. The raw workparts are generally placed in the heating furnace at random, and selecting the parts that are ready to be formed is an operator decision. Another problem is that different parts can require a different number of hammer strokes to form the final shape, a judgment that is made by the operator.

Each of these problem areas must be addressed in order for the robot forging application to be a success. Many of the problems are solved by making considerable use of interlocks and sensors. These devices permit the determination of such process variables as part temperature before processing, the presence of the workpart in the gripper, whether the part is stuck in the die, whether the robot arm is clear of the ram before operation of the drop hammer, and other factors.

Machining Operations

Machining is a metalworking process in which the shape of the part is changed by removing excess material with a cutting tool. It is considered to be a secondary process in which the final form and dimensions are given to the part after a process such as casting or forging has provided the basic shape of the part. There are a number of different categories of machining operations. The principal types include turning, drilling, milling, shaping, planing, and grinding. Commercially, machining is an important metalworking process and is widely used in many different products, ranging from those that are made in low quantities to those produced in very high numbers. In mid-volume and high-volume production, the operation is very repetitive with the same machining sequence being repeated on part after part.

The machine tools that perform machining operations have achieved a relatively high level of automation after many years of development. In particular, the use of computer control (e.g., computer numerical control and direct numerical control) permits this type of equipment to be interfaced with relative ease to similarly controlled equipment such as robots.

Robots have been successfully utilized to perform the loading and unloading functions in machining operations. The robot is typically used to load a raw workpart (a casting, forging, or other basic form) into the machine tool and to unload the finished part at the completion of the machining cycle. Figure 13-5 illustrates a machine tool loading and unloading operation in which the finished parts are palletized (lower left corner of the figure) after the machining cycle.

The following robot features generally contribute to the success of the machine tool load/unload application[2]:

Dual gripper. The use of a dual gripper permits the robot to handle the raw workpart and the finished part at the same time. This permits the production cycle time to be reduced.

Up to six joint motions. A large number of degrees of freedom of the arm and wrist are required to manipulate and position the part in the machine tool.

Good repeatability. A relatively high level of precision is required to properly position the part into the chuck or other workholding fixture in the machine tool.

Palletizing and depalletizing capability. In midvolume production, the raw parts are sometimes most conveniently presented to the workcell and delivered away from the workcell on pallets. The robot's controller and programming capabilities must be sufficient to accommodate this requirement.

Programming features. There are several desirable programming features that facilitate the use of robots in machining applications. In machine cells used for batch production of different parts, there is the need to perform some sort of changeover of the setup between batches. Part of this changeover procedure involves replacing the robot program for the previous batch

Figure 13-5 Unimate 2000 robot used to load and unload parts in a machine tool operation. (Photo courtesy of Unimation Inc)

with the program for the next batch. The robot should be able to accept disk, tape, or other storage medium for ease in changing programs. Another programming feature needed for machining is the capability to handle irregular elements, such as tool changes or pallet changes, in the program.

Example 13-2 This example illustrates some of the features of an automated machining cell consisting of a T3 robot, two turning centers, and an automated gauging station. Figure 13-6 shows the robot tending one of the turning centers. The raw workparts are castings which enter the workcell on a pallet delivered by means of a conveyor. The raw parts can be seen in the figure in the lower right foreground. The robot picks up a raw workpart and exchanges the raw part in the first turning center for the finished part. The finished part is then loaded into the gauging station, checked for the proper dimensions, and if determined to be within tolerance, is ready for loading into the second machine. The robot then exchanges the gauged part for the finished part on the second turning center (shown in the figure). That part is gauged in the automatic gauging station, and loaded onto the outgoing pallet if within tolerance.

Figure 13-6 Cincinnati Milacron T3 robot performing a machine tool load/unload operation. See Example 13-2. (Photo courtesy of Cincinnati Milacron)

The loading and unloading operations are accomplished from the rear of each turning center rather than the front of the machines. This means that the front of each machine is clear for operator access, tool replacement, and observation.

Stamping Press Operations

Stamping press operations are used to cut and form sheet metal parts. The process is performed by means of a die set held in a machine tool called a press (or stamping press). The sheet metal stock used as the raw material in the process comes in several forms, including coils, sheets, and individual flat blanks. When coil stock is fed into the press, the process can be made to operate in a highly automated manner at very high cycle rates. When the starting material consists of large flat sheets or individual blanks, automation becomes more difficult. These operations have traditionally been performed by human workers, who must expose themselves to considerable jeopardy by placing their hands inside the press in order to load the blanks. During the last decade, the Occupational Safety and Health Act (OSHA) has required certain alterations in the press in order to make its operation safer. The economics of the OSHA requirements have persuaded many manufacturers to consider the use of robots for press loading as alternatives to human operators. Noise is another factor which makes pressworking an unfriendly environment for humans.

Figure 13-7 GMF robot performing machine loading/unloading task at a stamping press. (Photo courtesy of GMF Robotics)

Robots are being used for handling parts in pressworking operations, largely as a result of the safety issue. The typical task performed by the robot is to load the flat blanks into the press for the stamping operation. There are variations in the way this can be done. In forming operations, the robot can be used to hold the blank during the cycle so that the formed part is readily removed from the press. In the case of many cutting operations, the robot loads the blank into the press, and the parts fall through the die during the press cycle. Another robot application in pressworking involves the transfer of parts from one press to another to form an integrated pressworking cell. Figure 13-7 shows a *TLL* configuration robot working in conjunction with a pressworking operation.

One of the limiting factors in using industrial robots for press loading is the cycle time of the press. Cycle times of less than a second are not uncommon in pressworking. These cycle rates are too fast for currently available commercial robots. There is generally a direct relationship between the physical size of the part and the press cycle time required to make the part. Bigger presses are needed to stamp bigger parts and bigger presses are inherently slower. Accordingly, robots are typically used in pressworking for larger parts.

REFERENCES

1. E. P. DeGarmo, *Materials and Processes in Manufacture*, 6th ed., Macmillan, New York, 1984, chaps. 9, 11, 14, 15, and 17.
2. J. F. Engelberger, *Robotics in Practice*, AMACOM (Division of American Management Associations), 1980, chaps. 10, 14, 15, 17, 19, and 22.
3. A. L. Giacobbe, "Diskette Labeling and Packaging System Features Sophisticated Robot Handling," *Robotics Today*, April 1984, pp. 73–75.
4. M. P. Groover and E. W. Zimmers, Jr., *CAD/CAM: Computer-Aided Design and Manufacturing*, Prentice-Hall, Englewood Cliffs, NJ, 1984, chap. 11.
5. R. Hinson, "Study Analyzes Utilizing Robots in Palletizing," *Industrial Engineering*, February 1984, pp. 21–24.
6. R. Hinson, "Case Study: Robots in Machine Load/Unload Operations," *Industrial Engineering*, March 1984, pp. 103–106.
7. R. A. Lindberg, *Processes and Materials of Manufacture*, 3rd ed., Allyn and Bacon, Boston, MA, 1983, chaps. 3, 4, and 5.
8. B. W. Niebel and A. B. Draper, *Product Design and Process Engineering*, McGraw-Hill, New York, 1974, chaps. 11, 12, 13, and 14.
9. R. N. Stauffer, "Palletizing/Depalletizing: Robots Make It Easy," *Robotics Today*, February 1984, pp. 43–46.
10. V. P. Valeri, "The Case for Gantry Robots in Machine Loading," *Robotics Today*, August 1984, pp. 23–26.
11. B. D. Wakefield, "Robot-Loaded Stamping Presses Keep Pace with Production," *Industrial Robots, Volume 2—Applications*, Soc. Mfg. Engrs., Dearborn, MI, 1979, pp. 33–36.
12. B. I. Zisk and W. E. Palmer, "FMS Robot-to-Material Transport Interfaces," *Robotics Today*, February 1984, pp. 38–41.

FOURTEEN

PROCESSING OPERATIONS

In addition to parts-handling applications, there is a large class of applications in which the robot actually performs work on the part. This work almost always requires that the robot's end effector is a tool rather than a gripper. Accordingly, the use of a tool to perform work is a distinguishing characteristic of this group of applications. The type of tool depends on the processing operation that is performed. We divide the processing operations that are performed by a robot into the following categories for purposes of organizing this chapter:

1. Spot welding
2. Continuous arc welding
3. Spray coating
4. Other processing operations

The two welding categories represent important application areas for robots. Spot welding is probably the single most common application for industrial robots in the United States today because they are widely used in automobile body assembly lines to weld the frames and panels together. Arc welding is an application that is expected to grow in use as we develop the technology required for using robots in this process. Spray coating usually means spray painting, an operation that is accompanied by an unhealthy work environment for humans, and therefore represents a good opportunity for robots. We use the term "spray coating" to indicate that there are additional applications beyond painting for a robot to spray a substance onto a surface. The final category in the listing above is a miscellaneous applications area. It includes certain machining operations, polishing, deburring, and other processing operations. These operations are usually, but not always, characterized by the use of a rotating spindle by the robot. We discuss the four categories of operations and how robots are used to accomplish these operations in the following sections.

14-1 SPOT WELDING

As the term suggests, spot welding is a process in which two sheet metal parts are fused together at localized points by passing a large electric current through the parts where the weld is to be made. The fusion is accomplished at relatively low voltage levels by using two copper (or copper alloy) electrodes to squeeze the parts together at the contact points and apply the current to the weld area. The electric current results in sufficient heat in the contact area to fuse the two metal parts, hence producing the weld.

The two electrodes have the general shape of a pincer. With the two halves of the pincer open, the electrodes are positioned at the point where the parts are to be fused. Prior clamping or fixturing of the parts is usually required to hold the pieces together for the process. The two electrodes are squeezed together against the mating parts, and the current is applied to cause heating and welding of the contacting surfaces. Then the electrodes are opened and allowed to cool for the next weld. A water circulation system is often used to accelerate the cooling of the electrodes. The actual welding portion of the sequence typically requires less than a second. Therefore, the rates of production in spot welding are largely dependent on the time required for positioning of the welding electrodes and the parts relative to each other. Another factor that affects production rate is the wear of the electrodes. Because of the heat involved in the process, the tips of the electrodes gradually lose their shape and build up a carbon deposit which affects their electric resistance. Both of these effects reduce the quality of the welds made. Therefore, the electrode tips must periodically be dressed to remove the deposits and restore the desired shape.

Spot welding has traditionally been performed manually by either of two methods. The first method uses a spot-welding machine in which the parts are inserted between the pair of electrodes that are maintained in a fixed position. This method is normally used for relatively small parts that can be easily handled.

The second method involves manipulating a portable spot-welding gun into position relative to the parts. This would be used for larger work such as automobile bodies. The word "portable" is perhaps an exaggeration. The welding gun consists of the pair of electrodes and a frame to open and close the electrodes. In addition, large electrical cables are used to deliver the current to the electrodes from a control panel located near the workstation. The welding gun with cables attached is quite heavy and can easily exceed 100 lb in weight. To assist the operator in manipulating the gun, the apparatus is suspended from an overhead hoist system. Even with this assistance, the spot-welding gun represents a heavy mass and is difficult to manipulate by a human worker at the high rates of production desired on a car body assembly line. There are often problems with the consistency of the welded products made on such a manual line as a consequence of this difficulty.

Robots in Spot Welding

As a result of these difficulties, robots have been employed with great success on this type of production line to perform some or all of the spot-welding operations. A welding gun is attached as the end effector to each robot's wrist, and the robot is programmed to perform a sequence of welds on the product as it arrives at the workstation. Some robot spot-welding lines operate with several dozen robots all programmed to perform different welding cycles on the product. Today, the automobile manufacturers make extensive use of robots for spot welding. In 1980 it was reported[3] that there were 1200 robots used in this application. Figure 14-1 shows an overview of an automobile body assembly line in which robots are used to perform the spot-welding operations. Figure 14-2 shows a close-up of a spot-welding gun mounted on a Cincinnati Milacron T3 robot performing its task inside the car body.

The robots used in spot welding must possess certain capabilities and features to perform the process. First, the robot must be relatively large. It

Figure 14-1 Robots performing spot welding operations on an automobile assembly line. (Photo courtesy of Unimation Inc., a Westinghouse Company)

NTRODUCTION

his chapter we seek to explain some of the terminology used in experimental thods and to show the generalized arrangement of an experimental system. We ll also discuss briefly the standards which are available and the importance of libration in any experimental measurement. A major portion of the discussion n experimental errors is deferred until Chap. 3, and only the definition of certain terms is given here.

2-2 DEFINITION OF TERMS

We are frequently concerned with the *readability* of an instrument. This term indicates the closeness with which the scale of the instrument may be read; an instrument with a 12-in scale would have a higher readability than an instrument with a 6-in scale and the same range. The *least count* is the smallest difference between two indications that can be detected on the instrument scale. Both readability and least count are dependent on scale length, spacing of graduations, size of pointer (or pen if a recorder is used), and parallax effects.

The *sensitivity* of an instrument is the ratio of the linear movement of the pointer on the instrument to the change in the measured variable causing this motion. For example, a 1-mV recorder might have a 25-cm scale length. Its sensitivity would be 25 cm/mV, assuming that the measurement was linear all across the scale. For a digital instrument readout the term "sensitivity" does not have the same meaning because different scale factors can be applied with the

push of a button. However, the manufacturer will usually specify the sensitivity for a certain scale setting, e.g., 100 nA on a 200-μA scale range for current measurement.

An instrument is said to exhibit *hysteresis* when there is a difference in readings depending on whether the value of the measured quantity is approached from above or below. Hysteresis may be the result of mechanical friction, magnetic effects, elastic deformation, or thermal effects.

The *accuracy* of an instrument indicates the deviation of the reading from a known input. Accuracy is usually expressed as a percentage of full-scale reading, so that a 100-kPa pressure gage having an accuracy of 1 percent would be accurate within ± 1 kPa over the entire range of the gage.

The *precision* of an instrument indicates its ability to reproduce a certain reading with a given accuracy. As an example of the distinction between precision and accuracy, consider the measurement of a known voltage of 100 volts (V) with a certain meter. Five readings are taken, and the indicated values are 104, 103, 105, 103, and 105 V. From these values it is seen that the instrument could not be depended on for an accuracy of better than 5 percent (5 V), while a precision of ± 1 percent is indicated, since the maximum deviation from the mean reading of 104 V is only 1 V. It may be noted that the instrument could be calibrated so that it could be used to dependably measure voltages within ± 1 V. This simple example illustrates an important point. Accuracy can be improved up to but not beyond the precision of the instrument by calibration. The precision of an instrument is usually subject to many complicated factors and requires special techniques of analysis, which will be discussed in Chap. 3.

We should alert the reader at this time to some data analysis terms which will appear in Chap. 3. Accuracy has already been mentioned as relating the deviation of an instrument reading from a *known* value. The deviation is called the *error*. In many experimental situations we may not have a known value with which to compare instrument readings, and yet we may feel fairly confident that the instrument is within a certain plus or minus range of the true value. In such cases, we say that the plus or minus range expresses the *uncertainty* of the instrument readings. Many experimentalists are not very careful in using the words "error" and "uncertainty." As we shall see in Chap. 3, uncertainty is the term that should be most often applied to instruments.

2-3 CALIBRATION

The calibration of all instruments is important, for it affords the opportunity to check the instrument against a known standard and subsequently to reduce errors in accuracy. Calibration procedures involve a comparison of the particular instrument with either (1) a primary standard, (2) a secondary standard with a higher accuracy than the instrument to be calibrated, or (3) a known input source. For example, a flowmeter might be calibrated by (1) comparing it with a standard flow-measurement facility of the National Bureau of Standards,

2-1 I

In t
me
sh
ca

TWO

BASIC CONCEPTS

2-1 INTRODUCTION

In this chapter we seek to explain some of the terminology used in experimental methods and to show the generalized arrangement of an experimental system. We shall also discuss briefly the standards which are available and the importance of calibration in any experimental measurement. A major portion of the discussion on experimental errors is deferred until Chap. 3, and only the definition of certain terms is given here.

2-2 DEFINITION OF TERMS

We are frequently concerned with the *readability* of an instrument. This term indicates the closeness with which the scale of the instrument may be read; an instrument with a 12-in scale would have a higher readability than an instrument with a 6-in scale and the same range. The *least count* is the smallest difference between two indications that can be detected on the instrument scale. Both readability and least count are dependent on scale length, spacing of graduations, size of pointer (or pen if a recorder is used), and parallax effects.

The *sensitivity* of an instrument is the ratio of the linear movement of the pointer on the instrument to the change in the measured variable causing this motion. For example, a 1-mV recorder might have a 25-cm scale length. Its sensitivity would be 25 cm/mV, assuming that the measurement was linear all across the scale. For a digital instrument readout the term "sensitivity" does not have the same meaning because different scale factors can be applied with the

push of a button. However, the manufacturer will usually specify the sensitivity for a certain scale setting, e.g., 100 nA on a 200-μA scale range for current measurement.

An instrument is said to exhibit *hysteresis* when there is a difference in readings depending on whether the value of the measured quantity is approached from above or below. Hysteresis may be the result of mechanical friction, magnetic effects, elastic deformation, or thermal effects.

The *accuracy* of an instrument indicates the deviation of the reading from a known input. Accuracy is usually expressed as a percentage of full-scale reading, so that a 100-kPa pressure gage having an accuracy of 1 percent would be accurate within ± 1 kPa over the entire range of the gage.

The *precision* of an instrument indicates its ability to reproduce a certain reading with a given accuracy. As an example of the distinction between precision and accuracy, consider the measurement of a known voltage of 100 volts (V) with a certain meter. Five readings are taken, and the indicated values are 104, 103, 105, 103, and 105 V. From these values it is seen that the instrument could not be depended on for an accuracy of better than 5 percent (5 V), while a precision of ± 1 percent is indicated, since the maximum deviation from the mean reading of 104 V is only 1 V. It may be noted that the instrument could be calibrated so that it could be used to dependably measure voltages within ± 1 V. This simple example illustrates an important point. Accuracy can be improved up to but not beyond the precision of the instrument by calibration. The precision of an instrument is usually subject to many complicated factors and requires special techniques of analysis, which will be discussed in Chap. 3.

We should alert the reader at this time to some data analysis terms which will appear in Chap. 3. Accuracy has already been mentioned as relating the deviation of an instrument reading from a *known* value. The deviation is called the *error*. In many experimental situations we may not have a known value with which to compare instrument readings, and yet we may feel fairly confident that the instrument is within a certain plus or minus range of the true value. In such cases, we say that the plus or minus range expresses the *uncertainty* of the instrument readings. Many experimentalists are not very careful in using the words "error" and "uncertainty." As we shall see in Chap. 3, uncertainty is the term that should be most often applied to instruments.

2-3 CALIBRATION

The calibration of all instruments is important, for it affords the opportunity to check the instrument against a known standard and subsequently to reduce errors in accuracy. Calibration procedures involve a comparison of the particular instrument with either (1) a primary standard, (2) a secondary standard with a higher accuracy than the instrument to be calibrated, or (3) a known input source. For example, a flowmeter might be calibrated by (1) comparing it with a standard flow-measurement facility of the National Bureau of Standards,

Figure 14-2 Close-up of a Cincinnati Milacron T3 robot performing a spot welding operation inside the car body. (Photo courtesy of Cincinnati Milacron)

must have sufficient payload capacity to readily manipulate the welding gun for the application. The work volume must be adequate for the size of the product. The robot must be able to position and orient the welding gun in places on the product that might be difficult to access. This might result in the need for an increased number of degrees of freedom. The controller memory must have enough capacity to accomplish the many positioning steps required for the spot-welding cycle. In some applications, the welding line is designed to produce several different models of the product. Accordingly, the robot must be able to switch from one programmed welding sequence to another as the models change. For welding lines in which there are multiple robots, programmable controllers are used to keep track of the different models at the various welding stations and to download the programs to the robots at individual workstations as needed.

The benefits that result from automation of the spot-welding process by means of robots are improved product quality, operator safety, and better control over the production operation. Improved quality is in the form of more consistent welds and better repeatability in the location of the welds. Even robots with relatively unimpressive repeatability specifications are able to locate the spot welds more accurately than human operators. Improved safety results simply because the human is removed from a work environment where there are hazards from electrical shocks and burns. The use of robots to automate the spot-welding process should also result in improvements in areas such as production scheduling and in-process inventory control. The maintenance of the robots and the welding equipment becomes an important factor in the successful operation of an automated spot-welding production line.

14-2 CONTINUOUS ARC WELDING

Arc welding is a continuous welding process as opposed to spot welding which might be called a discontinuous process. Continuous arc welding is used to make long welded joints in which an airtight seal is often required between the two pieces of metal being joined. The process uses an electrode in the form of a rod or wire of metal to supply the high electric current needed for establishing the arc. Currents are typically 100 to 300 A at voltages of 10 to 30 V. The arc between the welding rod and the metal parts to be joined produces temperatures that are sufficiently high to form a pool of molten metal to fuse the two pieces together. The electrode can also be used to contribute to the molten pool, depending on the type of welding process.

Arc welding is usually performed by a skilled human worker who is often assisted by a person called a fitter. The purpose of the fitter is to organize the work and to fixture the parts for the welder. The working conditions of the welder are typically unpleasant and hazardous. The arc from the welding process emits ultraviolet radiation which is injurious to human vision. As a result, welders are required to wear eye protection in the form of a welding helmet with a dark window. This dark window filters out the dangerous radiation, but it is so dark that the welder is virtually blind while wearing the helmet except when the arc is struck. Other aspects of the process are also hazardous. The high temperatures created in arc welding and the resulting molten metals are inherently dangerous. The high electrical current used to create the arc is also unsafe. Sparks and smoke are generated during the process and these are a potential threat to the operator.

There are a variety of arc-welding processes, and the reader is referred to the available manufacturing process texts for more details than we can include here. For robot applications, two types of arc welding seem the most practical: gas metal arc welding (GMAW) and gas tungsten arc welding (GTAW). GMA welding (also called MIG welding for metal inert gas welding) involves the use of a welding wire made of the same or similar metals as the parts being joined. The welding wire serves as the electrode in the arc-welding process. The wire is continuously fed from a coil and contributes to the molten metal pool used in the fusion process. GMA welding is typically used for welding steel. In GTA welding (also called TIG welding for tungsten inert gas welding), a tungsten rod is used as the electrode to establish the arc. The melting point of tungsten is relatively high, and therefore the electrode does not melt during the fusion process. If filler metal must be added to the weld, it must be added separately from the electrode. The GTA process is typically used for welding aluminum, copper, and stainless steel. In both GMA and GTA welding, inert gases such as helium or argon are used to surround the immediate vicinity of the welding arc to protect the fused surfaces from oxidation.

Problems for Robots in Arc Welding

Because of the hazards for human workers in continuous arc welding, it is logical to consider industrial robots for the process. However, there are significant technical and economic problems encountered in applying robots to arc welding. Continuous arc welding is commonly used in the fabrication industries where products consisting of many components are made in low quantities. It is difficult to justify automation of any form in these circumstances. A related problem is that arc welding is often performed in confined areas that are difficult to access, such as the insides of tanks, pressure vessels, and ship hulls. Humans can position themselves into these areas more readily than robots.

One of the most difficult technical problems for welding robots is the presence of variations in the components that are to be welded. These variations are manifested in two forms. One is the variation in the dimensions of the parts in a batch production job. This type of dimensional variation means that the arc-welding path to be followed will change slightly from part to part. The second variation is in the edges and surfaces to be welded together. Instead of being straight and regular, the edges are typically irregular. This causes variations in the gap between the parts and other problems in the way the pieces mate together prior to the welding process. Human welders are able to compensate for both of these variations by changing certain parameters in the welding process (e.g., adjusting the welding path, changing the speed at which the joint is traversed, depositing more filler metal where the gap is large, etc.). Industrial robots do not possess the sensing capabilities, skills, and judgment of human welders to make these compensations.

There are two approaches to compensate for these variations and irregularities in robot welding applications:

1. Correct the upstream production operations so that the variations are reduced to the point where they do not create a problem in the robot welding process.
2. Provide the robot with sensors to monitor the variations in the welding process and the control logic to compensate for part variations and weld gap irregularities.

Correction of the production operations that deliver parts to the arc-welding process is an attractive alternative because it tends to contribute to the overall quality of the product, and because it simplifies the welding robot project. The potential disadvantage of this approach is that it is likely to increase the cost of manufacturing the individual components because their dimensions must be held to closer tolerances. The second approach represents an area of intensive research and development activity in robotics. We will explore this area in a later subsection.

The Typical Robot Arc-Welding Application

Because of the technical and economic problems discussed above, the typical arc-welding application for a robot is one in which the quantities of production are medium or high. In these cases, a robot cell can be justified which consists of a welding robot and a part holder or part manipulator. The part holder is used to fixture the components and position them for welding. A part manipulator provides an additional capability in the form of 1 or 2 degrees of freedom to position and orient the components relative to the robot. The robot is equipped with a welding rod or wire feed system, and the required power source to provide the electric current for the operation. The workcell controller is used to coordinate the robot motion, the welding current, the wire feed, the part manipulator, and any other activities in the cell.

One possible organization of a robot welding cell is illustrated in Fig. 14-3.

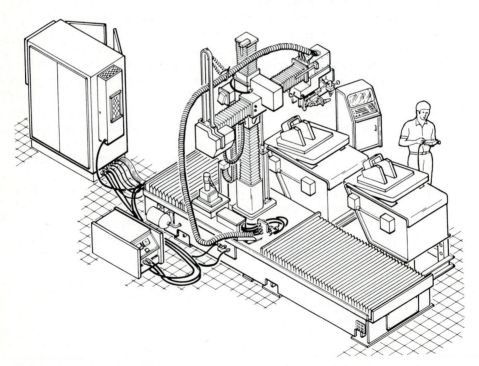

Figure 14-3 Robot arc welding cell. (Reprinted by permission of Advanced Robotics Corp.)

A human operator is typically used to load and unload the part fixtures. Two part fixtures are often used in the cell as shown in the figure. In this way, the robot can perform its welding cycle while the operator is unloading the previously welded assembly and loading the components into the holder for the next cycle. This arrangement permits higher utilization of the robot and greater productivity of the cell.

Features of the Welding Robot

An industrial robot that performs arc welding must possess certain features and capabilities. Some of the technical considerations in arc-welding applications are discussed in the following:

1. *Work volume and degrees of freedom.* The robot's work volume must be large enough for the sizes of the parts to be welded. A sufficient allowance must be made for manipulation of the welding torch. Also, if two part holders are included in the workstation, the robot must have adequate reach to perform the motion cycle at both holders. Five or six degrees of freedom are generally required for arc-welding robots. The number is influenced by the characteristics of the welding job and the motion capabilities of the parts manipulator. If the parts manipulator has 2 degrees of freedom, this tends to reduce the requirement on the number of degrees of freedom possessed by the robot.
2. *Motion control system.* Continuous-path control is required for arc welding. The robot must be capable of a smooth continuous motion in order to maintain uniformity of the welding seam. In addition, the welding cycle requires a dwell at the beginning of the movement in order to establish the welding puddle, and a dwell at the end of the movement to terminate the weld.
3. *Precision of motion.* The accuracy and repeatability of the robot determines to a large extent the quality of the welding job. The precision requirements of welding jobs vary according to size and industry practice, and these requirements should be defined by each individual user before selecting the most appropriate robot.
4. *Interface with other systems.* The robot must be provided with sufficient input/output and control capabilities to work with the other equipment in the cell. These other pieces of equipment are the welding unit and the parts positioners. The cell controller must coordinate the speed and path of the robot with the operation of the parts manipulator and the welding parameters such as wire feed rate and power level.
5. *Programming.* Programming the robot for continuous arc welding must be considered carefully. To facilitate the input of the program for welding paths with irregular shapes, it is convenient to use the walkthrough method in which the robot wrist is physically moved through its motion path. For straight welding paths, the robot should possess the capability for linear interpolation between two points in space. This permits the programmer to

define the beginning and end points of the path and the robot is capable of computing the straight line trajectory between the points.

Some welding applications require the robot to follow a weave pattern (back and forth motion across the welding seam) during the operation. Other applications require a series of passes along the same path, but each pass must be slightly offset from the previous one to allow for the welding bead that was laid down in the previous pass. Both of these requirements are generally associated with large welding jobs where the amount of material to be added is greater than what can be applied normally during a single welding pass. Robots intended specifically for arc welding are often provided with features to facilitate the programming of weave patterns and multiple welding passes. These programming capabilities permit the programmer to define the parameters needed to accomplish the special pass. For the weave pattern, the parameters are the amplitude of the weave, the number of weaves per inch of travel, and the dwell on either side of the weave. For multipass operation, the magnitude and direction of the offset between passes must be defined. In the appendix to Chap. Nine, we discussed the arc-welding programming features of the RAIL language, and the reader is invited to refer back to that section of the appendix.

The various programming features described above assume that the components to be welded possess regular edges along the intended weld path and that the components are uniform from part to part. Accordingly, the robot would be able to repeat the programmed motion path for each set of components to produce welded assemblies with great consistency. Unfortunately, some of the biggest technical difficulties in robot arc welding arise from the fact that these two assumptions are often not valid. To overcome the difficulties, robots are being equipped with smart sensor systems to track the welding path during the process and compensate for the irregularities in the path.

Sensors in Robotic Arc Welding

At present, a wide variety of arc-welding sensors are either commercially available or under development in various research and development laboratories. We will concentrate on the systems that are commercially available, leaving for the interested reader the task of exploring some of the research that is being done in this area. References 6 and 16 provide a more thorough treatment of these sensor systems than we can devote to the subject here.

The robotic arc-welding sensor systems considered here are all designed to track the welding seam and provide information to the robot controller to help guide the welding path. The approaches used for this purpose divide into two basic categories: contact and noncontact sensors.

Contact arc-welding sensors Contact arc-welding sensors make use of a mechanical tactile probe (some of the probe systems would better be described as electromechanical) to touch the sides of the groove ahead of the welding

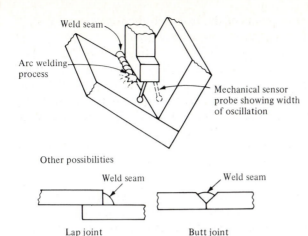

Figure 14-4 Diagram depicting the operation of the contact arc welding sensor and several of the types of joints in which it can be used.

torch and to feed back position data so that course corrections can be made by the robot controller. Some systems use a separate control unit designed to interpret the probe sensor measurements and transmit the data to the robot controller. To accomplish the position measurements, the probe must be oscillated from one side of the groove to the other by the sensor system. The nature of the operation of these sensor systems limits their application to certain weld geometries in which the side-to-side motion of the probe permits it to make contact with the edges or surfaces that are to be welded. Some of the weld geometries in this category include butt welds that have grooved joints, lap joints, and fillet welds. Figure 14-4 shows a diagram which depicts the operation of the contact arc welding sensor and several of the types of joints in which it can be used.

Another limitation of the contact arc weld sensor is that the probe must be maintained in the proper position ahead of the welding torch, and this makes these systems most effective on welds that are long and straight. These kinds of arc-welding applications do not make full use of most robots' capabilities for more complex path control.

Noncontact arc-welding sensors The second basic type of sensor system used to track the welding seam uses no tactile measurements. A variety of sensor schemes have been explored in this category, but our discussions will concentrate on arc-sensing systems and vision-based systems since these are the approaches used more in today's commercial systems.

Arc-sensing systems (sometimes called "through-the-arc" systems) rely on measurements taken of the arc itself, in the form of either electric current (in constant-voltage welding) or voltage (in constant-current welding). In order to interpret these signals, they must be varied during the arc-welding process. This is accomplished by causing the arc to weave back and forth across the joint as it moves down the path. The side-to-side motion along the joint can be achieved by programming the robot to perform the weave pattern, or by means of a servo system that attaches to the robot wrist and determines the position of the torch, or by other mechanisms. The weaving motion permits the electrical signals to be interpreted in terms of vertical and cross-seam position of the torch. The controller performs an adaptive positioning of the torch as it moves forward along the joint centerline so that the proper path trajectory can be maintained. As irregular edges are encountered along the weld path, the control system compensates by regulating either the arc length (for constant-current systems) or the distance between the torch tip and the work surface (for constant-voltage systems). Operation of the typical through-the-arc seam tracking system is illustrated in Fig. 14-5.

Limitations of the arc-sensing sensors are similar to the limitations in the operation of the tactile systems described above. The welding joint must possess a grooved geometry which permits the weaving motion to be effective. In spite of this limitation, through-the-arc systems constitute the largest number of noncontacting seam trackers commercially available at the time of this writing. Companies offering these systems include CRC Welding Systems (Nashville, Tennessee), Advanced Robotics Corp. (Columbus, Ohio), and Hobart Brothers Co. (Troy, Ohio).

Vision-based systems represent a promising technology for tracking the seam in arc-welding operations. These systems utilize a vision camera mounted on the robot near the welding torch to view the weld path. In some cases the camera is an integral component of the welding head. Highly structured light is usually required for the camera sensors to function reliably.

There are two approaches used with vision sensors for arc welding: two-pass systems and single-pass systems. In both types, the robot must be programmed for the welding path before the operation begins.

In the two-pass systems, the vision camera takes a preliminary pass over the seam before the welding operation begins. As indicated above, the robot must be programmed for the particular seam path before either pass is taken. Then the two passes are taken automatically by the robot. In the first pass, light is projected onto the seam and the camera scans the joint at high speed (speeds up to 1 m/s are claimed), checking for deviations from the anticipated

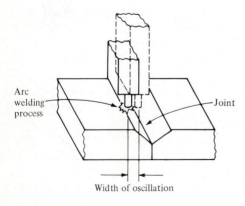

Arc welding process

Joint

Width of oscillation

Figure 14-5 Operation of a typical "through-the-arc" seam tracking sensor system, depicting oscillation from side to side in the welding groove.

seam path. These deviations are analyzed by the controller and remembered for the second pass. During the second pass, in which the welding process is performed, the controller makes adjustments in the seam path to correct for the deviations detected in the first pass. The first pass requires only about 10 percent of the time for the welding pass, and the advantage gained by using two passes is that the vision system can see a clean view of the welding path on the preliminary scanning pass, absent of the smoke and intense brightness encountered during actual welding. Examples of commercially available systems in this two-pass category include Unimation's Univision II system and the Robo Welder Series 1200 by Robotic Vision Systems Inc. (RVSI). Figure 14-6 shows the camera and the welding head of the RVSI seam tracking vision system, both components mounted as the end effector to the robot wrist.

In the single-pass system, the vision camera is aimed at the welding seam just ahead of the torch. Deviations from the programmed seam location are detected and corrections are made in the weld path. The obvious advantage of the single-pass systems, compared to the two-pass systems, is that time is saved by eliminating the need for a second pass along the weld path. Another advantage is that the single-pass systems are able to compensate for thermal distortions in the weld path caused by the welding process.

Examples of commercial vision systems in the single-pass category are the Robovision II from Automatix Inc., and WeldVision from General Electric. In the Automatix system, the camera is focused about 4 cm in front of the weld. The observed image is analyzed to extract the location of the center of the seam, the seam width, and the distance of the seam from the camera. In the General Electric system, the vision sensor is incorporated into the design of the welding torch. The image observed by the camera includes the weld puddle and the seam ahead of it. By analyzing both the weld puddle and the seam, the controller is able to make adjustments in the process to automatically track the seam.

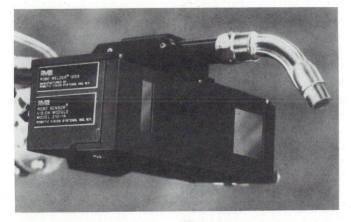

Figure 14-6 Robo Welder Series 1200 seam tracking vision system. (Photo courtesy of Robotic Vision Systems Inc.)

Advantages and Benefits of Robot Arc Welding

A robot arc-welding cell for batch production has the potential for achieving a number of advantages over a similar manual operation. These advantages include the following:

1. Higher productivity
2. Improved safety and quality-of-work life
3. Greater quality of product
4. Process rationalization

The productivity of a manual arc-welding operation is characteristically quite low. The productivity is often measured by the "arc-on" time. This gives the proportion of time during the shift that the welding process is occurring, and therefore production is taking place. Typical values of arc-on time range between 10 and 30 percent. The lower value corresponds to one-of-a-kind welding jobs, and the higher value corresponds to batch type production. One of the reasons why the arc-on time is low in manual welding is the fatigue factor. The hand–eye coordination required and the generally uncomfortable working environment tend to be tiring to the human welder and frequent rest periods must be taken. With robot welding cells for batch production, a 50 to 70 percent arc-on time can be realized. There are several factors that contribute to the increased arc-on time when robots are used in batch production. Certainly one factor is the elimination of the fatigue factor. Robots do not experience fatigue in the sense that human workers do. A robot can continue to operate during the entire shift without the need for periodic rest breaks. Another contributing factor is the presence of two parts positioners in the cell. The robot can be performing the welding operation at one positioner while the human operator is unloading the previous assembly and loading new components at the other positioner.

Improved safety and quality-of-work environment result from removing the human operator from an uncomfortable, fatiguing, and potentially dangerous work situation. As described above, the welding environment contains a number of serious hazards for human beings.

Greater product quality in robot arc welding results from the capability of the robot to perform the welding cycle with greater accuracy and repeatability than its human counterpart. This translates into a more consistent welding seam, one that is free of the start-and-stop buildup of filler metal in the seam that is characteristic of many welds accomplished by human welders.

The term process rationalization refers to the systematic organization of the work and the material flow in the factory. The design and installation of a robot welding cell forces the user company to consider such issues as the delivery of materials to the cell, the methods required to perform the welding process, the design of the fixtures, and the problems of production and inventory control related to the operation of the cell. Typically, these issues are not adequately addressed when the company relies on human welding stations.

14-3 SPRAY COATING

Most products manufactured from metallic materials require some form of painted finish before delivery to the customer. The technology for applying these finishes varies in complexity from simple manual methods to highly sophisticated automatic techniques. We divide the common industrial coating methods into two categories:

1. Immersion and flow-coating methods
2. Spray-coating methods

Immersion and flow-coating methods are generally considered to be low-technology methods of applying paint to the product. Immersion involves simply dipping the part or product into a tank of liquid paint. When the object is removed, the excess paint drains back into the tank. The tanks used in the process can range in size from 1 or 2 gallons for small objects to thousands of gallons for large fabricated metal products. Closely related to immersion is the flow-painting method. Instead of dipping the parts into the tank, they are positioned above the tank and a stream of paint is directed to flow over the object. Both of these methods are relatively inefficient in terms of the amount of paint deposited onto the object. Although dipping and flow coating are relatively simple processes, the methods for delivering the product to the painting operation may involve considerable mechanization. For example, conveyor systems are often used in high production to carry the parts down into the dipping tanks to apply the coating.

A more advanced immersion method is electrodeposition. This is a process in which a conductive object (the part or product) is given a negative electrical charge and dipped into a water suspension containing particles of paint. The paint particles are given a positive electrical charge, and consequently they are attracted to the negatively charged object (the cathode). The electrodeposition coating method is a highly sophisticated technique and requires close control over the process parameters (e.g., current, voltage, concentration of paint in suspension) in order to ensure the success of the operation. Its advantage is that it does not waste nearly as much paint as conventional immersion methods.

The second major category of industrial painting is spray coating. This method involves the use of spray guns to apply the paint or other coating to the object. Spray painting is typically accomplished by human workers who manually direct the spray at the object so as to cover the desired areas. The paint spray systems come in various designs, including conventional air spray, airless spray, and electrostatic spray. The conventional air spray uses compressed air mixed with the paint to atomize it into a high velocity stream. The stream of air and paint is directed through a nozzle at the object to be painted. The airless spray does not use compressed air. Instead the liquid paint flows under high fluid pressure through a nozzle. This causes the liquid to break up into fine droplets due to the sudden decrease in pressure in front of the nozzle.

The electrostatic spray method makes use of either conventional air spray or airless spray guns. The feature which distinguishes the electrostatic method is that the object to be sprayed is electrically grounded and the paint droplets are given a negative electrical charge to cause the paint to adhere to the object better.

The spray-coating methods, when accomplished manually, result in many health hazards to the human operators. These hazards include[5]:

Fumes and mist in the air. These result naturally from the spraying operation. Not all of the paint droplets become attached to the surface of the object. Some remain suspended in the atmosphere of the spray painting booth. To protect the human operators, ventilation systems must be installed in the booth and protective clothing and breathing masks must be worn. Even with this protection, the environment is uncomfortable and sometimes toxic for humans.

Noise from the nozzle. The spray gun nozzle produces a loud shrill noise. Prolonged exposure by humans can result in hearing impairments.

Fire hazards. Flammable paint, atomized into a fine mist and mixed with air, can result in flash fires in the spray painting booths.

Potential cancer hazards. Certain of the ingredients used in modern paints are believed to be carcinogenic, with potentially unsafe health consequences to humans.

Robots in Spray Coating

Because of these hazards to humans, the use of industrial robots has developed as an alternative means of performing spray-coating operations. Spray-coating operations to which robots have been applied include painting of car bodies, engines, and other components in the automotive industry, spraying of paint and sound absorbing coatings on appliances, application of porcelain coatings in bathroom fixtures, and spray staining of wood products. Some of the applications have consisted of a stand-alone robot spraying a stationary workpart that has been positioned in a paint booth by a human worker. However, these applications are generally less successful because they rely heavily on the human worker and the utilization of the robot is relatively low.

In most robot spray-coating applications, the robots are usually part of a system that includes a conveyor for presenting the parts to the robot, and a spray booth for shielding the spraying operation from the factory environment. Figure 14-7 illustrates a robot spray painting a part. When a conveyor–robot system is used, the operation of the robot and the conveyor must be closely synchronized. In the case of an intermittent conveyor system, interlocks are used to coordinate the start and finish of the robot program with the movement of the conveyor. With a continuously moving conveyor, some form of baseline tracking system (discussed in Chap. Eleven) is required in order to synchronize the robot's motions with the movement of the conveyor.

Another feature of many robot spray-coating applications is that the

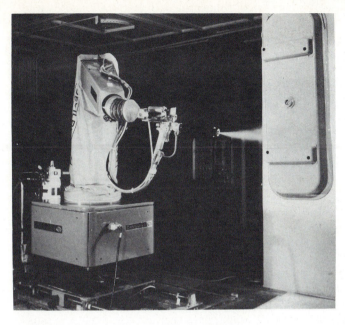

Figure 14-7 DeVilbiss/Trallfa robot performing a spray coating operation. (Photo courtesy of DeVilbiss Company)

system must be designed to process a variety of part styles, each with its own unique configuration. This is usually accomplished by providing the workcell with a parts identification system. Once the part has been properly identified, the robot can then apply the correct spray cycle for that part.

In general, the requirements of the robot for spray-coating applications are the following:

1. *Continuous-path control.* In order to emulate the smooth movement of a human spray paint operator, the robot must possess many degrees of freedom in its manipulator and it must have continuous path capability.
2. *Hydraulic drive.* Hydraulic drive is preferred over electric or pneumatic drive in spray-painting applications. In electric drive there is danger that a spark in the electric motor system may ignite the paint fumes in the spray booth environment. The motions generated in pneumatic drive are generally too jerky to be suitable for spray-coating applications.
3. *Manual leadthrough programming.* In most spray-coating applications, the most convenient method of teaching the robot involves leadthrough programming in which the robot arm is manually pulled through the desired motion pattern by a human operator who is skilled in the techniques of spray painting. During the programming procedure, a "teach arm" which is light and maneuverable, is often substituted for the actual robot arm, which tends to be heavy and difficult to manipulate smoothly.
4. *Multiple program storage.* The need for multiple program storage arises in

paint production lines in which more than one part style are presented to the robot for spraying. The capability to quickly access the program for the current part is a requirement for these lines. Either the robot itself must have sufficient memory for the programs required or it must be interfaced to the cell controller (computer or programmable controller) for random access to this memory capacity.

In robot spray-coating operations, the spray gun is the robot's end effector. Control over the operation of the spray gun system must be accomplished by the robot during program execution. In addition to on–off control over the spray gun nozzle, some of the important process variables that must be regulated during the spray cycle include paint flow rate, fluid and/or air pressure, and atomization. These variables are regulated through the output interlock functions of the robot controller. The operation of the interlock functions is established during the programming procedure. Other parameters that must be controlled during the spray-coating process are related to the coating fluid (e.g., the paint). Viscosity, specific gravity, temperature, and other characteristics of the fluid must be maintained at consistent levels in order for the results of the finishing operation to be acceptable.

Another requirement for consistent quality in the finishing operation is that the spray gun must be periodically cleaned. This can be accomplished without significant loss of production time by programming a cleaning cycle into the workcell operation at regular intervals. The cleaning operation takes only a few seconds to complete and consists of the robot placing the spray nozzle under cleaning jets which spray solvent into the nozzle opening. Incorporating the cleaning operation into the work cycle should be planned to minimize the impact on the productive portion of the cycle.

Benefits of Robot Spray Coating

Use of robots to perform spray-finishing operations provide a number of important advantages. These advantages include:

1. Removal of operators from hazardous environment
2. Lower energy consumption
3. Consistency of finish
4. Reduced coating material usage
5. Greater productivity

Removing the human workers from the kinds of hazards which characterize the manual spray-finishing environment (i.e., fumes, fire hazards, etc.) is a significant health benefit of using robots. Also, because humans are not in the spray booth, the ventilation requirements are reduced below the levels needed when humans are present. Therefore less energy is needed to control the environment.

Other advantages include better quality and fewer rejects. Because the robot performs the same spraying cycle on every workpart, the quality of the finishing job is more consistent compared to a human worker. As an added benefit, the amount of paint required to coat the parts is typically reduced by 10 to 50 percent when robots are used. These various features of the robot spray-coating cell result in substantial labor savings and improved productivity in the process.

Example 14-1 An example of a modern high-technology robotic spray-painting cell design is presented in a paper by Akeel.[1] The cell design is the result of the development efforts of the General Motors Manufacturing Staff directed at the problem of automating the paint shops in an automobile production plant. A typical paint cell for producing 60 jobs per hour consists of eight robots (four pairs, each pair servicing the two sides of the automobile) and is illustrated in Fig. 14-8. Other features of the cell include:

A machine vision system for identifying the body style so that the proper robot program can be used.

Backup robots so that if one of the production robots breaks down, they can be quickly replaced by the backups. An overhead crane system serves to replace the robots when needed.

Automatic two-axis door openers so that the internal surfaces of the car can be sprayed.

Supervisory computer control of the cell, including a backup computer

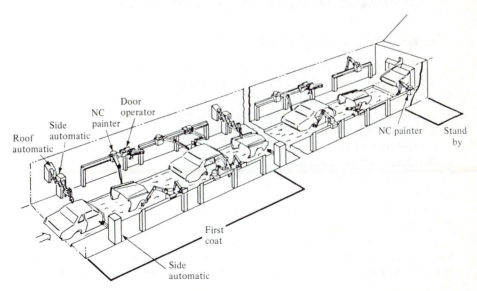

Figure 14-8 Typical configuration of GM paint spray cell. (Reprinted with permission from Akeel [1])

that is operated in parallel with the primary system. The backup can be switched into service should the primary computer fail.

The individual painting robots (GM calls them paint spray machines) are seven-axis manipulators that are operated under computer control. The supervisory computer communicates the correct work cycles to the individual machines on the line. Each spray paint machine is equipped with two spray guns especially designed by GM engineers for automatic operation in the cell. Each spray gun is provided with its own paint supply system so that one system can be in the process of being changed over (purged, cleaned, and refilled with the next paint) while the other system is in production. The parallel paint supply lines are contained inside the manipulator arm.

14-4 OTHER PROCESSING OPERATIONS USING ROBOTS

In addition to spot welding, arc welding, and spray coating, there are a number of other robot applications which utilize some form of specialized tool as the

Figure 14-9 PUMA 500 robot performing a wire brush operation to deburr a workpart. (Photo courtesy of Unimation Inc.)

end effector. Operations which are in this category include:

Drilling, routing, and other machining operations
Grinding, polishing, deburring, wire brushing, and similar operations
Riveting
Waterjet cutting
Laser drilling and cutting

We are excluding from this category applications in assembly, inspection, and nonmanufacturing operations which might employ a tool as end effector. Assembly and inspection applications are discussed in the following chapter, and nonmanufacturing applications are covered in Chap. Twenty. From the preceding list, it can be seen that a typical end effector in this category is a powered spindle attached to the robot's wrist. The spindle is used to rotate a tool such as a drill or grinding wheel. The purpose of the robot is to position the rotating tool against a stationary workpart in order to accomplish the desired processing operation. In the other examples given in the above list (riveting, waterjet cutting, and laser operations), the end effector is not a powered spindle, but the job of the robot is still to position the tool relative to the part. Requirements of these applications vary, but one of the inherent disadvantages of robots in some of these operations is their relative lack of accurcy as compared to a regular machine tool. Small robots tend to be more accurate than large robots, but large robots are more likely to possess the strength and rigidity necessary to withstand the forces involved to hold the powered spindle against the part during the process. Figure 14-9 shows a PUMA 500 robot performing a wire brush deburring operation.

Example 14-2 This example resulted from an Air Force ICAM (Integrated Computer-Aided Manufacturing) sponsored project,[4] and represents an important project in the development of processing applications. The application involves drilling and routing operations of different aircraft components. The components are various sheet metal fuselage panels for the F-16 fighter aircraft which is fabricated by General Dynamics Corporation in Fort Worth, Texas. A diagram of the cell concept is illustrated in Fig. 14-10. The principal components of the cell are the robot and a part fixture. The robot used in the application is a Cincinnati Milacron T3 robot which has a repeatability of + or −0.050 in. and a relatively large work envelope. The application requirement was that the robot should be able to reach over a 3 by 4 ft part area. The fixture has two positions so that the robot can be processing one part while loading and unloading is being accomplished at the opposite position. A human worker performs the loading and unloading tasks. The cell works in a batch production mode and was first placed in operation in October, 1979. Some of the important features about this robot application are:

Supervisory computer control to coordinate the activities of the robot and the part fixture.

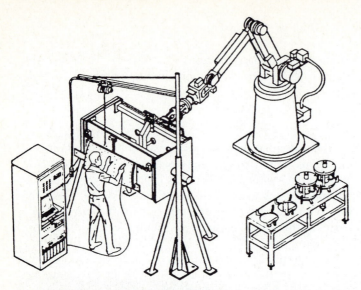

Figure 14-10 Diagram of the drilling and routing cell developed under Air Force ICAM sponsorship. (Reprinted with permission from Golden et al. [4])

Multiple program storage so that the robot executes the correct program on each different fuselage component.

Automatic workpart identification system employing an optical character reader to recognize which part is to be processed next. This allows the correct program to be called from multiple program storage.

Automatic tool changing of drills and routing tools. A tool rack is used to hold the different tools required in the sequence of processing operations. The robot is programmed to select the proper tool for the particular operation to be performed.

Templates are used to define hole patterns and router paths. Also, compliant end effectors are used to permit the tools to line up with the templates. These measures were adopted in order to overcome the inherent accuracy and repeatability limitations of the T3 robot.

Example 14-3 A more recent application, similar to the one described in Example 14-2, is the installation at Grumman Aerospace Corp. The robot cell is used to accomplish light machining of complex contour sheet metal parts for aircraft skins and other components. An ASEA IRB-60 robot performs routing, trimming, and drilling of the aircraft parts. The workcell consists of the robot mounted on a linear track system (this is a mobile robot cell), the part holders, tool-changing module, and cell controller. The cell layout is illustrated in Fig. 14-11. The track system provides approximately 20 ft of linear movement for the robot. This dramatically increases the robot's work volume.

The tool-changing module is shown in Fig. 14-12. Several air-powered

Figure 14-11 Cell layout for Example 14-3. (Photo courtesy of Grumman Aerospace Corp.)

Figure 14-12 Tool-change set up for Example 14-3. (Photo courtesy of Grumman Aerospace Corp.)

413

drilling and routing tools are used as the robot's end effector. Human operators load the parts into the fixtures and notify the cell controller which parts are to be processed.

REFERENCES

1. H. A. Akeel, "Expanding the Capabilities of Spray Painting Robots," *Robotics Today*, April 1982, pp. 50–53.
2. T. J. Bublick, "Guidelines for Applying Finishing Robots," *Robotics Today*, April 1984, pp. 61–64.
3. J. F. Engelberger, *Robotics in Practice*, AMACOM (American Management Association), New York, 1980, chaps. 11, 12, and 16.
4. H. D. Golden et al., "ICAM Robotics System for Aerospace Batch Manufacturing—Task A," *Technical Report AFWAL-TR-80-4142*, Vol. I, Materials Laboratory, Air Force Wright Aeronautical Laboratories, OH, 1980.
5. M. P. Groover and E. W. Zimmers, Jr., *CAD/CAM: Computer-Aided Design and Manufacturing*, Prentice-Hall, Englewood Cliffs, NJ, 1984, chap. 11.
6. J. Jablonowski, "Robots that Weld," *American Machinist*, Special Report 753, April 1983, pp. 113–128.
7. R. J. McCluskey, "Robotic System Cuts Airplane Parts," *American Machinist*, August 1984, pp. 71–73.
8. S. Muller, "Spot Welding: The Classic Case for the Quality Robot," *Decade of Robotics*, IFS Publications, Bedford, England, 1983, pp. 34–39.
9. G. M. Nally, "Robotic Arc Welding: At What State is the Art?" *Robotics Today*, August 1983, pp. 37–40.
10. K. Ostby, "Robots Automate Routing and Water Jet Cutting," *Robotics Today*, June 1984, pp. 42–43.
11. P. F. Rogers, "The Economics of Robotic Arc Welding Workcells," *Robotics Today*, June 1984, pp. 46–48.
12. R. N. Stauffer, "Robogate and Unimates Team Up to Improve Quality and Efficiency," *Robotics Today*, Summer 1980, pp. 24–30.
13. R. N. Stauffer, "Anatomy of a Successful Arc Welding Installation," *Robotics Today*, April 1982, pp. 41–42.
14. R. N. Stauffer, "Automated Body Assembly—Circa 1972," *Robotics Today*, April 1982, pp. 58–60.
15. R. N. Stauffer, "Welding Robots: The Practical Approach," *Robotics Today*, August 1983, pp. 43–44.
16. R. N. Stauffer, "Update on Noncontact Seam Tracking Systems," *Robotics Today*, August 1983, pp. 29–34.
17. J. A. Vaccari, "Robots that Paint Can Create Jobs," *American Machinist*, January 1982, pp. 131–134.
18. J. Weston, "Arc Welding: A Difficult Path for Robots to Tread," *Decade of Robotics*, IFS Publications, Bedford, England, 1983, pp. 40–43.

PROBLEMS

14-1 A robotic arc welding cell is proposed to replace a high-production manual welding operation. The manual operation requires two operators; a fitter and a welder. The method in the current operation is for the fitter to place and clamp the components into a rudimentary fixture and then for the welder to perform the welding operation on the parts to form the assembly. The fitting time per cycle is 8.0 min, and the welding time is 9.0 min, which includes 4.0 min for repositioning the welding rod during the welding. The repositioning involves moving the welding rod to a new location on the parts without actually welding. It is therefore nonproductive time. The two operators are paid for 8 hours per day (250 days per year), although 1 hour is lost per day for rest breaks, cleanup, and so on. The fitter is paid at the rate of $12.00 per hour (including fringes) and the welder is paid $13.00 per hour. Overhead rates will not be considered in the analysis.

The proposed robot cell would replace the manual cell and would use one operator at the $12.00 rate. The cell would utilize a manipulator/fixture with two positions so that the human operator would load and unload components into the fixture at one position while the robot was doing the welding operation at the opposite position. With this arrangement, the welding and fitting could be accomplished simultaneously. The fitting time would be reduced to 6.0 min per cycle because of improvements in the fixturing. The actual welding time would remain the same because of technological process considerations, but the repositioning time (previously 4.0 min with the human welder) would be reduced by one-half with the robot welder. Time required for indexing positions with the manipulator/fixture would be 15 s. It is anticipated that the robot cell would operate for 7.5 hours per day although the operator would be paid for 8 hours.

The investment cost of the welding robot would be $95,000 including the welding apparatus. The cost of the manipulator/fixture would be $40,000, and the installation cost for the cell would be $30,000. A 5-year project life will be used with the salvage value of the robot and manipulator/fixture equal to one-third their combined initial values at the end of the project. Annual maintenance and operating costs are anticipated to be $10,000. The rate of return criterion is 20 percent.

(*a*) What is the hourly production rate for the current manual operation and what is the anticipated hourly production rate for the proposed robot cell?

(*b*) What is the arc-on time for the current manual operation and what is the anticipated arc-on time for the proposed robot cell?

14-2 Should the proposed robot cell in Prob. 14-1 be installed? Support your answer using the economic analysis methods of Chap. Twelve.

FIFTEEN

ASSEMBLY AND INSPECTION

There is a growing interest in automated assembly because of the high manual labor content of most assembly operations today. Automated assembly systems have traditionally been applied to high-volume products in which a large investment is made in custom-engineered equipment, designed to perform the specific operations required for those products. However, there are a large number of products, representing a majority of the assembly operations performed in the United States, where the volume of production is low or medium. In these cases, it is not economically feasible to make large investments in specialized assembly equipment. Programmable and flexible systems, including robotics, must be applied to these low- and medium-volume assembly operations if automation is to be successfully achieved.

Inspection is another area of factory operations in which there is a significant interest in automation. Machine vision and other sensor technologies are being investigated for this purpose. Today, inspection is commonly performed manually according to the sampling procedures of statistical quality control. Even with sampling, the human work involved in most inspection tasks is tedious and boring. With greater importance being placed on product quality in manufacturing, inspection is an area that is ripe for automation. Instead of accomplishing the inspection process manually on a sampling basis, it will be accomplished automatically on a 100 percent basis. Robotics technology figures to play a significant role in automating certain aspects of inspection.

This chapter will deal with both assembly and inspection operations and how robotics can be applied to these operations. We cover the assembly application first.

15-1 ASSEMBLY AND ROBOTIC ASSEMBLY AUTOMATION

The term assembly is defined here to mean the fitting together of two or more discrete parts to form a new subassembly. The process usually consists of the sequential addition of components to a base part or existing subassembly to create a more complex subassembly or a complete product. As such, assembly operations involve a considerable amount of handling and orienting of parts to mate them together properly. The difference between assembly tasks and other material-handling tasks is that value is added to the product through the assembly operation. Also, there are often interactions that take place between the two parts being assembled, between the gripper and the part, and between other elements of the workcell. When parts are fastened together (called parts joining), there are often additional interactions with the medium used to join the components (e.g., adhesive). All of these potential interactions can make assembly operations considerably more complex compared to the simpler task of moving a part from one location to another.

There are a variety of assembly processes used in industry today. These include mechanical fastening operations (using screws, nuts, bolts, rivets, swaging, etc.), welding, brazing, and bonding by adhesives. Some of these processes are more adaptable to automatic assembly. We will discuss the various assembly methods in more detail in a later section.

It was mentioned in the introduction that assembly operations can be performed manually, or by high-speed automatic assembly machines, or by robots and other programmable systems. In addition, combinations of these techniques can be used in the design of an assembly system for a particular application.

In our coverage of the application of robotics to assembly, we will divide the subject into three areas as suggested by the preceding discussion:

Parts presentation methods
Assembly tasks
Assembly cell designs

The following four sections will examine these three areas and some of the particular problems associated with them. We also present a discussion of a major development project devoted to the application of robots to assembly called the Adaptable-Programmable Assembly System (APAS). Our discussion of assembly will conclude with a section devoted to the topic of product design for automated assembly.

15-2 PARTS PRESENTATION METHODS

In order for a robot to perform an assembly task, the part that is to be assembled must be presented to the robot. There are several ways to accomplish this presentation function, involving various levels of structure in the workplace:

Parts located within a specific area (parts not positioned or oriented)
Parts located at a known position (parts not oriented)
Parts located in a known position and orientation

In the first case, the robot is required to use some form of sensory input to guide it to the part location and to pick up the part. A vision system could be used as the sensory input system for this purpose. In the second case, the robot would know where to go to get the part, but would then have to solve the orientation problem. This might require the robot to perform an additional handling operation to orient the part. The third way of presenting the part to the robot (known position and orientation) is the most common method currently used, and is in fact the method used in automatic assembly that precedes the advent of robotics. This approach requires the least effort from the robot and sensor system, but it places the largest requirement on the parts feeding system.

There are a number of methods for presenting parts in a known position and orientation, including the use of bowl feeders, magazine feeders, trays, and pallets. These and other techniques of parts presentation are discussed in detail in some of the references at the end of the chapter.[1,3,11] Our discussion of these methods will emphasize the issues that relate to robotics applications.

Bowl Feeders

Bowl feeders are the most commonly used devices for feeding and orienting small parts in automated assembly operations. They are made by numerous companies and have been used to feed everything from delicate electronic parts to rugged metal castings. A bowl feeder consists of two main components: the bowl and the vibrating base. A track rising in a spiral up the sides of the bowl is used to deliver parts in the bottom of the bowl to an outlet point. This track is commonly located on the inside of the bowl. A typical bowl feeder is shown in Fig. 15-1.

The base of the bowl feeder is constructed of leaf springs and an oscillating electromagnet which causes the bowl and track to vibrate. The vibratory motion causes the parts to be driven up the spiral track until they reach the outlet point.

As the parts are driven up the track and approach the outlet point, they are oriented randomly and must be placed in the proper orientation for delivery out of the bowl feeder. This can be done by either of two methods, called selection and orientation. Selection (sometimes called passive orientation) involves taking parts that are not properly oriented and rejecting them from the track back into the bottom of the bowl, thus permitting parts that are properly oriented to pass through to the outlet point. Orientation involves taking parts that are not oriented properly and physically reorienting them as

Figure 15-1 Vibratory bowl feeder (Photo courtesy of FMC Corp., Material Handling Equipment Dev.)

desired. Both methods are usually accomplished by means of a series of obstacles located along the track. These obstacles allow the parts to pass through only if they meet certain orientation criteria. In the case of part reorientation, obstacles or other mechanisms are used to physically change the orientation of the part as it moves along the track. Some of the techniques that are used in selection and orientation devices are pictured in Fig. 15-2. By providing a sufficient number of obstacles along the track, we can ensure that only parts that possess the desired orientation will successfully reach the outlet point.

Parts exiting the bowl feeder usually travel down a track or chute to some type of holding fixture. This fixture is located at an elevation which is below the outlet point of the bowl feeder so that gravity can be used to deliver the parts from the outlet to the holding fixture. The fixture is isolated from the vibration of the bowl feeder so the robot or other device that retrieves the part will not have to contend with the problems of a vibrating or moving target. The holding fixture maintains the desired position and orientation of the part until it is removed for the assembly operation. If a robot is to perform the assembly operation, the fixture must be designed with enough clearance for the robot gripper to grasp the part.

Another issue that must be addressed is the "back pressure" caused by the parts along the track leading to the holding fixture. Back pressure is the result of two forces: the force imparted on the parts by vibration in the track, and the force generated by the weight of all the parts in the track ahead of the fixture.

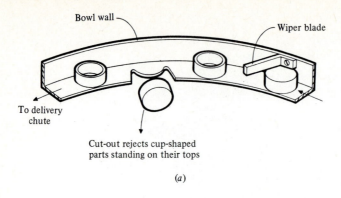

Bowl wall

Wiper blade

To delivery
chute

Cut-out rejects cup-shaped
parts standing on their tops

(a)

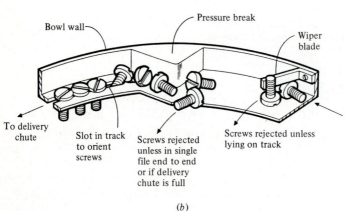

Pressure break

Bowl wall

Wiper
blade

To delivery
chute

Slot in track
to orient
screws

Screws rejected
unless in single
file end to end
or if delivery
chute is full

Screws rejected unless
lying on track

(b)

Figure 15-2 Part selection and orientation methods used in vibratory bowl feeders. (a) Selection and orientation of cup-shaped parts, (b) selection and orientation of screws. (Reprinted with permission from Boothroyd and Redford [3])

If the back pressure is large enough, it can inhibit the robot from successfully removing the part from the fixture. The following relation describes when the robot will be unable to remove a part from the fixture:

$$u_f F_g < u_p [F_b + nW(\sin \theta)]$$

where u_f = coefficient of friction between the gripper and the part
F_g = gripping force of the gripper
u_p = coefficient of friction between the parts in the track
F_b = back pressure force due to track vibration
n = number of parts in the track leading up to the fixture
W = weight of each part
θ = angle that the track makes with the horizontal (the angle is assumed constant in the portion of the track containing parts leading to the fixture)

There are several ways to limit the back pressure at the holding fixture.

From feeder

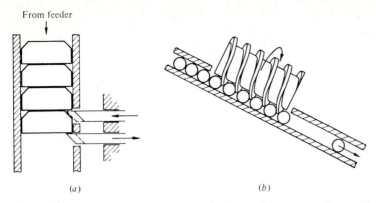

(a)   (b)

Figure 15-3 Several types of escapement devices used in automated assembly. (a) Linear motion escapement device for disk-shaped parts, (b) worm escapement device. (Reprinted with permission from Boothroyd and Redford [3])

The first is to turn off the vibration of the bowl whenever there are a large enough number of parts in the track that the back pressure would reach an undesirable level. On–off control of the bowl operation can easily be accomplished by providing a sensor in the track that detects the presence of the parts. A simple limit switch can be used for this purpose. When parts back up along the track to the point where the sensor is located, the bowl would be turned off. By properly positioning the sensor, the back pressure can be controlled to allow smooth pickup of parts by the robot from the holding fixture. A second way to reduce the back pressure is to make the angle of the track leading into the fixture relatively small. The risk here is that the parts will not properly slide down the track. A third way is to use an "escapement" device at the end of the track to individually place the parts into the holding fixture. This avoids the weight of many parts stacked against each other in the track. Figure 15-3 illustrates a number of escapement devices.

Magazine Feeders

Bowl feeders are generally used to handle parts that are received at the workstation in bulk. Parts supplied in bulk are randomly oriented and one of the functions served by the bowl feeder is to deliver the parts to the track in proper orientation. An alternative to the use of a bowl feeder is to receive the parts at the workstation in a preoriented orderly fashion.

The use of magazine feeders is one technique in which preoriented parts can be received at the workstation. Of course, this does not eliminate the problem of orienting and loading the parts; it simply transfers the problem away from the workstation. The most convenient way to load parts into the magazine in a proper orientation is to perform the loading operation as an integral element of the production process that makes the parts. What makes this possible is that the parts come out of the production process already oriented. For example, in a sheet metal stamping operation, the parts always come out of the press in the same way. It is therefore possible to load the stamped parts one on top of the next in the same orientation, into some kind of tube or other container. This container would constitute the magazine. The tube filled with parts could subsequently be attached to a mechanism (e.g., an escapement) designed to withdraw the parts and present them to the assembly workhead or the robot.

On the other hand, if the parts cannot be loaded into the magazine directly from the production operation and must be loaded instead manually, then the parts magazine loses much of its appeal. One of the disadvantages associated with the use of a parts magazine is that it typically holds fewer parts than a bowl feeder. Consequently, it must be replaced and refilled more frequently requiring a greater level of human attention at the workstation.

Trays and Pallets

Sometimes it is too expensive to use bowl feeders or parts magazines. In those cases, trays or pallets can be used. A specific advantage of using trays, pallets, and other similar storage containers is that they can be used for a variety of different part geometries. Bowl feeders and magazines must usually be custom engineered for a particular part configuration. However, there are certain conditions that generally must be satisfied in using trays and pallets in robotics. Namely, the parts must be located in known positions and orientations with respect to certain reference points on the device, usually the edges of the containers. This allows the trays to be registered correctly at the workstation and for the robot to be programmed to go to the known positions in the trays to retrieve the parts.

If the cycle time of the operation is relatively long, and the tray capacity is large, then the trays could be presented to the workstation by a human operator as required. If the cycle rate is fast, and a more automated operation is desirable, then some type of materials handling system must be devised to present the trays to the workstation automatically. In either case, an issue of great importance in the design of this type of container system (in robotics or any other form of automation) is that the containers must be positioned accurately at the workstation and the parts must be positioned precisely in the containers. If different part styles use the same basic container and material-handling system, the information system supporting the operation must be sophisticated enough to handle the differences.

The alternative to the approach of precise part location is for the parts to be randomly oriented in the trays, and for the robot to perform some kind of "bin-picking" procedure in order to pick out the parts one at a time. The bin-picking problem requires the use of machine vision and is explained in Chap. Seven.

15-3 ASSEMBLY OPERATIONS

Assembly operations can be divided into two basic categories: parts mating and parts joining. In parts mating, two (or more) parts are brought into contact with each other. In parts joining, two (or more) parts are mated and then additional steps are taken to ensure that the parts will maintain their relationship with each other. In this section we discuss a number of assembly tasks that fall into these two categories, along with their implications for a robot system's capabilities.

Parts Mating

The variety of parts mating operations include the following assembly situations:

   1. Peg-in-hole. This operation involves the insertion of one part (the peg) into another part (the hole). It represents the most common assembly task. Peg-in-hole tasks can be divided into two types: the round peg-in-hole and the square peg-in-hole. The two types are illustrated in Fig. 15-4. With the round peg-in-hole, the robot needs only 5 degrees of freedom to insert the peg since there is no requirement to align the peg about its own axis. With the square peg-in-hole case, a full 6 degrees of freedom are needed in order to mate the corners of the square peg with the corners of the hole.

   2. Hole-on-peg. This is a variation of the peg-in-hole task. Similar problems exist in defining the degrees of freedom needed to execute the mating of the two parts. A typical example of the hole-on-peg task would be the placement of a bearing or gear onto a shaft.

   3. Multiple peg-in-hole. This is another variation on case 1 except that

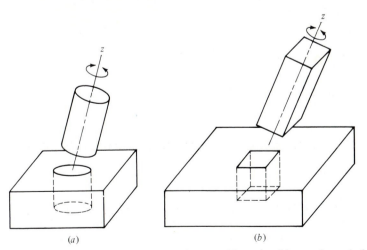

(a)   (b)

Figure 15-4 Two types of peg-in-hole assembly tasks. (a) round peg-in-hole and (b) square peg-in-hole—orientation about z-axis required.

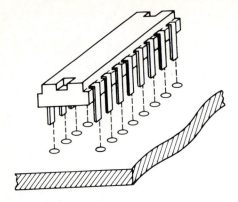

Figure 15-5 Multiple peg-in-hole assembly task: insertion of a semiconductor chip module into a circuit card.

one part has multiple pegs and the other part has corresponding multiple holes. Consequently, the assembly task always requires the ability of the assembly system to orient the parts in all directions. An example would be the assembly of a microelectronic chip module with multiple pins into a circuit card with corresponding holes, as illustrated in Fig. 15-5. This example represents a common assembly problem in the electronics industry.

4. Stacking. In this type of assembly, several components are placed one on top of the next, with no pins or other devices for locating the parts relative to each other. In a subsequent assembly operation, the group of parts would be joined together. An example of the stacking assembly operation would be a motor armature or a transformer in which the individual laminations are stacked.

Parts-Joining Tasks

In parts joining, not only must the two (or more) components be mated, but also some type of fastening procedure is required to hold the parts together. The possible joining operations include the following:

1. Fastening screws. The use of screws is a very common method of joining parts together in manual assembly. Self-tapping screws are often used and this eliminates the need to perform the extra operation of tapping the threaded hole in the mating part. There are two ways in which a robot can perform the screw-fastening operation: it can drive the screw by advancing and simultaneously rotating its wrist, or it can manipulate a special end effector consisting of a power screwdriver. Power screwdrivers are available that not only drive the screws but also feed them automatically to the bit. Figure 15-6 shows a power screwdriver that can be attached to a robot wrist. Screw fastening without the aid of a power screwdriver turns out to be a relatively difficult task for a robot because of the complexity of the motions involved to rotate the screw and advance it into the hole at the same time. Also, when a screw is to be fastened into a threaded hole (in other words, a self-tapping screw is not used), there is the possibility for binding to occur between the screw threads if the mating hole threads are not properly aligned.

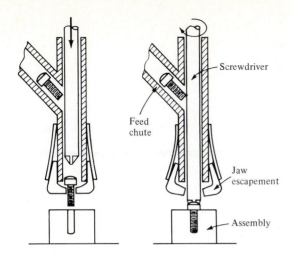

Screwdriver

Feed
chute

Jaw
escapement

Assembly

Figure 15-6 Operation of a power screwdriver. (Reprinted with permission from Boothroyd and Redford [3])

2. Retainers. Retainers can take a number of alternative configurations. They can be pins inserted through several parts in order to maintain the relationship among the parts. Another form of retainer is a ring that clamps onto one part to establish its relationship with another part. Common ring retainers are snap rings and C-rings.

3. Press fits. This is another variation of the peg-in-hole task except that the parts to be mated have an interference fit. This simply means that the peg is slightly larger than the hole into which it is to be inserted. Press-fitted parts can form a very strong assembly. However, a substantial force is required to accomplish the insertion operation. In most force-fit operations, the robot will not be able to provide the necessary force to press the parts together, and therefore the application will be designed so that the robot loads the parts into a power press which performs the actual press-fitting operation.

4. Snap fits. This joining technique has features of both the retainer and the press-fitting methods. A snap fit involves the joining of two parts in which the mating elements of the parts possess a temporary interference that only occurs during the joining process. When the parts are pressed together, one (or both) of the parts elastically deforms to accommodate the interference, then catches into the mating element of the part. The parts are usually designed so that a slight interference fit exists between the two parts even after they are snapped together. Figure 15-7 illustrates the snap fit assembly. This joining method turns out to be an ideal method for automatic assembly methods including robotics.

5. Welding and related joining methods. Continuous arc welding and spot welding are two common welding operations used to joint parts together. We have discussed these two joining techniques in the preceding chapter. In addition there are other similar joining techniques requiring heat energy that are used in assembly operations. These include brazing, soldering, and

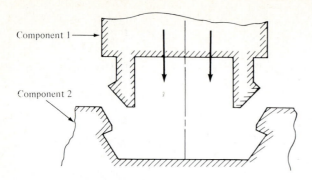

Figure 15-7 Snap fit assembly showing cross-section of tab-hook configuration for two mating parts.

ultrasonic welding. All of these joining methods can be implemented by means of robots.

6. **Adhesives.** Glue and similar adhesives can be applied to join parts together by using a dispenser to lay down a bead of the adhesive along a defined path (for a robot, the motion cycle is similar to arc welding) or at a series of points (similar to spot welding). In most applications the adhesive dispenser is attached to the robot's wrist, while in other cases the robot manipulates the part and presents it to the dispenser.

7. **Crimping.** The term crimping, in the context of assembly, refers to the process of deforming a portion of one part (often a sheet metal part) to fasten it to another part. A common example of crimping is when an electrical connector is crimped (squeezed) onto a wire. To perform a crimping process, the robot requires a special tool or pressing device attached to its wrist. Staking operations and riveting operations are similar to crimping in that they involve deformation of one part to attach it to another. Staking usually refers to the use of metal tabs on one part that are bent over the joining part. Riveting involves specially designed fasteners (screws without threads) whose ends are flattened over the joining part.

8. **Sewing.** Although not typically considered as a robot application, this is a common joining technique for soft, flexible parts (e.g., cloth, leather).

15-4 COMPLIANCE AND THE REMOTE CENTER COMPLIANCE (RCC) DEVICE

Let us examine the peg-in-hole assembly task and consider the potential problems that are encountered during insertion. When a peg is inserted into a hole there are two possible positioning errors for the peg: a lateral position error and an angular error. These possibilities are illustrated in Fig. 15-8. When the parts are chamfered and there is a position error small enough to allow insertion to begin, it is still likely that an angular error will result during chamfer crossing as the peg rotates about the grip point at the top of the peg. Figure 15-9 shows how this problem can occur. The angular error allowable on the peg is a function of the clearance of the hole and the depth of the

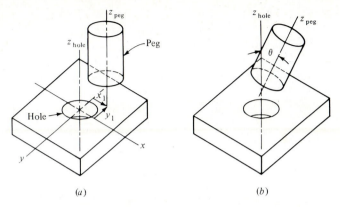

Figure 15-8 Two possible positioning errors for the peg-in-hole insertion task. (a) lateral position error, and (b) angular error.

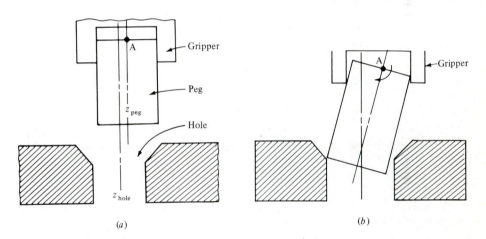

Figure 15-9 Peg-in-hole insertion task. (a) effect of small position error with chamfered parts. (b) small angular error results during chamfer crossing as the peg rotates about the grip point at the top of the peg.

insertion. That is, the deeper the part is inserted into the hole, the less angular error can be tolerated; likewise, the smaller the clearance between the parts and the hole, the smaller the angular error. If the angular error is greater than the tolerable error, the parts will wedge into place much in the same way that a dresser drawer gets wedged when it becomes cocked in the drawer slide. Conceptually, what must be done to perform a successful peg-in-hole insertion task is to correct for the lateral and angular errors during assembly. A common solution to the problem makes use of the Remote Center Compliance device.

The Remote Center Compliance device, or RCC, was developed during research on assembly at the Charles Stark Draper Laboratory in Cambridge, Massachusetts. Today, RCC products are commercially available and one such product is pictured in Fig. 15-10. The RCC device is typically mounted between the wrist of the robot and its gripper. Figure 15-11 shows a mechanical gripper attached to the RCC device.

The RCC device is capable of accommodating the lateral errors and angular errors encountered in an insertion operation and in other tasks requiring limited compliance. In the peg-in-hole insertion process, the RCC operates as though the part (peg) were being pulled into the hole by the tip, rather than being pushed from the top. The action is accomplished as illustrated in Fig. 15-12.[7] Parts (a) and (b) of the figure show the accommodation of the lateral forces. Suppose there is a lateral error in position between the peg and the hole as shown in part (a). Because of the chamfer on the hole, the error will cause a lateral force on the peg. This force causes the RCC to translate the peg so that it can be inserted into the hole. Next, consider the possibility of an angular error. Suppose the axis of the hole is not parallel with

Figure 15-10 Remote center compliance (RCC) device. (Photo courtesy of Lord Corporation)

Figure 15-11 Mechanical gripper attached to the remote center compliance device. (Photo courtesy of Lord Corporation)

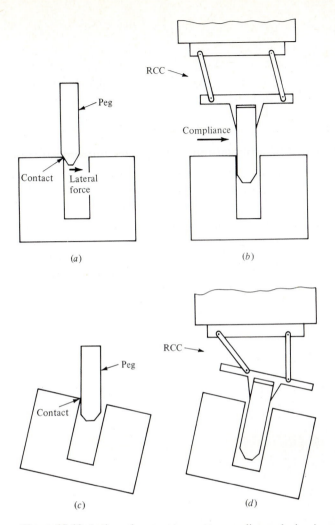

RCC

Peg

Contact Lateral
force

Compliance

(a)

(b)

Peg

RCC

Contact

(c)

(d)

Figure 15-12 Action of a remote center compliance device in peg-in-hole insertion task. (a) action of RCC for lateral displacement, (b) action of RCC for angular displacement. (Reprinted by permission courtesy of Lord Corporation [7])

the axis of the peg. The peg will enter the hole (assuming the necessary lateral compliance occurs), but its leading edge will contact one side of the hole, while the edge of the hole will contact the other side of the shaft as shown in Fig. 15-12(c). This will cause a moment on the peg. The RCC will accommodate the moment by means of a rotation about the compliant center as illustrated in part (d) of the figure.

RCCs are typically constructed using elastomer springs rather than the mechanical linkages shown in Fig. 15-12. This has resulted in designs that are simple, small, and lightweight. The parameters to be considered when selecting a remote center compliance device include the following:

Remote center distance. This is also called the center of compliance dimension. It is the distance between the RCC bottom surface and the compliant center of the RCC device. The compliant center (also called the elastic center) is the point about which the forces acting on the object being inserted are minimum. The remote center distance should be selected on the basis of the length of the part and the gripper.

Axial force capacity. This is the maximum force in the axial direction which the RCC device is designed to withstand and still function properly.

Compressive stiffness. This is also called the axial stiffness. It is the force per unit distance or spring constant required to compress the RCC device in the direction of insertion. Generally, the compressive stiffness is relatively high to allow for press fitting of parts together.

Lateral stiffness. This is the spring constant relating to the force required to deflect the RCC laterally (perpendicular to the direction of insertion). This parameter should be determined according to the stiffness of the robot and the delicacy of the parts being assembled.

Angular stiffness. This is also called the cocking stiffness. It is the rotational spring constant that relates to the force required to rotate the part about the elastic center by a certain amount.

Torsional stiffness. This is the torsional spring constant which relates to the moment required to rotate the part about the axis of insertion. This parameter becomes important when the insertion task requires orientation relative to the axis of insertion.

Other parameters to be considered in the specification of the remote center compliance device are the maximum allowable lateral and angular errors. These errors are generally determined by the relative size of the product and by its design (e.g., design of the chamfers). They must be sufficiently large to compensate for errors in the workcell due to parts, robot, and fixturing.

A second approach to provide compliance would be to measure the forces and moments encountered by the part and to servo the robot to compensate for these forces. In our discussion in Sec. 6-3 (Chap. Six on robotic sensors) we examined the capability of the force-sensing wrist to accomplish this type of compensation. Also, the Instrumented Remote Center Compliance (IRCC) device is a possible approach to this type of problem. The IRCC is an RCC device that has been instrumented to measure deflections. These deflections provide an indication of the forces and moments being applied to the wrist. Whereas most force sensors are very rigid and deflect very little under load, the IRCC is compliant in certain directions. This permits high-speed part insertion owing to the compliance of the RCC while allowing monitoring and data collection of forces during operation of the system.

Finally, a third approach to provide compliance for assembly tasks is the class of robot manipulators known by the acronym SCARA robots, for Selective Compliance Arm for Robotic Assembly. SCARA robots are horizontally articulated manipulators with a vertical insertion axis at the wrist end. The SCARA type robot was illustrated in Fig. 2-8 (Chap. Two). The arm is very stiff in the vertical direction, but is relatively compliant laterally. This is very convenient for a variety of assembly tasks in which components are stacked from the vertical direction.

15-5 ASSEMBLY SYSTEM CONFIGURATIONS

There are two basic configurations of assembly systems, a single workstation, and a series of workstations (an assembly line). Combinations of these two basic types are also possible. For example, it is sometimes advantageous to design a series configuration with certain stations in parallel. The following subsections will cover the various possibilities.

Single-Workstation Assembly

In this configuration all of the parts which are required to complete the desired assembly are presented to the operator or robot at a single workstation. All of the parts mating and joining tasks for the assembly are accomplished at the single workstation. In manual assembly, this configuration is generally used for low-volume products (e.g., custom-engineered machinery). In robotic assembly, the conditions warranting the use of this configuration are different from those for manual assembly. A single-station robotic assembly system would typically be used for low- and medium-volume work in which there were a limited number of assembly tasks and parts to be handled. This means that the product is of low to medium complexity. The features and problems of this configuration are illustrated by means of an example.

Example 15-1 The workcell is illustrated in Fig. 15-13 and is designed to assemble an electric motor consisting of the following components: rear endbell, front endbell, rotor, stator, two brushes, two bearings, two screws. There are many alternative ways to define the assembly sequence. In this example, we will begin with the procedure shown in Table 15-1.

Some of the problems of robotic assembly become apparent in the sequence. Can the robot handle all of the different shaped parts? Can it reach all of the points on the assembly? It is quite possible that a single gripper will be inadequate to perform all of the handling and assembly tasks. For example, the robot will probably have to use a special powered screwdriver in order to accomplish the screw-fastening operations. It is also likely that the subassembly will have to be repositioned at some point during the assembly sequence. In our analysis of the work cycle, we will assume that one gripper can be designed to handle the rotor, stator, endbells, and completed motor. It will also be assumed that the same gripper can be used to retrieve and grasp a powered screwdriver to

Table 15-1 Sequence of steps to assemble electric motor of Example 15-1

1. Place rear endbell into fixture
2. Set first bearing into endbell
3. Set rotor into bearing–endbell
4. Set stator around armature
5. Set second bearing on top of rotor shaft
6. Set front endbell over bearing–rotor–stator
7. Insert first screw
8. Insert second screw
9. Drive both screws
10. Insert first brush holder
11. Insert second brush holder
12. Press both brush holders
13. Off-load completed motor

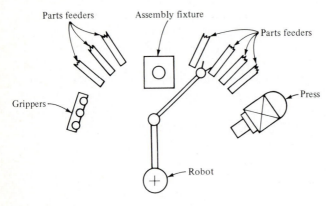

Figure 15-13 Single station robotic workcell for Example 15-1.

perform the screw-fastening operations. A second gripper will be designed to handle the screws, brush holders, and bearings.

We will assume that the RTM method of Chap. Eleven is used to analyze the work cycle and to define the following element times for certain general categories of tasks shown in Table 15-1:

Assembly task	8 s
Gripper change	12 s
Reorientation of work	7 s
Tool cycle (screwdriver and press)	4 s

Given these times and the gripper limitations, the sequence of operations and the corresponding element times would be as shown in Table 15-2. The total assembly time is 234 s. If it were possible to design a single gripper to handle all of the different part configurations in the assembly sequence, then a total of 96 s (elements 2, 4, 7, 9, 11, 13, 18, and 20) could be saved, thus reducing the cycle time to 138 s. If the powered screwdriver could somehow be built into the gripper design, and the screwdriver could be designed to feed the screws as part of its process rather than merely drive them, then an additional 48 s (elements 12, 14, 15, and 16) could be saved. This would decrease the cycle time to 90 s.

The production rate of the assembly workstation can be determined from the cycle time analysis. Using the 234-s cycle time the production

Table 15-2 Sequence of assembly steps for robot in Example 15-1

Step	Time, s
1. Load first endbell (gripper 1)	8
2. Change gripper	12
3. Install bearing (gripper 2)	8
4. Change gripper	12
5. Install rotor (gripper 1)	8
6. Install stator (gripper 1)	8
7. Change gripper	12
8. Install bearing (gripper 2)	8
9. Change gripper	12
10. Install endbell (gripper 1)	8
11. Change gripper	12
12. Install two screws (gripper 2)	16
13. Change gripper	12
14. Retrieve screwdriver (gripper 1)	12
15. Drive two screws (gripper 1)	8
16. Replace screwdriver (gripper 1)	12
17. Reorient motor (gripper 1)	7
18. Change gripper	12
19. Install two brush holders (gripper 2)	16
20. Change gripper	12
21. Load motor into press (gripper 1)	8
22. Cycle press	4
23. Unload motor (gripper 1)	7
Total	234

rate would be

$$R = 60/234$$

$$= 0.2564 \text{ motors/min or } 15.38 \text{ motors/hour}$$

If the 106-s cycle time is used (assuming the improvements in the work cycle could be made), the production rate would be increased to

$$R = 3600/90 = 40.0 \text{ motors/hour}$$

It is obvious that changing grippers can become a time-consuming portion of the cycle. The reader should also note that reorienting parts (in the example there was a need to reorient the motor in element 17) results in lost time. This can often be avoided by designing a workholding fixture which presents the assembly to the robot in a proper orientation.

Several conclusions can be drawn from this example about single-station assembly cells. First, the single-workstation configuration is not very fast, even for assemblies of relatively low complexity (a total of 10 components in our example). Second, other things being equal, the production rate of the cell is inversely proportional to the number of parts in the assembly. Also, it is reasonable to infer, and experience bears this out, that the more parts that must be assembled by the system, the less reliable it will be. Third, a larger number of distinct parts that must be assembled will present a more difficult problem to the designer of the gripper and other tooling (e.g., workholding fixture). In the case of the gripper, either the design must be more complicated to handle the variety of different parts; or the gripper must be instrumented with sensors (usually tactile sensing) to accommodate the part differences; or a tool-changing mechanism must be devised for exchanging the various specialized grippers and other end effector tooling.

The single-workstation assembly system possesses one attractive merit, and that is it requires the least capital expense for low-volume automated production of the alternatives in robotics.

Series Assembly Systems

The manual assembly line is the most familiar assembly configuration. It constitutes the series assembly system. This configuration is used in many medium- and high-production assembly situations, such as automobiles, household appliances, small power tools, and other products made in large quantities. The assembly line consists of a series of workstations at which only a few operations are performed on the product at each station. Each station is working on a different product, and the products are gradually built up as they move down the line. The principle behind this form of assembly is labor specialization: each worker becomes an expert at performing a limited number of tasks and can therefore do them in minimum possible time. Consequently the production rate for this type of system tends to be significantly higher than for the single-station system.

The series assembly configuration can be applied to the design of a robotic assembly system. Instead of workstations where humans perform the assembly operations, the workstations are manned by robots. We have previously referred to this type of workcell in Chap. Eleven as an in-line configuration. The transfer systems used in an in-line robot cell were discussed in Sec. 11-2. Typical hardware used for a series assembly system includes indexing conveyors (intermittent transfer) and power-and-free systems (nonsynchronous transfer). In both of these transfer systems, the work travels in a straight line. Another configuration commonly used in high-speed automated assembly systems, and a possible configuration for a series-type robotic assembly system, is the dial indexing machine. The dial indexing machine consists of a round table with a number of workstations evenly spaced around the periphery. At the end of each work cycle, the table indexes the work to the next station (intermittent transfer). Figure 15-14 shows an assembly robot loading parts in a dial indexing machine.

Current applications experience suggests that the robotic assembly line seems to be most appropriate for medium-volume and even moderately high-volume production. In the medium-volume case, the assembly line would take advantage of the robot's ability to be reprogrammed to perform different tasks, therefore allowing the same assembly line to be used to work on different product styles, probably in a batch mode. In principle it should be possible to implement a flexible automation system in robotic assembly, where the robots would be able to deal with different product styles all being produced simultaneously on the same assembly line.

In the moderately high-production category, the robot's capacity to per-

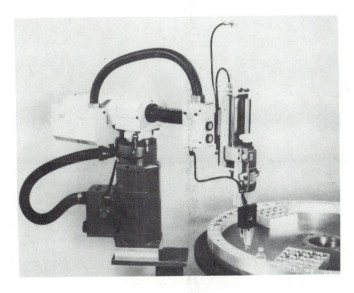

Figure 15-14 GMF robot loading a dial indexing machine. (Photo courtesy of GMF Robotics)

form complex motion patterns or changes in motion pattern would have to be utilized in order for a robot system to be economically feasible. This case would be represented by the assembly of a product in high volumes where there are certain differences in product style requiring alternative programs for the various styles. The spot welding of automobile bodies would be an example of this situation (welding is, after all, an assembly process). If these features of a robot could not be used to advantage, a fixed automation system could probably be designed that would be capable of substantially greater production rates.

Example 15-2 Figure 15-15 shows a series assembly cell laid out for the motor assembly job of Example 15-1. We are assuming that a separate

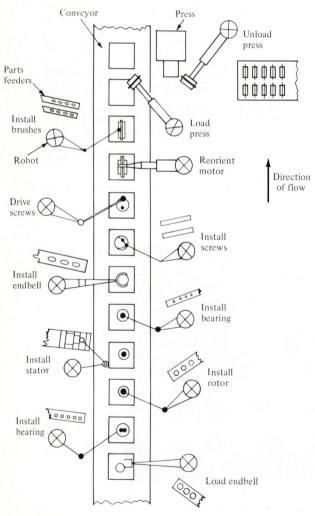

Figure 15-15 Series assembly cell in Example 15-2.

robot at each individual station would be equipped to perform each of the assembly and handling tasks identified in the previous example (elements 1, 3, 5, 6, 8, 10, 12, 15, 17, 19, 21, and 23). Using the RTM data from Example 15-1, it would be possible to design a series system that could work at a cycle time of 12 s plus the time required to move the parts between stations. We are assuming that elements 21 and 22 would be performed at one station (the sum of their element times is 12 s) and that this station would be the limiting cycle on the line. Assuming the transfer could be accomplished in 4 s, this would lead to a cycle time of 16 s, with a corresponding production rate of 225 motors/hour.

Of course, this high production rate did not come cheaply. The series system requires a dozen robots plus a parts transfer system to perform the same basic assembly sequence as the single robot of Example 15-1. Also, the reliability of the system is adversely affected by the large number of inter-dependent components in the system. In the case of an intermittent transfer system, the failure of a single component will cause the entire system to either stop or produce incomplete motors. The use of a nonsynchronous transfer system would reduce the interdependence of the stations on one another.

Example 15-3 This case study describes a series system that was installed to assemble a residential smoke detector. The assembly is made up of four components: base part, cover, printed circuit board, and battery. The components and a completed assembly are shown in Fig. 15-16. An overall layout of the assembly system is illustrated in Fig. 15-17 and an

Figure 15-16 Smoke detector and components in Example 15-3. (Photo courtesy of United States Robots, subsidiary of Square D Company)

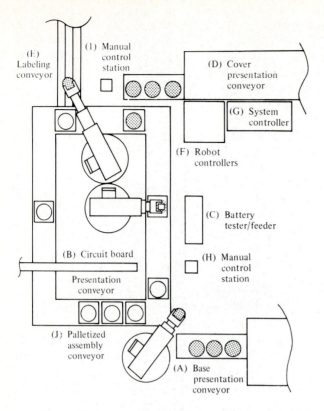

Figure 15-17 Layout of assembly system for Example 15-3.

overview of the cell is pictured in Fig. 15-18. The assembly sequence is as follows:

1. Load base onto assembly fixture.
2. Insert printed circuit board into base.
3. Load battery into base.
4. Snap cover onto base.

The assembly buildup is performed on pallet fixtures which travel to three workstations along the system conveyor. Photoelectric sensors are used almost exclusively to satisfy interlock requirements of the system. Let us examine the operation of each station and some of the features of the workcell control.

Station 1: Load Base. The function of the first station is to properly orient and load the base of the smoke detector into an assembly fixture. An operator loads a full box of bases into a feeder which delivers the parts one at a time to the pickup location near the first robot. A sensor at the pickup location confirms the presence of the part. The robot grasps the

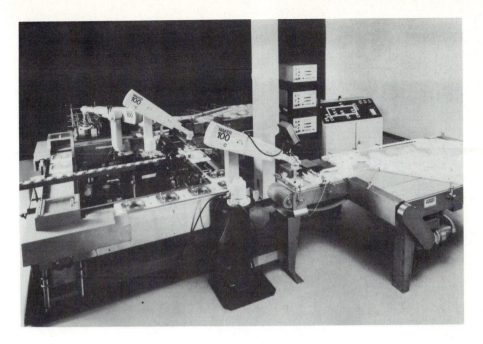

Figure 15-18 Overview of the smoke detector assembly cell in Example 15-3. (Photo courtesy of United States Robots, subsidiary of Square D Company)

part using a vacuum cup gripper and loads it into the fixture. The presence of the fixture at the workstation is also detected by an interlock sensor. The robot must reorient the part onto the fixture by rotating it until a sensor detects a specific feature on the base which indicates the correct orientation. The robot then releases the part, clears its arm from the part by a sufficient distance, and signals the controller that the loading task has been completed. The pallet fixture is then transferred to the second workstation.

Station 2: Battery and Printed Circuit Insertion. Two tasks are performed at the second workstation: insertion of the battery and insertion of the printed circuit board (PCB). A special gripper was designed to grasp both parts using two sets of fingers. This gripper is shown in Fig. 15-19.

The batteries are presented to the robot by a magazine feeder. As each battery is escaped from the feeder, it is tested to ensure that it meets a minimum power output requirement. If the battery fails the test, the workcell controller informs the robot to reject the part into a reject bin. If the battery passes the test, it is inserted into the base.

After completing the first task, the robot grips a PCB from a feeding conveyor. The conveyor delivers the PCB directly from the circuit assembly area and is designed to minimize possible damage from static electricity to the circuit. At the end of the PCB conveyor, an escapement is used to present the boards one at a time to the robot. An input interlock

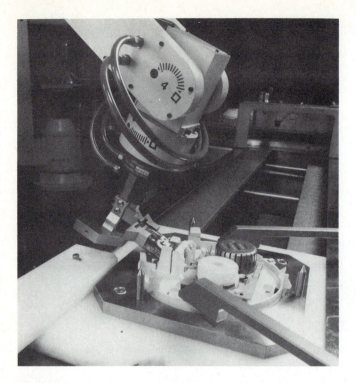

Figure 15-19 Special gripper used at station 2 in Example 15.3. (Photo courtesy of United States Robots, subsidiary of Square D Company)

is used to indicate the presence of a board to the workcell controller. The PCB is grasped by the other side of the gripper used to hold the battery and inserted into the base at the same time that special tooling pins are used to back bend the retaining snaps on the base. The tooling pins are operated by a pneumatic cylinder located beneath the conveyor. As the robot completes the PCB insertion, it signals the workcell controller and the assembly fixture is transferred to the next workstation.

Station 3: Cover Assembly and Final Test. At this station, the robot assembles the cover to the smoke detector base subassembly and performs a functional test of the completed product. The covers are presented to the robot on a conveyor coming from the cover-and-test-button assembly area. The robot orients the cover in a fixture at the pickup point and snaps it onto the base. After the cover is snapped into place, a small plunger in the gripper is activated to push the test button on the completed assembly. A microphone listens for the alarm and the system controller determines if the device is operating properly. If the smoke detector passes the test, it is loaded into an automatic labeler where the product identification label is applied. If the product fails the test, it is dropped into a reject bin. When the sequence of tasks at the third station is completed, the robot signals the cell controller to return the empty pallet fixture to the first station to be reloaded.

Workcell controller. A programmable controller was used as the workcell controller for this application. The cell control functions included:

Interlocking to detect part presence at each station
Controlling the movement of the pallet fixtures through the system
Confirming the correctness of the assemblies
Controlling part escapement at station 2
Synchronizing robot operations
Selecting robot functions
Providing system status display
Providing an operator input station
Providing for manual operation of the system if that becomes necessary
Controlling tests and inspections
Informing the operator when to refill the magazines and feeders

Parallel Assembly Systems

The concept of a parallel arrangement in a robotic workcell is pictured in Fig. 15-20. In essence the work can take either of two (or more) routes to have the same operations performed. There are two conditions under which parallel workstations would normally be considered. The first is the situation where production cycle times at a particular workstation or group of workstations are too long to keep up with the other sections of the line. The other stations are forced to wait for the slow workstations. In this case, the use of two parallel stations effectively halves the cycle time (doubles the production rate) for the stations. This permits the workload to be more evenly distributed among the workstations. The second reason for considering a parallel configuration is when reliability of a certain station (or group of stations) is a problem. Production requirements may be such that a shutdown of the line cannot be tolerated. For example, in the motor assembly line it is likely that certain stations will have greater chance of breakdown, and those are the stations that would be provided with a duplicate station in parallel. This would improve the

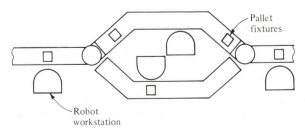

Robot
workstation

Pallet
fixtures

Figure 15-20 Parallel workstations on an assembly system.

proportion of uptime of the line. Again, nothing comes free, and the duplication of equipment adds cost to the system.

Other System Configurations

In addition to the three traditional configurations presented above, it is possible to develop other systems that take advantage of the robot's programmability and adaptability. Based on our discussion of the previous configurations, we have seen how increasing the number of robots in the system increases throughput but decreases reliability. We have also seen how redundancy (e.g., parallel stations) adds to the reliability. A desirable objective in the design of any robotic production system is to provide the greatest throughput with the highest reliability at the lowest possible cost.

An advantage that robotic systems have over most other forms of automation is their capability to adapt to changing conditions in the workplace. If required, the robot can alter its programmed cycle in response to the changes. One cell concept which utilizes the adaptability of the robot is the series layout with overlapping work envelopes, illustrated in Fig. 15-21. As the name implies, this is basically a series configuration in which the robots' work envelopes overlap each other significantly. The tooling of each robot is configured so that they can perform the tasks of the robots in adjacent stations. The reason for this redundancy is to allow the system to tolerate failures of some of the workstations and to continue to operate, although at a reduced production rate. When a station failure occurs, the workcell controller commands the robots on either side of the failed station to share its work cycle. Each robot on the line would be programmed to perform its primary tasks (the work at its station); in addition, when a failure occurs at an adjacent station, a secondary program would be downloaded, instructing it to perform a portion of the work of the broken robot next to it. Although the throughput of the system would be decreased, the line would remain operational in this mode

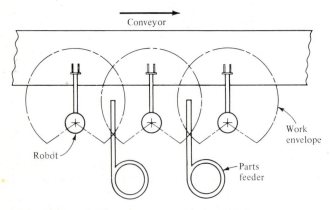

Figure 15-21 Series layout with overlapping work envelopes.

long enough to repair the failure. The advantage of this concept is that it provides the redundancy of the parallel system without the expense.

15-6 ADAPTABLE-PROGRAMMABLE ASSEMBLY SYSTEM

A research and development project in robotic assembly automation which merits special mention in this chapter is the APAS project.[4,12] APAS stands for Adaptable-Programmable Assembly System. The project was sponsored jointly by the National Science Foundation and the Westinghouse Electric Corporation. The actual work was done at Westinghouse and used the assembly of endbells for electric motors as the object of the study. The purpose of the development effort was to advance the state of the art in automated batch assembly. Batch assembly refers to the assembly of various styles of a product in small to medium batch sizes. Because the size of each batch is relatively small and there is a need to change over the production system to accommodate the differences between the product styles, automation of batch assembly has been a difficult technological problem. Yet, it has been estimated that batch assembly represents approximately 75 percent of all manufacturing done in the United States, and constitutes roughly 22 percent of the U.S. Gross National Product.[4] As it is commonly performed it is a highly labor intensive production activity.

The assembly of small electric motors was selected as the basis for the APAS development because it is representative of batch manufacturing. A factory that produces small motors may ship several hundred different styles during a given year. Since the customer orders are generally small, the plant may have to change over from one style to the next several times each day. From the 450 motor types produced by Westinghouse, five representative styles were selected for the study. An experimental demonstration line was constructed with six workstations, four of which utilized robots to perform the assembly tasks. A conveyor system is used to transfer pallet fixtures holding the partially completed assemblies between workstations. The APAS line is illustrated in Fig. 15-22. One end of the line is pictured in Fig. 15-23.

In the first station, the endbell is presented manually to the robot. A vision system is used to locate the endbell with sufficient accuracy for the pickup. The vision system also confirms that the endbell is the one required to satisfy the desired production schedule. If it is the correct endbell, it is placed into a fixture which is transferred to the second workstation. The second station is merely a buffer and transfer station.

At the third workstation, a robot picks up a bearing cap, a felt washer, and a thrust washer. All three parts are inserted at once into the endbell. Feeding of the felt washers and thrust washers proved to be difficult using conventional feeding mechanisms. Special feeders were developed at the University of Massachusetts to feed these kinds of components. Photoelectric sensing was utilized to assure correct feeding of the parts.

The fourth station performs several operations on the top, side, and

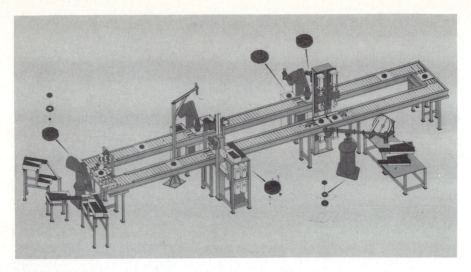

Figure 15-22 Overview of the Adaptable-Programming Assembly System (APAS) for assembling different styles of electric motors.

bottom of the endbell using special-purpose tooling. Automatic screwdrivers were used to drive screws into the tops and bottoms of the endbells, while a specially engineered device presses the caps into the sides of the endbells.

At the fifth station, the feeders are arranged to feed all of the parts needed for assembly to one position for pickup by the robot. Another feature of this station is that all of the tooling is changed for the different parts automatically, without operator assistance. This allows for changeovers with minimum time loss.

The sixth and final station consists of a machine vision system to inspect the entire endbell assembly for missing or misplaced components. The system is integrated with a robot which sorts completed assemblies into appropriate holding bins.

Control of the system is accomplished by a master supervisory computer which communicates with local computers at each workstation. The local computers are interfaced to the workcell using programmable controllers. The master computer is responsible for product scheduling and coordination, and informs the local computers which tasks must be performed at each station.

Some of the conclusions drawn from the APAS project about automated batch assembly include the following:

1. The concept of adaptable-programmable assembly is feasible.
2. Future APAS type assembly systems will probably be a combination of fixed automation and robotics and other programmable automation.
3. Inspection will have to be done as the assembly operations are performed in order to be certain that the assembly is built up as intended at each stage of the overall process.
4. It is important to design the product to facilitate the assembly processes.

Figure 15-23 View of one end of the APAS. (Photo courtesy of Unimation, Inc., Subsidiary of Westinghouse Corp.)

Certain assembly operations are easier to automate than others and this should be taken into account in product design.

Considering this final conclusion, let us devote the last section in our coverage of robotic assembly systems to the problem of product design for automated assembly.

15-7 DESIGNING FOR ROBOTIC ASSEMBLY

Certain assembly tasks are more difficult for a robot to perform than others. If possible, this difficulty factor should be considered in the design of the product. As an example, for a robot to accomplish the screw-fastening operation without the use of an automatic screwdriver is difficult. Even with a powered device to perform the operation, the process of turning the screw into the part requires time. If the objective of using a threaded fastener is to allow for subsequent disassembly (e.g., for service of the product), then the use of screws may be an appropriate design decision. However, if the particular assembly or subassembly is designed to be permanent, then perhaps a better choice than screw fastening would be a press-fit or adhesive bonding of the parts.

Another consideration in the design of an assembly is the direction in which the parts are to be added in the assembly operation. If the parts can all be added without reorienting the partially completed subassembly, then time and money can be saved. On the other hand, if the subassembly requires many reorientations, then handling time is being spent without adding any real value. Similarly, if all the components can be added along the same axis direction, a robot with fewer degrees of freedom can perform the assembly tasks. This suggests that stacking of parts during the construction of an assembly is advantageous.

Today's robots are typically one-armed machines. Coordination of more than one arm at a time is difficult with current control technology. Interpreting this limitation in terms of limitations on the assembly process, an automated mating or joining operation should require the robot to handle no more than one part at a time. Assemblies that require the robot to manipulate two parts simultaneously or to maintain the relationship between two parts while adding a third may require a significant amount of fixturing. The solution to these problems is to design the parts so that they maintain their relationships with each other by designing such features into the components as locating bosses, grooves, and other mating elements.

In order to facilitate automatic assembly, it is often appropriate to add certain features to the components. For example, breaking edges and corners on parts, and adding chamfers to the holes will make it easier to accomplish part insertion tasks. These design features added to the components will minimize the robot's accuracy requirements and should allow faster operating cycles. Also, the design of distinct alignment features into the parts, or purposely making an otherwise symmetric part into an asymmetric part, makes it easier to feed and mate the parts in the proper orientation. These added part features will probably necessitate extra processing operations which may increase manufacturing costs of the components. The increases in part costs must be justified by corresponding reductions in assembly costs.

There are many other considerations in designing for automated assembly, and this section is not intended to be a thorough treatment of the subject. We leave it to the interested reader to explore the literature, in particular Refs. 1, 3, and 11.

15-8 INSPECTION AUTOMATION

Inspection is a quality control operation that involves the checking of parts, assemblies, or products for conformance to certain criteria generally specified by the design engineering department. The inspection function is commonly done for incoming raw materials at various stages of the production process, and at the completion of manufacturing prior to shipping the product. Testing is another quality control operation often associated with inspection. The distinction between the two terms is that testing normally involves the functional aspects of the product, such as testing to ensure that the product operates properly, fatigue testing, environmental testing, and similar procedures. Inspection is limited to checking the product in relation to nonfunctional design standards. For example, a mechanical component would be inspected to verify the physical dimensions (e.g., length, diameter, etc.) that have been established by the design engineer.

This section will consider how robotics can be used to perform inspection operations. It turns out that fully automated manufacturing systems have a significant need for inspection to be incorporated as part of the production operation. We encountered this need in our discussion of the APAS project. While automation may reduce inconsistency and errors in manufacturing, it also removes the sensory capabilities and judgment that the human operator can bring to the production process. These capabilities (or some of them, at least) must somehow be replaced in fully automated production operations.

Robotics can be used to accomplish inspection or testing operations for mechanical dimensions and other physical characteristics, and product function and performance. Generally the robot must work with other pieces of equipment in order to perform the operations. Examples include machine vision systems, robot manipulated inspection and/or testing equipment, and robot loading and unloading operation with automatic test equipment. The following subsections will discuss these three categories of robotic inspection systems.

Vision Inspection Systems

Machine vision as a sensor in robotics was discussed in Chap. Seven. Some of the robotic applications of vision systems include part location, parts identification, and bin picking. Machine vision can also be used to implement a robotic inspection system. Typical robotic vision systems are capable of analyzing two-dimensional scenes by extracting certain features from the images. Examples of inspection tasks carried out by this procedure include dimensional accuracy, surface finish, and completeness and correctness of an assembly or product. The robot's role in the inspection process would be either to present the parts to the vision system in the proper position and orientation, or to manipulate the vision system over the portions of the parts or assemblies that must be inspected.

In the design of a machine vision inspection system, there are a number of factors that must be considered in order for the system to operate reliably. These factors include:

The required resolution of the vision camera
The field-of-view of the camera relative to the object being inspected

Any special lighting requirements
The required throughput of the inspection system

A representative example of a machine vision inspection system will illustrate this robotic application.

Example 15-4 Inspection of a PCB provides a good example of a potential vision system application. The robot is used to present the PCB to the vision system in the proper position. If the board is relatively large or there is a need to inspect both sides of the board, the robot would be used to reposition as needed. The PCB consists of a number of electronic components mounted on an epoxy board. The process of fabricating the assembly is complex and there are many opportunities for error. Yet the nature of the product makes it difficult to inspect by human eye.

There are a number of features of the board assembly that would be of interest to the manufacturer. These include:

The size of the epoxy board and the location of certain mounting holes
The number and location of the electronic components on the board
The identification of the components on the board

For instance, the PCB may consist of an epoxy board measuring 250 by 450 nm, with a certain number of integrated circuit modules and other components each with its own unique identification (size, color, and part number). The vision inspection system would be programmed to perform the following operations:

1. Check the dimensions of the circuit board to ensure that it is within specification.
2. Search for the components mounted on the board and count them to verify that the correct quantity are present.
3. Identify the components by type to make certain that the correct component was located in the designated position on the board.

Robot-Manipulated Inspection or Test Equipment

This method of robotic inspection involves the robot moving an inspection or testing device around the part or product. An example would be for a robot to manipulate an electronic inspection probe or a laser probe along the surface of the object to be measured. As long as the accuracy of the measurement is not required to exceed the repeatability of the robot, the approach is feasible.

Example 15-5 Manual methods of inspecting a car body often involve building a fixture large enough to surround the body, and measuring the desired dimensions on the body relative to gauge points on the fixture. Since robots are capable of positioning an end effector to precise locations in space, the car body dimensions can be measured relative to the robot's coordinate system. There are a number of examples in the automotive industry where robots are used to inspect certain dimensions on car bodies.

In one application reported for one of the Ford Motor Company assembly plants,[8] four Cincinnati Milacron T3 robots perform a series of dimensional checks on automobile bodies at the rate of five or six bodies per hour. This compared with the previous manual inspection method in which only one or two inspections were completed each shift. The robots manipulate electromechanical probes to make approximately 150 dimensional checks around windshield, door, and window openings.

At a General Motors plant, an application was developed in which several robots were fitted with laser-ranging equipment. As the automobile body passed through the robotic inspection station, the laser gauges were positioned at critical locations around the car body and measurements were taken. Using this system, GM was able to test every car body that came down the line, rather than one per shift using the traditional manual techniques.

Robot-Loaded Test Equipment

The third application area in robotic inspection is loading and unloading inspection and testing equipment. This application is very similar to machine tool loading/unloading. There are various types of inspection and testing equipment that can be loaded by a robot. These include mechanical, electrical, and pneumatic gauges, and functional testing devices.

A robotic inspection application would logically be incorporated into the manufacturing method. The robot would be used to unload the finished part from the production machine and to load it into an inspection gauge which would determine if the part was acceptable. If the part were within tolerance, it would be passed to the next step in the production process. If it did not meet the tolerance specification, the part would be rejected. Example 13-2 (Chap. Thirteen) illustrated this type of production-inspection cell.

Taking this inspection system one logical step further, the robotic system would act as a feedback control system by making adjustments for tool wear and other sources of variation in the metal cutting process. The automatic inspection system would be programmed to determine not only whether the parts were within tolerance, but also to perform a trend analysis and to input this information into the machine tool control system so that compensating adjustments could be made to the tool path.

Functional testing is commonly used in the electronics industry. The quality and performance of the product cannot be determined by visual inspection alone. Accordingly, the parts must be loaded into functional testers which measure the response of the system to certain controlled inputs. For example, if a voltage is placed across the coil of a relay, one would expect the contacts to close and the resistance of the contacts to decrease. A tester can be built to perform this kind of performance test. A more elaborate functional

Figure 15-24 Automatic test cell of Example 15-6. (Photo courtesy of United States Robots, subsidiary of Square D Company)

tester can be designed to test the response of all the circuits in a printed circuit board.

Example 15-6 Figure 15-24 shows a robot cell for testing printed circuit boards. The PC boards are presented to the robot in a tote bin. The robot unloads the board from the tote bin and places it in one of the available testers. It then signals the tester to perform the test. When the tester completes the functional testing, it informs the robot and indicates whether the board has passed the testing procedure. The robot then sorts the PCB accordingly into the appropriate output tote bins.

Integrating Inspection into the Manufacturing Process

As we have discussed earlier in this chapter, inspection is a vital component of the automated assembly process. This is true not only in assembly but also in other automated manufacturing methods as well. As the human operator is removed from the workstation, the function of checking the work must be taken over by other means. One of the features of a robotic workcell is that the inspection can usually be added for a nominal capital cost. The inspection process can often be accomplished on the finished part at the same time as the production process is working on the next part. Therefore, the added time to inspect can be minimized. It is likely that the automated factory of the future will be characterized by a very high level of integration between the manufacturing process and the inspection process. The earlier that a defect is discovered in the automated manufacturing process, the less expensive the part repair or scrap will be. In fact, automatic part inspection may even eliminate certain breakdowns in the production equipment due to jamming of the parts or similar failures.

Automating a production operation requires more than merely mechanizing the process itself. Using a robot to perform the production or assembly process and then to inspect its work, and even correct for errors, is a fundamental step toward achieving the fully automated factory of tomorrow.

REFERENCES

1. M. M. Andreasen, S. Kahler, and T. Lund, *Design for Assembly*, IFS Publications Ltd. (U.K.), and Springer-Verlag (Heidelberg), 1983.
2. F. L. Bracken, "Design of Data Processing Equipment for Automated Assembly," *IBM Technical Report*, Endicott, NY, June 29, 1983.
3. G. Boothroyd and A. H. Redford, *Mechanized Assembly*, McGraw-Hill, London, U.K., 1968.
4. N. Captor, B. Miller, B. D. Ottinger, A. J. Riggs, L. M. Tomko, and M. C. Culver, "Adaptable-Programmable Assembly Research Technology Transfer to Industry," *Final Report*, NSF Grant ISP 78-18773, January 1983.
5. M. P. Groover, *Automation, Production Systems, and Computer-Aided Manufacturing*, Prentice-Hall, Englewood Cliffs, NJ, 1980, chaps. 4, 5, and 6.
6. M. P. Groover and E. W. Zimmers, Jr., *CAD/CAM: Computer-Aided Design and Manufacturing*, Prentice-Hall, Englewood Cliffs, NJ, 1984, chaps. 11 and 19.
7. Lord Corporation, *Robowrist—Remote Center Compliance Devices (RCC Series)*, Publication PC-8034b, Erie, PA, 1983.
8. G. C. Macri and C. S. Calengor, "Robots Combine Speed and Accuracy in Dimensional Checks of Automotive Bodies," *Robotics Today*, Summer 1980, pp. 16–19.
9. W. E. McIntosh, "Automating the Inspection of Printed Circuit Boards," *Robotics Today*, June 1983, pp. 75–78.
10. J. L. Nevins and D. E. Whitney, "Computer-controlled Assembly," *Scientific American*, February 1978, pp. 62–74.
11. F. J. Riley, *Assembly Automation—A Management Handbook*, Industrial Press, New York, 1983.
12. R. N. Stauffer, "Westinghouse Advances the Art of Assembly," *Robotics Today*, February 1983, pp. 33–36.
13. R. N. Stauffer, "IBM Advances Robotic Assembly in Building Word Processor," *Robotics Today*, October 1982, pp. 19–23.
14. R. N. Stauffer, "Robotic Assembly," Special Report, *Robotics Today*, October 1984, pp. 45–52.
15. D. E. Whitney, "Quasi-static Assembly of Compliantly Supported Rigid Parts," *Journal of Dynamic Systems, Measurement, and Control*, March 1982, pp. 65–77.

SIX

IMPLEMENTATION PRINCIPLES AND ISSUES

The preceding chapters on applications engineering for manufacturing were concerned with the technical problems that must be solved in order to apply robotics in specific factory operations. In addition to the technical problems, there are other problem areas and issues that must be considered. These problems and issues are more management-oriented than the engineering issues considered in Part Four. The implementation of robotics in an organization requires management involvement as well as engineering expertise. In the final analysis, the introduction of a successful robotics program in a firm is the responsibility of management.

This part of the book is concerned with the management issues that must be addressed in the implementation of robotics. We divide these issues into two chapters. The first chapter (Chap. Sixteen) provides a step-by-step approach that a company can apply to introduce industrial robots into its operations. The approach presumes a minimum starting knowledge about robotics on the part of the company. It includes a procedure for making a plant survey, selecting a proper application, and selecting the best robot for the application. This approach takes the reader up to the installation of the robot cell, relying on analysis methods presented in Chaps. Eleven and Twelve.

Chapter Seventeen describes several additional problem areas that management must deal with during and after installation. The title of this chapter indicates the specific areas that are discussed: operator safety, training, maintenance, and quality issues associated with robotics.

SIXTEEN

AN APPROACH FOR IMPLEMENTING ROBOTICS

Robotics is a sophisticated technology and the successful implementation of this technology in industry is a formidable management problem as well as a technical problem. The purpose of this chapter is to describe a logical approach that we propose for introducing a robotics program into an organization. Some of the steps described in the approach relate closely to the applications engineering methods previously discussed in Chaps. Eleven and Twelve. Other aspects of the approach go beyond the engineering techniques required to implement robotics.

We describe the approach for implementing robotics in terms of a logical sequence of steps that a company would want to follow in order to implement a robotics program in its operations. The steps in the approach are the following:

1. Initial familiarization with the technology
2. Plant survey to identify potential applications
3. Selection of the application
4. Selection of the robot
5. Detailed economic analysis and capital authorization
6. Planning and engineering the installation
7. Installation

We describe these seven implementation steps in the following seven sections of this chapter. The sections contain a number of tables and checklists that might be useful to a firm in its implementation program.

16-1 INITIAL FAMILIARIZATION WITH ROBOTICS TECHNOLOGY

Many companies are in the following situation: None of their personnel have any expertise in robotics, but they believe that there are potential applications for robots in their plants. In order to become involved with the technology and be capable of making rational decisions on robot projects, these personnel are faced with the problem of quickly becoming knowledgeable with the field.

The sources of information on robotics include books, technical magazines and trade journals, robot manufacturing companies, consulting firms, technical seminars, conferences, and trade shows. At the time of this writing, there are several dozen books on robotics available. There are more than half a dozen magazines, trade journals, and newsletters devoted specifically to the field of robotics. In addition, many of the other trade publications include feature articles on topics in robotics. There are well over 250 organizations that provide products and services related to robotics. These organizations include manufacturers, suppliers, consultants, schools, and research institutions working in robotics or fields related to robotics.

In addition to these materials, there are also seminars, conferences, and trade shows devoted to the technology and application of robotics. At Lehigh University we offer a workshop seminar on "robotics in manufacturing." Similar seminars are offered by organizations like the Institute of Industrial Engineers (IIE) and Robotics International of the Society of Manufacturing Engineers (SME). More specialized seminars are also offered by a number of organizations on topics such as end effector design, developing robot work-cells, and arc-welding applications. The annual Robot Conference and Trade Show, cosponsored by the Robotic Industries Association and Robotics International of SME, provides an opportunity for the robot industry to show its products and exchange ideas at the various technical paper presentations included in the conference.

In the process of introducing robotics into a firm, the importance of management support should not be underestimated. Many of the reports of successful robot installations point to this as a critical factor. The implementation of robotics in a firm is usually a long process, perhaps spanning several years before the first application project is completed. It is important that management provide continuing support and encouragement during this startup period. Some companies have lost valuable time by turning on and off their support to their manufacturing staff functions as attempts were made to implement robotics.

Another management issue involves the approval process required to install a robot project. One might assume that the authorization to implement a robot project would constitute management support for the project. However, the approval process in the firm might be such that the manager who decided to invest in the robot system is not the same as the manager who will use it in the plant. For the project to be successful, it is important that the manager responsible for using the system also be committed to its success.

Checklist:

Item	Points to be Distributed	Driving	Restraining
1. Can workers be openly assured of job retention?	20		
2. Can workers displaced, but retained, be placed in equally rated jobs?	15		

3. Will the installations benefit the workers in terms of:
  a. Health?
  b. Safety?
  c. Relief from dehumanizing jobs?
  d. Relief from dirty, overly hot, back-breaking, onerous tasks?
	15		

4. Is the present union–management climate favorable to open exchange? Disclosure of economic conditions? Labor unrest and frequent grievances? Usually distrustful? (If no union, assign points on similar issues for management–work force relations.)	15		
5. Is the present economic condition of the organization sufficiently healthy to guarantee that promises are kept?	5		
6. Have manufacturing engineering and other management units shown the ability to establish rapport with workers or does inordinate "social distance" exist?	5		
7. Is there management recognition and concern for the dehumanizing aspects of jobs to be performed by a robot? Or is the concern solely economic?	5		
8. Is there a plan to select and upgrade workers who will supervise or perform setups for the robot?	5		
9. Will workers on incentive rates be penalized by new rates or robot downtime that is not attributable to operator?	5		
10. Has management in the past demonstrated respect and regard for the talents, skills, and intelligence of the workers?	3		
11. Is the organization willing to share the results of this checklist with the work force and/or union?	3		
12. Will robot training be on organization time? Is there willingness to send the workers (if required) to the robot vendor's training school?	2		

Figure 16-1 General Electric work force acceptance checklist. (*Adapted from General Electric source materials by permission.*)

Fig. 16-1 (*continued*)

Item	Points to be Distributed	Driving	Restraining
13. Can workers express their concerns, apprehensions, and fears, without riducule?	2		

Scoring:

    1. Total Driving Points = _____

    2. Total Restraining Points = _____

        Net Score (1–2) = _____

Interpretation:

Range of Net Score	Probability of Acceptance
80–100	High. Implementation may proceed, assuming management acceptance conditions are equally high rated.
60–80	Proceed with Caution. After examination of the feasibility of changing strength of existing forces.
40–60	Insufficient. Reexamination of forces and management action required to increase probability.
Below 40	Failure More Than Likely. A score in this range indicates a poor probability of even modifying the forces.

Another key to successful implementation of robotics in a company is to include production personnel in the project. Production operators are likely to know the manufacturing operations better than anyone since they perform these operations. It makes sense to try to use their knowledge in developing and implementing the robot application. Also, production people will be more likely to accept the robot if they have participated in its installation in some way. The General Electric Company has developed a work force acceptance checklist to assess the potential for successfully implementing robotics in a plant. Figure 16-1 presents an adapted version of this checklist. Many of the issues considered in this checklist are contingent upon management commitment and action. Without the acceptance of the workers in the plant, the problem of installing and operating a robot cell in the plant becomes much more difficult.

16-2 PLANT SURVEY TO IDENTIFY POTENTIAL APPLICATIONS

As described in Chap. Twelve (Sec. 12-1), there are two general categories of robot applications that must be distinguished. The first category is where the robot project involves the design of a new plant or a new facility within an existing plant. Here, the applications engineer has greater flexibility in the design of the project. Although the same general application selection criteria apply to both cases, the new facility offers the opportunity for designing the application to achieve the greatest benefit from robotics technology. The manufacturing operation itself can be examined to determine the best method for accomplishing the process using available robotics technology.

The second category is the robot project in an existing facility. The problem here is to substitute a robot in place of the human operator in an existing production operation. The applications engineer has fewer options to select from in this case because the robot must be adapted to existing equipment. The least expensive robot installation often requires that the robot be used in the same way that the human operator performed the job. In this second case, the potential applications for robots are most conveniently identified by means of a tour through the plant to survey the existing operations.

In performing a plant survey, the objective is to determine those existing operations which are susceptible to automation by robotics. Opportunities for the application of today's robotics technology have certain characteristics in common. These general characteristics will usually make a potential robot application technically practical and economically feasible. These general characteristics are[3]:

1. *Hazardous or uncomfortable working conditions.* Work situations possessing potential hazards to a human operator are often ideal situations to install a robot. The potential hazards include physical dangers and health hazards from heat, sparks, radiation, toxicity, or the use of carcinogenic materials. Even if the job situation is not actually hazardous, but the workplace is considered uncomfortable, unpleasant, and undesirable by humans, this represents a good potential robot application. Hazardous or uncomfortable work situations for humans have a high probability for worker acceptance of a robot application. The workers do not like to work under these conditions, so they can more readily accept the automation of their jobs so long as it does not lead to unemployment. There are many examples of current-day robot applications in this category, including spot welding, arc welding, die casting, and spray painting.

2. *Repetitive operations.* Repetitive operations are very common in high-volume and medium-volume production. The operation consists of a sequence of work elements that are performed over and over. Human operators usually perform this work and they generally find it boring and degrading. Industrial robots are ideally suited to many operations in this category because of their capability to repeat a fixed-motion pattern without deviation from one cycle to the next. The basic requirements are that the robot must be provided with the proper end effector to accomplish the particular task, and its work volume must be sufficient to include the workspace needed for the operation. Examples of repetitive operations where robots have been successfully used include pick-and-place operations, machine loading and unloading, and spot welding.

3. *Difficult handling jobs*. A third general characteristic of jobs where robots are applied is in the handling of difficult-to-hold objects. The objects can be either workparts or tools, and the reasons why they are difficult to hold is that they are heavy, hot, or possess a shape that makes them awkward for a human to grasp. Plate glass is an example of this last characteristic. Workers would probably need some form of mechanical assistance to hold and manipulate these kinds of parts, such as a crane or a hoist. A robot with enough lift capacity and equipped with an appropriate end effector should be considered for these difficult handling tasks.

4. *Multishift operation*. Many manufacturing operations run two or three shifts in order to meet the demand for the product. In some cases, the nature of the process requires that it be operated 24 hours per day. Plastic molding and many other high-temperature operations need startup periods that make it economical only to run the process continuously rather than intermittently. When these processes are operated using human workers, the labor cost is a variable cost that continues during the second and third shifts at the same or slightly higher rate. When industrial robots can be used to substitute for the human workers, there is a high fixed cost and a relatively low variable cost associated with the installation. The advantage over the use of human labor is that the fixed cost can be spread over the number of shifts, thus reducing the total operating costs for the process.

During the plant tour, an attempt is made to identify the operations that possess these characteristics. The characteristics can be found by looking for: operations in which some form of protective clothing is worn by the workers (e.g., welding masks and helmets, safety clothing, etc.); operations requiring special equipment to protect the workers (e.g., ventilating systems); operations which have a repetitive work cycle (e.g., large and medium quantities of product); and operations where the operator needs some form of mechanical assistance in handling the workparts or tools (e.g., hoists and cranes). The plant supervision will know which operations are carried out on more than one shift; this is not always clear from a plant survey carried out during the day shift.

16-3 SELECTION OF THE BEST APPLICATION

By surveying the plant operations, a number of potential robot applications can usually be identified. The problem is then to determine which potential application(s) to pursue. An obvious criterion in selecting the best application(s) is economic. It is appropriate to perform a preliminary economic analysis on the alternative potential applications, using the methods of Chap. Twelve, to determine which alternative seems to offer the best financial payback and return on investment. To accomplish this analysis, the existing operations would have to be studied to determine current production rates and costs, and a robot method would have to be proposed in order to estimate its investment costs and operating costs.

In addition to the economic criteria, the potential applications must be subjected to certain technical criteria as well. The General Electric Company has been quite successful at finding good robot applications by applying the following criteria to the industrial operations during the survey[1]:

The operation is simple and repetitive.
Cycle time for the operation is greater than 5 s.
Parts can be delivered to the operation in proper location and orientation.
Part weight is suitable (1100 lb is typically used as the upper weight limit).
No inspection is required for the operation.
One or two persons can be replaced in a 24-hour period.
Setups and changeovers are infrequent.

If the potential application satisfies all of these criteria, it is considered to be an attractive candidate for a robot application. It should be noted that these General Electric criteria include not only technical issues, but economic issues also.

For a company engaged in its first robot project, an additional piece of advice often given by experienced applications engineers is to start with a simple application, one that does not require a high level of sophistication in the workcell layout, workstation control, and end effector design. It is important for the company's initial application to be a success so that the new technology will be accepted by the personnel in the plant. Even for companies that have experienced a number of prior robot installations, it is a good idea to accomplish the straightforward applications and minimize the risk of a failure. There is a tendency among technical personnel to become enthralled with the technology of robotics, and this may cause the company to attempt an application whose probability of success is relatively low. The application may work adequately under pilot conditions in the laboratory, but its chances of success in the factory are much lower.

In the case of a new facility being planned but not yet in operation, the same general criteria for deciding on appropriate robot applications can be applied. However, in this case, the applications engineer is not necessarily bound by existing equipment limitations. The project can be designed from the ground up, and a wider variety of alternatives can therefore be considered.

16-4 SELECTION OF THE ROBOT

After the application has been selected, the applications engineer must choose which robot to use for the job from the many commercial models available. Indeed, selection of the application must often be done with consideration of available robots in mind.

The robot selected should possess an appropriate combination of technical features (e.g., number of axes, type of control system, work volume, ease of programming, precision of motions, load carrying capacity, etc.) for the application being considered. If a deviation from the needed specifications is made, it should be made in the direction of greater robot capabilities rather than less. It might turn out that when the current application is completed and another application is being sought for the robot, the new application may require greater technological capabilities than the current application.

Table 16-1 presents a listing of the kinds of typical robot technical features needed in common applications. It should be emphasized that the data included in this table are considered to be representative of current robot application practice, but exceptions to the recommendations in the table can be found in successful installations in industry today. However, the use of this table might help the applications engineer to reduce the number of alternative robot models to select from.

Table 16-1 Technical features required of robots for selected applications

Application	Typical technical features required
Material transfer	Number of axes: 3 to 5 Control system: limited sequence or point-to-point playback Drive system: pneumatic or hydraulic (for heavy loads) Programming: manual, powered leadthrough
Machine loading	Anatomy: Polar, cylindrical, jointed arm Number of axes: 4 or 5 Drive system: electric or hydraulic (for heavy loads) Programming: powered leadthrough Control system: limited sequence or point-to-point playback
Spot welding	Anatomy: polar, jointed arm Number of axes: 5 or 6 Drive system: hydraulic or electric Programming: powered leadthrough Control system: point-to-point playback
Arc welding	Anatomy: polar, jointed arm, cartesian Number of axes: 5 or 6 Drive system: electric or hydraulic Programming: manual or powered leadthrough Control system: continuous-path playback
Spray coating	Anatomy: jointed arm Number of axes: 6 or more Drive system: hydraulic Programming: manual leadthrough Control system: continuous-path playback
Assembly	Anatomy: jointed arm, cartesian (box), SCARA Number of axes: 3 to 6 Drive system: electric Programming: powered leadthrough, textual language Control system: playback: point-to-point or continuous path Accuracy and repeatability: high

To make the final selection of the robot, the following decision procedure is suggested. The procedure consists of preparing a detailed listing of the technical features for the particular application and then systematically comparing these features against the specifications of the alternative models under consideration. It is advantageous to divide the list of technical features into two categories: "must" and "desirable." The "must" features are ones that must be satisfied by the robot in order to perform the application. If any of the candidates do not satisfy the "must" features, then that robot model is excluded from further consideration.

The "desirable" features are ones that are not necessarily required to accomplish the application but would be highly beneficial during installation and/or operation. The specifications of each robot candidate would be compared to each of the desirable features, and a rating score would be assigned to the candidate to indicate how well the robot satisfies the particular feature. There may be differences in relative importance among the various features, and this would be taken into account by giving each feature a maximum possible point score. Determination of the rating score for the different robot models in each feature category would be a judgment call that the applications engineer would have to make based on the relative merits of each candidate.

Example 16-1 Figure 16-2 illustrates a possible format for making the comparison of the application features against the available robot specifications. This is based on an actual decision table developed by a company for selecting a robot for an arc-welding application. The organization of the form has been changed and some of the features have been stated differently here for clarity. The company, of course, will remain anonymous. The robots are identified in Fig. 16-2 as A, B, C and D, rather than using the actual company and model number.

The features are divided into the two categories: "must" and "desirable," according to our suggested procedure. The must features were considered essential for the welding application. It turned out that one of the models being considered has only five axes, whereas six axes were considered a requirement. Therefore, that model (Model C) was eliminated from further consideration.

The desirable features were each evaluated as to its relative priority by giving it a possible range of point score values (the numbers in parentheses in Fig. 16-2). The applications engineer made judgments to determine how each of the remaining three models rated for the given feature. It should be noted that some of the desirable features listed in the form included nontechnical considerations as well as technical considerations. Price, delivery, and vendor evaluation were considered important by the company in selecting the robot model. Based on the sums of the point scores, Model D was selected as the best robot for this application.

Technical Feature	Robot Model Candidates			
	Model A	*Model B*	*Model C*	*Model D*
"Must" Features				
Continuous-path control	OK	OK	OK	OK
6 axes	OK	OK	X	OK
Walkthrough programming	OK	OK	OK	OK
"Desirable" Features				
Ease of programming (0–9)	6	4	—	6
Ability to edit program (0–5)	4	2	—	5
Multipass features (0–4)	2	2	—	2
Work volume (0–9)	5	8	—	6
Repeatability (0–5)	5	2	—	4
Lowest price (0–5)	4	5	—	3
Delivery (0–3)	1	1	—	3
Evaluation of vendor (0–9)	6	5	—	8
Totals	33	29	—	37

Conclusions:

1. Model C is eliminated from consideration because it does not satisfy all of the "must" features.

2. Model D would be selected because it satisfies all of the "must" features and it has the highest point score among the "desirable" features.

Figure 16-2 Sample form used to compare application features against robot technical specifications.

16-5 DETAILED ECONOMIC ANALYSIS AND CAPITAL AUTHORIZATION

With reference to our seven-step approach, the company has now selected an appropriate robot application and has also decided which robot model would be best for that application. In order to implement the project, authorization must be given by upper management. The procedures for management authorization of an investment project vary from company to company. What is usually required is that a detailed economic and technical analysis be documented to justify the proposed project. The economic analysis would estimate the probable financial benefits to be derived from the project. These benefits are often reduced to measures such as the payback period and the return on investment, using the methods discussed in Chap. Twelve.

The technical analysis would detail the engineering and technical feasibility of the project. The engineer must describe the project in terms of its application features, required change to existing equipment, new equipment that must be acquired, fixtures and tooling, anticipated production rates, effects on labor, potential problem areas, and other similar characteristics. Many of these details are required in order to accomplish the economic analysis.

Based on the documentation of the economic and technical analysis, management must decide whether to proceed with the project or not. If the decision is to implement the project, a capital authorization is provided to spend money. The funds would be used to accomplish the detailed planning and engineering work and to purchase and install the equipment for the project.

16-6 PLANNING AND ENGINEERING THE INSTALLATION

The planning and engineering of the installation involves many of the analysis and design considerations discussed in Chap. Eleven. Table 16-2 provides a checklist of these considerations in the approximate order in which the applications engineer would have to deal with them in the implementation of a robotics project.

The initial item in the list is a careful study of the operation and the way it should best be performed using a robot. Some prior thought has undoubtedly been given to this issue during the selection of the application, the selection of the robot, and the detailed economic analysis. Now that approval has been given to proceed with the project, a fresh look at the problem is appropriate.

The study would consider the basic purpose and function of the operation. What task must be accomplished during the operation? A common shortcoming is to limit one's thinking to consideration of how a human would perform the operation. There are likely to be differences between the most appropriate method for a robot and the best method for a human. For example, a human operator can readily perceive many types of defects in a workpart and discard those parts that are defective. A robot, without some form of sensing capability, is unable to detect even the most obvious flaws in the part. Attempting to process a defective part could cause damage to the tools or the equipment used in the operation. The limitations of the robot must somehow be taken into account in the design of the method. It might be necessary to inspect the parts before they are delivered to the workstation, or to incorporate a manual or automatic inspection procedure into the robot cell operation.

Another issue dealing with the operation method is the relationship of the part and the tooling used to process the part. The most natural method for a human might be to fixture the workpart in a stationary position and to manipulate the tool relative to the part. The applications engineer should consider alternatives to the best human method when planning the robot method. It might be more appropriate to reverse the positions of the tool and workpart if a robot performs the task. The RTM (Robot Time and Motion)

Table 16-2 Checklist of considerations and problem areas to be addressed during planning and engineering of the robot installation

---

1. Study of the operation method
   What is the basic purpose and function of the operation?
   What is the best method for a robot to perform the operation?

2. Design of robot workcell
   Which of the three basic types of cell layout should be used?
   a. Robot-centered cell.
   b. In-line robot cell.
   c. Mobile robot cell.
   What changes to other equipment must be made to accommodate operation and control in the robot cell?
   Consideration of part positioning and orientation coming into the cell and leaving the cell.
   Consideration of part identification methods if more than one part style is processed through the cell.
   Protection of the robot from its environment.
   Provision for utilities and other services required for the cell.

3. Workcell control
   What are the basic functions that must be performed by the workcell controller in this operation?
   What interfaces to the human operator must be included?
   What interlocks must be designed into the cell?
   What sensors must be used to accomplish the interlocks?
   Are there additional sensor requirements that must be satisfied?
   Type of workcell controller. Does the robot have sufficient control capacity or must an additional cell controller be incorporated into the cell?

4. Safety considerations designed into the cell.

5. End effector design.

6. Design of other tools and fixtures for the cell.

---

method discussed in Chap. Eleven might be useful in assessing which method would be the least time consuming.

An issue closely related to the operation method is the workcell configuration. The method used for the operation will largely determine the configuration. Again, the best cell layout for a human operator might not be what is best for a robot. The applications engineer should attempt to develop a workplace layout that is best suited to the robot rather than to a human. A human is very mobile and capable of walking from one location in the workplace to another. Most robots are stationary. For a stationary robot to perform the task, the cell would have to be designed so that the work is within the robot's reach. This would probably result in a much more compact workcell than what would be considered safe or desirable for a human. As an extreme example of the contrast between the robot cell configuration and the human cell configuration, the best mounting for the robot might be in an inverted position from an overhead stand. For machine loading/unloading applications, many robots today are mounted directly to the machine tool. The applications engineer should not restrict the possibilities for locating the robot to floor mountings.

In order to perform the operation, the robot will require some form of end effector. Additional tooling and fixturing may also be required for the application. Chapter Five discussed the various types of end effectors and the calculations involved in the design of a gripper.

16-7 INSTALLATION

After approval has been given for the detailed plans and designs of the robot project, installation begins. Basically, installation involves the implementation of the detailed plans that have been prepared. Table 16-3 presents a checklist of the kinds of activities included in the installation phase.

Other aspects of the installation phase include startup, debugging, trial production runs, and fine tuning of the workcell. It usually turns out that there are minor problems with the workcell immediately after setup. Examples of these problems might include software bugs, sensor problems, improperly located components in the cell, and mechanical difficulties with the equipment. These minor problems must be solved before full-scale production can be achieved.

The time required to complete the installation is typically from three months to a year. Critical items in the installation process are the lead times between order and receipt of the robot and other equipment. Depending on business conditions and the backlog of orders at the robot company, these lead times can cause significant delays in the installation procedure.

There are several additional issues related to installation that should be

Table 16-3 Checklist of activities included in the installation phase

Purchase of the robot(s) and other equipment and supplies needed to install the workcell.

Preparation of the physical site in the plant where the robot cell is to be located. This might include altering the foundation to support heavy machine tools in the cell and to fix the relative positions of the robot and other equipment. Also included would be any provisions for protection of the robot from its environment (e.g., high temperatures, dangerous fumes or mist in the atmosphere, electrical noise, fire hazards, etc.)

Provision of electrical, pneumatic, and other utilities for the cell.

Adaptation of standard pieces of equipment for use in the cell.

Placement of robot and other equipment: installation of conveyors and other materials-handling systems for delivery of parts into and out of the cell.

Installation, checkout, and programming of the workcell controller.

Installation of interlocks and sensors, and integration with the workcell controller.

Installation of safety systems.

Fabrication of end effectors and other tooling.

considered. These issues are safety, training, maintenance, and quality control. They are concerned with activities for which some planning must take place in advance of the installation, but which occur during and after installation. The following chapter deals with these management issues.

REFERENCES

1. J. A. Behuniak, "Planning the Successful Robot Installation," *Robotics Today*, Summer 1981, pp. 36–37.
2. J. F. Engelberger, *Robotics in Practice*, AMACOM (American Management Association), New York, 1980, chaps. 6, 7.
3. M. P. Groover and E. W. Zimmers, Jr., *CAD/CAM: Computer-Aided Design and Manufacturing*, Prentice-Hall, Englewood Cliffs, NJ, 1984, chaps. 10, 11.
4. R. N. Stauffer, "Equipment Acquisition for the Automatic Factory," *Robotics Today*, April 1983, pp. 37–40.
5. W. R. Tanner, "A User's Guide to Robot Applications," *Industrial Robots, Volume I—Fundamentals*, Society of Manufacturing Engineers, Dearborn, MI, 1979, pp. 13–22.
6. L. L. Toepperwein, M. T. Blackman, et al., "ICAM Robotics Application Guide," Technical Report AFWAL-TR-80-4142, Vol. II, Materials Laboratory, Air Force Wright Aeronautical Laboratories, OH, April 1980.
7. J. P. Van Blois, "Strategic Robot Justification: A Fresh Approach," *Robotics Today*, April 1983, pp. 44–48.

SEVENTEEN

SAFETY, TRAINING, MAINTENANCE, AND QUALITY

In addition to the implementation issues discussed in Chap. Sixteen, which deal mainly with the problems of getting a company started in robotics, there are other issues that are concerned with the ongoing operation and management of the robot installations. Primary among these are operator safety, technical training, robot maintenance and repair, and quality benefits that derive from robotics. In this chapter we discuss these management-oriented issues.

17-1 SAFETY IN ROBOTICS

There are two aspects of the safety issue in robotics. The first deals with the justification of robots. Historically, one of the fundamental reasons for using robots in industrial applications is to remove human operators from potentially hazardous work environments. The hazards in the workplace include heat, noise, fumes, and other discomforts, physical dangers (potential injuries or even loss of limbs), radiation, toxic atmospheres, and other health hazards. The problem of removing or reducing these hazards from the workplace has provided one of the important justifications for industrial robots in applications such as welding, forging, spray painting, and die casting. Since the Occupational Safety and Health Act (OSHA) was enacted in 1971, worker safety has become a significant factor in promoting the substitution of robots for human labor in these kinds of dangerous jobs.

The second aspect of the safety issue involves the potential hazards to humans posed by the robot itself. The use of robots presents a new set of possible dangers to the human worker for which precautions must be taken. It is this second aspect of the safety issue which we will address in this chapter.

In considering the potential hazards that are encountered in the use of a robot, it seems reasonable to question when during the use of the robot are humans in contact or close proximity to it. There are three occasions when humans are close enough to the machine to be exposed to danger.[6] These are:

During programming of the robot
During operation of the robot cell when humans work in the cell
During maintenance of the robot

The types of risks encountered during these times include physical injury from collision between the human and the robot, electrical shock, objects (parts or tools) dropped from the robot gripper, and loose power cables or hydraulic lines on the floor. Some of these risks can be reduced with straightforward safety measures such as proper grounding of electrical cables to prevent shock, and raised floor platforms to cover power cables and hydraulic lines. In other cases, operator safety is improved by requiring certain common sense procedures to be followed. For example, when the robot is being programmed, the speed of the arm should be set at a low level during teaching and testing of the program. Another example would be when the maintenance personnel are servicing the robot. During maintenance, the power to the machine should be turned off under normal circumstances.

More extensive safety measures must be taken to guard against hazards that arise during robot operation. Many of these safety measures must be designed into the workcell, either as part of the workplace design or as part of the workcell control system.

Workplace Design Considerations for Safety

Certain safety features can be designed into the robot workcell. These include physical barriers to limit intrusion into the cell, emergency stop buttons to halt the cell operation, and laying out the equipment in the cell for maximum safety.

The most common approach is to construct a physical barrier around the periphery of the robot workcell. The periphery of the robot cell must be defined to be outside the farthest reach of the robot in all directions with end effector attached to the wrist. The workcell would also include any equipment in the cell which operates with the robot. The barrier should not be designed only for the programmed work cycle envelope because a malfunction of the controller may cause the robot to follow a trajectory different from its normal program. The barrier has the effect of preventing human intruders from entering the vicinity of the robot while it is operating. The barrier often consists of a fence with a gate for access to the workcell. The gate is equipped with an interlock device so that the work cycle is interrupted when the gate is open. A positive restart procedure is designed into the cell for resumption of the cycle, rather than using the gate closure for restarting. Other possible physical barriers include safety rails and chains, although these are not as effective as a full fence.

Another approach that has been used in industry as a physical barrier is a steel post in the floor at the limits of the programmed motion cycle, so that an out-of-control robot arm crashes into the post rather than go beyond its allowed space. However, there are certain undesirable aspects to this type of barrier that make it unsuitable. First, if the robot arm crashes into the steel post at high speed, it will no doubt be damaged or destroyed. This kind of mishap could occur during programming of the robot rather than as a result of a malfunction of the robot itself. Thus, although the steel post is intended to protect against a robot control failure, it is possible that the robot arm could be ruined due to human error instead. A second feature which makes the post unsatisfactory is that it does not prevent intruders from entering the robot cell. A third undesirable aspect of the steel post is that a human could be pinned between the robot arm and the post. This would probably result in greater injury to the human than by simply being struck in open space by the robot arm.

In a robot cell designed to operate with humans as coworkers in the production process, certain features must be designed into the cell layout to protect the workers. This is typically encountered when human workers are employed to load and unload workparts in the cell and the robot is used to perform a processing operation such as arc welding or grinding. In these cases, some form of two-position parts manipulator can be used to exchange parts between the robot and the worker. One possible workcell configuration for accomplishing this exchange is illustrated in Fig. 17-1. It uses a rotary indexing table to move raw workparts from the human operator's position to the robot position for processing, and simultaneously moves completed parts from the robot position to the operator's position for unloading. This arrangement prevents inadvertent collision between the worker and the robot. Some form of operator interface is required to permit the worker to index the

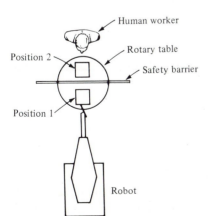

Figure 17-1 Cell layout using part manipulator to separate human worker from robot for safety and production efficiency.

table for the next cycle, and to perform other functions such as emergency stop. In addition to its safety features, this type of layout has the additional advantage of production efficiency since it allows two operations to be performed simultaneously rather than sequentially. The loading and unloading of parts by the human operator takes place simultaneously with the processing of parts by the robot.

Safety Sensors and Safety Monitoring

In addition to these various approaches for designing safety into the robotic workcell, other safety provisions can be made as well. We will describe some of the possible safety monitoring schemes that can be utilized in robot workcells in this subsection, and other measures, including emergency stop buttons and "deadman switches."

Safety monitoring, as previously defined in Chap. Eleven, involves the use of sensors to indicate conditions or events that are unsafe or potentially unsafe. The objectives of safety monitoring include not only the protection of humans who happen to be in the cell, but also the protection of the equipment in the cell. The sensors used in safety monitoring range from simple limit switches to make sure that certain steps in the sequence control have been carried out, to sophisticated vision systems that are able to scan the workplace for intruders and other deviations from normal operating conditions. We have discussed some of the possible sensors that are used in robotic workcells in Chaps. Six and Seven. An important point that should be made in the context of this discussion on safety monitoring is that the workcell controller is limited in its monitoring capability to irregularities that have been foreseen by the designer of the cell control system. If the designer has not anticipated a particular hazard, and consequently has not provided the robot with the sensing capacity to monitor that hazard, the workcell controller will not be able to respond to the event. Great care must be taken in workcell design to anticipate all of the possible mishaps that might occur during the operation of the cell, and to design safeguards to prevent or limit the damage resulting from these mishaps.

The National Bureau of Standards defines three levels of safety sensor systems in robotics[4]:

Level 1—Perimeter penetration detection
Level 2—Intruder detection inside the workcell
Level 3—Intruder detection in the immediate vicinity of the robot

The first level systems are intended to detect that an intruder has crossed the perimeter boundary of the workcell without regard to the location of the intruder. In effect this would operate much the same as the fence surrounding the cell. Level 2 systems are designed to detect the presence of an intruder in the region between the workcell boundary and the limit of the robot work volume. The exact definition of this region would depend on the cell layout and the strategy used to ensure the safety of the intruder. Level 3 systems provide intruder detection inside the work volume of the robot. These sensor systems are intended to protect workers who must be in close proximity to the robot during operation of the robot (e.g., during programming of the robot). This third category must be capable of detecting an imminent collision between the worker and the robot, and of executing a strategy for avoiding the collision. Figure 17-2 illustrates the three sensor levels.

Two common means of implementing a robot safety sensing system are pressure sensitive floor mats and light curtains. Pressure sensitive mats are area pads placed on the floor around the workcell which sense the weight of someone standing on the mat. Light curtains consist of light beams and photosensitive devices placed around the workcell that sense the presence of an intruder by an interruption of the light beam. Pressure sensitive floor pads can be used for either level 1 or level 2 sensing systems. The use of light curtains would be more appropriate as level 1 systems. Proximity sensors located on the robot arm could be utilized as level 3 sensors.

The safety monitoring strategies that might be followed by the workcell controller would include the following schemes. Some of the strategies would be more appropriate for certain levels of sensor detection systems than for others.

Complete shutdown of the robot upon detection of an intruder
Activation of warning alarms
Reduction in the speed of the robot to a "safe" level
Directing the robot to move its arm away from the intruder to avoid collision.
   This is sometimes referred to as "obstacle avoidance."
Directing the robot to perform tasks in another region of the workcell away
   from the intruder.

A more sophisticated system used in safety monitoring is called a "fail–safe hazard detector".[7] The concept of this detector is based on the recog-

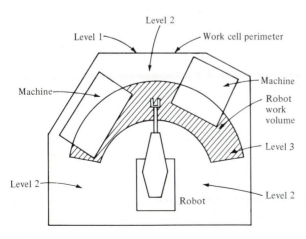

Figure 17-2 Three levels of safety sensor systems: Level 1—perimeter penetration; Level 2—intruder detection in the workcell; Level 3—intruder detection inside the robot work volume.

nition that some component of the basic hazard sensor system might fail and that this failure might not be found out until some safety emergency occurred. The fail–safe hazard detector is designed to overcome this problem. The detector consists of the usual sensor subsystem for monitoring some potential hazard in the cell, but it also possesses the capability to periodically and automatically check the sensor subsystem to make certain it is operating properly. This capability is achieved by means of a challenge subsystem which simulates the hazard that the sensors are supposed to detect, and then checking for any interruption in the anticipated response of the sensor subsystem. In effect, the fail–safe hazard detector monitors for possible failure of the basic sensor subsystem. This fail–safe arrangement provides for a much more reliable safety monitoring system.

Other Safety Measures

Emergency stop buttons (also referred to by the term "panic button") are usually located on both the main control panel and the robot teach pendant. They are designed to be easily identifiable in case of an emergency situation. The stop control should be capable of stopping not only the robot itself, but also the other moving equipment (e.g., conveyors) in the cell. The obvious purpose of the panic button is to interrupt the work cycle in the event of an emergency situation that may cause harm to either equipment or humans.

A "deadman switch" is a useful control feature during leadthrough programming. It is a trigger or toggle switch device generally located on the teach pendant which requires active pressure to be applied to the device in order to drive the manipulator. If the pressure is removed from the trigger or toggle switch, then the device springs back to its neutral position which stops all robot movement. This type of teach control avoids the possibility of the programmer inadvertently leaving the power drive to the arm in the "on" position.

Safety is an important issue in robotics as it should be in any automated production system. As previously indicated, one of the important social objectives of robots and other forms of automation is to remove humans from unsafe and uncomfortable working conditions. It would be counterproductive to this objective if the robot workcell itself proved to be unsafe during those infrequent occasions when humans must enter.

17-2 TRAINING

Training is an important factor in the successful implementation of any advanced technology in a company. In robotics, training is especially important because this technology impacts on so many different areas of a company's operations. The training must include the management and engineering staff as well as the operating and maintenance personnel in the plant. Hinson[2] divides the robotics training that should take place in a company into five categories:

Awareness
Justification
Application
Operations and maintenance
Safety

Awareness training provides a survey of robotics, including technology, applications, economics, and social implications. It also explores the future trends and research developments taking place in robotics. This category of training is typically presented to managers and engineers to encourage the implementation of robotics and to examine the opportunities for applying this technology. Training programs designed for production and maintenance personnel are designed to dispel the mystery surrounding robots by explaining how the machines operate and by presenting examples of successful applications.

Training in the justification of robots is intended for engineers and managers who are responsible for implementing robot projects in the company. Justification training deals with economic issues and the unique problems that arise in the justification of robots as compared to other investment projects. It is often the case that conventional company justification criteria do not include some of the benefits that accrue from the use of robots. The purpose of the justification training is to examine these benefits and to incorporate consideration of them in the justification procedures.

Training in applications is designed principally for technical people (engineers, engineering managers, production managers, and foremen) who must select the applications and plan the installations. Coverage in this category would include technical areas such as basic technology (Chap. Two), robot programming (Chaps. Eight and Nine), and applications engineering issues (Chaps. Eleven through Fifteen).

Training in robot operation and maintenance must be provided for production and maintenance personnel. It is designed to give these persons the detailed technical skills and knowledge required to use and service the equipment. Most robot manufacturers provide training programs as part of the sales contract with the customer. These programs are held either at the robot company's facilities or, in some cases, at the customer's plant where the robot is to be installed. A typical training course would cover areas such as programming, operation, maintenance, and repair of the robot. A typical training course offered by the robot manufacturer may last 1 or 2 days for a relatively simple robot and from 1 to 2 weeks for a more complex robot. This type of training should be timed to coincide with the installation of the robot. Because of employee turnover and the need to periodically upgrade operating personnel, the user company must plan on regular training programs for its people.

The construction and operation of robots are generally similar to other existing machinery in the plant. As a result, the electrical and mechanical trade skills of the maintenance staff already in the plant are usually good enough to build any additional competence that is required for robotics. The new skills that might need to be developed for robot maintenance would include more electronics, computers, microprocessors, programmable controllers, and robot programming. In addition, the maintenance staff might have to become familiar with the use of special diagnostics equipment designed for analyzing robot equipment problems.

Because the robot training courses provide instruction in the programming of the robot, the following question arises: Who should do the robot programming for the user company? Direct labor operating personnel? Maintenance staff? Manufacturing or industrial engineering? It seems consistent with other production technologies such as numerical control that direct labor should not be given the overall responsibility for programming the robot. Neither should the maintenance staff be given this responsibility. Robot programming should be considered a management planning function similar to numerical control part programming or methods and workplace design. Accordingly, the primary responsibility for programming should be assigned to a suitably trained staff person who acts as an agent for management. The actual input of the program to the robot may be performed by a direct labor operator under the supervision of the staff person.

The final category of training is concerned with safety. It should be the goal of the company to give all personnel who are involved with robots in the plant an awareness of the potential dangers associated with this technology. Production supervision, engineers, operating personnel, and maintenance staff must all be provided with appropriate safety training in robotics.

17-3 MAINTENANCE

Robots are sophisticated electronic/mechanical systems whose reliability is generally good. Nevertheless, the complexity of these machines means that occasional equipment failures do occur and periodic maintenance service is required. The essential ingredients of an effective robot maintenance program for a company are the following:

A maintenance staff of highly skilled, highly trained personnel
An appropriate preventive maintenance program
A rational spare parts policy

In the subsections below, we will discuss these areas, and attempt to define several basic concepts and methods of analysis that are applicable to robot maintenance.

Maintenance Staff

Perhaps the most important single ingredient in the maintenance program is the existence of a good maintenance staff in the company, trained and skilled specifically for robotics.

The maintenance staff is responsible for two types of maintenance activity: preventive maintenance (PM) and emergency maintenance. Preventive maintenance involves the planned servicing of equipment at periodic intervals. We will discuss PM in the next subsection. Emergency maintenance (also called remedial maintenance) is the case when the maintenance crew is called in to repair a robot that malfunctions or breaks down during regular operation.

In the case of emergency maintenance, the maintenance staff must respond as quickly as possible to the problem. Because of the cost of lost production time and the possible dependency of other equipment on the robot, it is imperative that the downtime be minimized. If a crew is available (i.e., it is not occupied servicing another piece of equipment elsewhere in the plant), then the crew can respond immediately to the service call. If the crew is engaged in repairing another machine, then existing company maintenance policy should dictate whether the robot has a higher priority than the equipment currently being serviced. The determination of priority level should be based on parameters such as safety and cost. If the crew cannot respond immediately to the robot repair call, then the total downtime will consist of waiting time plus service time at the robot. Economic decisions for determining the optimum number of repair crews can be derived from queueing theory. These decisions reflect an attempt to balance the cost of equipment downtime against the maintenance labor cost.

The service time at the robot in an emergency situation divides into three categories: diagnosing the problem, actual repair, and checkout of the equipment to verify that the problem has been corrected. It often turns out that the most time-consuming category is the diagnosis of the problem. This is the area where the skill level of the maintenance crew is probably the most critical. The actual repair and checkout of the robot are usually routine activities once the problem has been identified.

The capabilities of the user company's staff in diagnosing maintenance problems can be augmented in several ways. Some robot manufacturers sell diagnostics equipment that connects to the robot controller to help identify the probable cause of the breakdown and the proper remedy to apply. The economics of the situation are such that the user company would want to have more than a few robots of the particular model in order to justify the investment in the diagnostics equipment for that model. Another aid to the user's maintenance crew is the robot manufacturer's service staff. The problem can often be properly diagnosed by means of a telephone call to the manufacturer in which the nature of the problem is explained, and the service engineer attempts to identify the repair which seems most appropriate. If the preceding approaches do not work, the last resort is to have the manufacturer's service engineer make a service call to the user's plant to repair the robot. The cost of this alternative is high, with rates at the time of this writing averaging about $600 per day plus travel expenses. The likelihood of avoiding these expenses can be increased by improving the qualifications of the user's own maintenance personnel.

Preventive Maintenance

In a plant that is highly mechanized and automated, emergency maintenance problems are unavoidable. One way of reducing their frequency is for the company to have an appropriate preventive maintenance (PM) program. The objective of PM is to service the equipment at periodic intervals to reduce the occurrence of emergency breakdown incidents. By servicing the machines in a planned and systematic fashion, it is expected that the number of equipment failures will be reduced, and that those which do occur will be less severe. In addition, PM can be accomplished more conveniently during times when the production equipment is not in regular operation. For example, the PM on the machinery might be performed during model changeovers or on the third shift in a two-shift plant.

In robotics, PM consists of checking, cleaning, and possibly replacing certain mechanical and electrical components of the robot at regular time intervals. The typical components include "O" rings, seals, bearings, bushings, valves, and other parts that are subject to wear. The robot manufacturers usually include a recommended maintenance program in their operating manuals, indicating which components should be periodically serviced. The preventive maintenance programs vary greatly depending on the manufacturer and the complexity of the robot.

One of the measures used to assess the reliability of a piece of machinery is the mean time between failure (MTBF). This measure indicates how long, on average, the machinery will operate between breakdowns. Engelberger[1] reports that one of the design criteria used at Unimation, Inc. for its robots is a MTBF of 400 hours. When a breakdown occurs, a certain amount of time is required to service the robot. The mean time to repair (MTTR) is the measure used to indicate how much time, on average, is spent repairing the robot for each breakdown. These two measures (MTBF and MTTR) can be combined to indicate the proportion of time that the robot is available for operation. This measure is called the availability:

$$\text{Availability} = \frac{\text{MTBF} - \text{MTTR}}{\text{MTBF}}$$

The effect of good preventive maintenance should be to increase the MTBF and to reduce the MTTR for an emergency breakdown situation. This would result in an increase in the availability of the equipment as the following example illustrates.

Example 17-1 Suppose the present MTBF of a particular robot is 200 hours and the MTTR when breakdowns occur is 8 hours. A preventive maintenance program is to be initiated in the plant which is expected to increase the MTBF to 300 hours and reduce the MTTR to 6 hours. Determine the effect of the PM program on the availability of the robot.

Before the PM program is introduced the availability is

$$\text{Availability} = \frac{200 - 8}{200} = 0.96 \text{ or } 96\%$$

As a result of preventive maintenance, the expected availability of the robot will become

$$\text{Availability} = \frac{300 - 6}{300} = 0.98 \text{ or } 98\%$$

In order to operate a successful PM program, it is important to maintain a record of the maintenance performance of each piece of equipment in the plant. This record should include data on the times between failures of the equipment, the times to repair, the nature of the maintenance problem, and the remedial action taken including the components repaired or replaced. By keeping this kind of maintenance log for each robot, statistics such as MTBF and MTTR can be calculated so that the most appropriate PM can be planned and scheduled for each robot. Another reason for keeping a record of the maintenance performance for each robot is to help develop an effective spare parts policy.

Spare Parts Policy

A robot might consist of several hundred to several thousand components. Certain of these components are subject to gradual wear or sudden failure which could disable the entire robot. It is important that the user company consider the problem of keeping an inventory of parts on hand to replace those on the robot that wear or fail. Ottinger[5] suggest a budgetary estimate for spare parts of 10 percent of the robot base price. Many of the robot manufacturers provide a list of recommended spare components that should be stocked by the user.

The user's spare parts policy can cover the full range of inventory levels. At the one end is the policy of keeping no spare parts on hand, except miscellaneous items such as fuses that the company would probably have anyway. At the opposite end of the range, the company might elect to keep a complete duplicate robot available to replace a robot that has failed. The decision depends on finding an appropriate balance between the cost of robot downtime and the cost of spare parts inventory. For an automobile manufacturer with several dozen spot-welding robots on its car body assembly line, it might be worthwhile to have one or more entire robots on hand as spares to replace robots on the line which periodically fail. The broken robot would be quickly removed from its position in the line and replaced by the spare robot, and the program for that station would be entered into the memory of the replacement robot. The broken-down robot could then be repaired and used as one of the spares. This kind of policy allows the downtime on the assembly line to be minimized. For most companies, the optimum spare parts policy would require a certain level of spare parts to be maintained rather than a complete spare robot.

As indicated above, deciding on the best spare parts policy is a matter of balancing the inventory cost of the spare parts against the cost of robot downtime. The inventory cost would include the cost of storage space plus the cost of interest on the capital tied up in the inventory. The purchase cost of the actual parts is not included in the inventory cost since these components would have to be purchased in any case, either as spare parts or when the robot breaks.

When a breakdown occurs, there will be production time lost while the robot is being repaired. If the spare parts needed to fix the robot are available in the plant, it is expected that the repair time and cost will be minimized. On the other hand, if the spare parts are not available in the plant, the time to repair the robot will be delayed while the required spare parts are ordered and delivered from the manufacturer. The corresponding cost of lost production time will be greater for this case.

The user's problem is to decide which spare parts to keep in inventory. From a statistical viewpoint, there are certain components of the robot that are more likely to fail. There are other components that are quite unlikely to fail. The user company would want to select its spare parts inventory so as to emphasize those components that are more likely to fail. There would be little merit in keeping spares of components for which there is no chance that they will be needed. As the company considers what level of spare parts inventory to maintain, it is faced with the law of diminishing returns. As it increases its level of inventory, a point will be reached at which the cost of keeping the part is greater than its value as a spare. The nature of the problem can perhaps best be illustrated by means of an example.

Example 17-2 Suppose that the cost of the robot is $80,000, and the user company wants to establish an optimum level of spare parts inventory. From previous experience and manufacturer's recommendations, the company knows which components of the robot are most likely to fail, and it wants to build its spare parts to a stock level at which the total cost of inventory plus lost production time is minimized. When a breakdown occurs, the repair time will average 2 hours if the required spare part is available. If it is not available, the company figures that it will lose an average of 10 hours of production while the part is being ordered and delivered. In addition, rush delivery of the part(s) will cost $50 per order. The company estimates that the cost of downtime is $100 per hour. Therefore, the cost per downtime incident will be $2 \times \$100 = \$200$ if the spare part is available, and $10 \times \$100 + \$50 = \$1050$ per occurrence if it is not available.

The robot operates one shift per day for 250 days per year, a total of

2000 hours per year. The mean time between failure of the robot is 200 hours. It is therefore anticipated that the robot will break down an average of $2000/200 = 10$ times per year.

The company is considering alternative levels of spare parts inventory, from no spare parts all the way to a spare of every component in the robot. As the level of inventory gets larger, the probability increases that the company will have the needed spare part to fix a broken-down robot. The following table indicates the various alternative spare parts levels being considered and the corresponding probability that the robot can be repaired using available parts in stock. We refer to this probability as the Pr(coverage) and the probability that the needed spare is not available in stock as the Pr(no coverage):

Spare parts level	Pr(coverage)	Pr(no coverage)
0	0	1.00
$ 5,000	0.25	0.75
10,000	0.40	0.60
20,000	0.65	0.35
30,000	0.80	0.20
40,000	0.90	0.10
80,000	1.00	0

The cost of inventory is 25 percent per year of the value of the spare parts in stock. The total annual cost of inventory plus downtime is an expected cost. The expected cost is equal to the sum of three components: (1) the inventory cost, (2) 10 breakdowns per year times the 2-hour downtime cost times the Pr(coverage), and (3) 10 breakdowns times the 10 hour downtime cost times the Pr(no coverage). For example, the total annual cost of keeping the $20,000 spare parts level would be

$$E(\text{cost}) = 0.25(\$20,000) + 10(0.65 \times \$200) + 10(0.35 \times \$1050) = \$9975$$

The results of the computations for the seven levels of spare parts inventory are as follows:

Spare parts level	E(cost)—Annual
0	$10,500
$ 5,000	9,625
10,000	9,600
20,000	9,975
30,000	11,200
40,000	12,850
80,000	22,000

The minimum expected cost occurs at a spare parts inventory level of $10,000.

The difficulty in this problem is not in the computations. It is in determining which spare parts to keep in inventory and the relationship between the level of spare parts and the probability that the robot can be repaired using the available parts. We referred to this probability as the Pr(coverage).

The example illustrates the potential value of keeping an inventory of spare parts, if the spares are properly selected. Generally, the optimum level of spare parts inventory will increase as any of the following factors increases: the cost of robot downtime, the number of shifts of operation of the robot cell, and the difference in downtime between whether the spare part is available in the plant or it needs to be specially ordered.

17-4 QUALITY IMPROVEMENT

One of the benefits often claimed for robotics is that the quality of the manufactured product will improve when a robot performs the operation. In this section we examine this claim and suggest the possible sources of quality improvement that might result from the use of a robot as a substitute for human labor.

It turns out that the robotics quality benefit derives not so much from the robot itself as from the requirements that are imposed on the operation when a robot is used. These requirements include certain aspects of the workcell design, and the prerequisite quality of the workparts received from upstream operations. In the workcell design category, certain demands are made by the robot when it is used to accomplish a manufacturing operation. In a typical application, the robot must receive the raw workpart in a known orientation and location; other equipment in the cell must be in an exact position relative to the robot; and the delivery of the finished parts must be made to a consistent location after the operation. A greater level of automation and mechanization is required in the robot installation than for a manual operation. Each of the above demands must be satisfied in the design of the workcell. The benefits derived from these workcell features include consistency in handling and processing of parts, and less damage to the workparts due to rough handling and part-to-part contact. These benefits contribute to better quality control in manufacturing.

The second requirement imposed by a typical robot application is better consistency of workparts coming into the cell. The usual robot is not capable of dealing with part variations and defects. A human operator can recognize a defective part and is capable of making adjustments in the process, or discarding the defect, or taking other steps to compensate for the defect. (It should be noted, however, that in spite of their capabilities, human workers cannot always be relied on to take these compensating actions in the face of poor incoming quality.) A dumb robot is incapable of any of these responses unless special sensors and program subroutines are designed into the cell. Consequently, most robot cells require that the parts delivered into the cell are of higher uniformity and quality than for a typical manual operation. This requirement places additional demands on the supply sources and the previous processes in the manufacturing sequence.

Both of the preceding improvements in manufacturing quality result indirectly from the use of robotics. In the first case, mechanized parts-handling techniques and better workcell design are likely to improve quality control whether or not robots are used. And in the second case, there is every reason to believe that better parts quality going into the operation will result in better outgoing quality in both a manual cell and a robot cell.

A third area of quality improvement, more closely related to the use of the robot, is the regularity and repeatability of the robot cycle. The robot is capable of performing the work cycle with less variation in the operation time and in the motion pattern. In processes that require a minimum of variation in cycle times (e.g., plastic injection molding and die casting), robots are able to perform these operations with greater uniformity than human workers. The processes can be fine tuned to operate at optimum conditions, and these conditions contribute to higher quality. For processes that require consistency in the motion pattern (e.g., spray painting, spot welding, and many other robot processing operations), robots can accomplish these tasks with more repeatability than humans. Given that the parts presented to the robot are uniform and that the programming has been properly carried out, the resulting quality is generally higher on average than when human workers perform the operation.

REFERENCES

1. J. F. Engelberger, *Robotics in Practice*, AMACOM (American Management Association), New York, 1980, chaps. 5, 6.
2. R. Hinson, "Training Programs Are Essential for Robotics Success", *Industrial Engineering*, September 1983, pp. 26–30.
3. R. Hinson, "Robots Provide Improved Quality in Manufacturing," *Industrial Engineering*, January 1984, pp. 45–46.
4. R. D. Kilmer, "Safety Sensor Systems for Industrial Robots," *Technical Paper MS82-221*, Society of Manufacturing Engineers, Dearborn, MI, 1982.
5. L. V. Ottinger, "Robot System's Success Based on Maintenance," *Industrial Engineering*, June 1983, pp. 38–43.
6. R. D. Potter, "Requirements for Developing Safety in Robot Systems," *Industrial Engineering*, June 1983, pp. 21–24.
7. L. L. Toepperwein, M. T. Blackman, et al., "ICAM Robotics Application Guide," Technical Report AFWAL-TR-80-4142, Vol. II, Materials Laboratory, Air Force Wright Aeronautical Laboratories, OH, April 1980.

PROBLEMS

17-1 Data have been collected for a certain robot model on times between breakdowns and it has been determined that the mean time between failures is 324 hours. Repairing the robot has required an average of 6.5 hours, according to the records of the maintenance department. Calculate the robot's availability.

17-2 Four robots of the same manufacturer and model are used in a robot-automated manufacturing line. The line is fully integrated and if one robot fails, the entire line must be stopped until repairs are completed. Data have been collected for each robot on times between failures and repair times. The data were taken after the normal break-in problems were solved and cover the first 1200 hours of operation of the line. The data are as follows:

Robot	Time between failures, hours	Repair times, hours
1	196	4.7
	115	2.1
	280	10.8
	304	5.2
	237	6.6
2	76	1.1
	404	7.4
	282	3.9
	126	6.2
	205	9.3
3	165	3.4
	358	8.2
	260	5.5
	329	7.6
4	45	3.3
	124	5.5
	236	9.4
	288	6.1
	301	7.0
	201	3.2

(a) Compute the mean time between failure (MTBF) and the mean time to repair (MTTR) for this robot model according to the four samples given.

(b) Determine the availability for this robot model from your computed values of MTBF and MTTR in part (a).

(c) What is the effective availability of the manufacturing line as a result of the fact that there are four robots?

17-3 The cost of a certain robot is $90,000. It is desired to establish the appropriate level of spare parts inventory for this machine. Based on previous experience, it is known which components of the robot are most likely to fail. The spare parts inventory level will be established so that the total cost of inventory plus lost production time is minimized. The cost of inventory is 35 percent per year of the value of the spare parts in stock.

The robot operates one shift per day (8 hours per day) for a total of 2000 hours per year. The mean time between failure (MTBF) of the robot is 250 hours. When a breakdown occurs, the repair time averages 4 hours if the required spare part is available. If the part is not available, the company loses the 4 hours repair time plus the time for the needed part to be delivered. The delivery time results in the loss of 2 working days. For an 8-hour shift, an average of 20 hours of production time is lost while the part is being delivered and repairs are being made. In addition, there is a delivery charge of $25 per order for the needed parts. The company figures that the cost of downtime is $75 per hour. The following table indicates the various alternative spare parts levels being considered and the corresponding probability that the robot can be repaired using available parts in stock [we have called this the Pr(coverage)]:

Spare parts level	Pr(coverage)	Pr(no coverage)
0	0	1.00
$ 5,000	0.20	0.80
10,000	0.35	0.65
15,000	0.45	0.55
20,000	0.55	0.45
30,000	0.65	0.35
40,000	0.80	0.20
50,000	0.90	0.10
60,000	0.95	0.05

(*a*) Based on the data provided, determine the optimum level of spare parts inventory to maintain.

(*b*) At the level of inventory determined in part (*a*), determine the "availability" of the robot.

17-4 Solve Prob. 17-3 only assume that the robot will operate for two shifts per day (16 hours per day) instead of 8 hours per day.

SEVEN

SOCIAL ISSUES AND THE FUTURE OF ROBOTICS

The final part of the book is concerned with the social aspects of robotics, and with the future technology and applications of robotics. In the social area, we focus mainly on the labor issues that are raised by robotics. How will the labor market be affected by robotics? How many workers are likely to be displaced? What are the impacts on the professional and semiprofessional workers who are employed in manufacturing? These, in addition to other social issues (e.g., productivity, international economic competition, and education), will be analyzed in Chap. Eighteen.

In Chap. Nineteen, we attempt to assess where the technology of robotics is heading. What are the research and development projects currently being performed in areas related to robotics? These areas include control systems, improved sensors, advances in the mechanical design of the manipulator, mobile robots, and anthropomorphic hands. By examining the work in these areas, we are able to construct a profile of what the future robot will probably be like.

In Chap. Twenty, we discuss the possible applications that are likely to result from these technological developments. The applications can be divided into manufacturing and nonmanufacturing types. In manufacturing, we consider how the applications will change in the future. In the nonmanufacturing category, robots of the future are likely to be exploring outer space, fighting fires, mining coal, and performing other jobs that are considered hazardous to humans. In addition to these kinds of jobs, future robots are also likely to be used in the service industries working in restaurants, banks, and hospitals. Perhaps by the year 2000, we will see a mass-produced household robot performing chores around the home.

EIGHTEEN
SOCIAL AND LABOR ISSUES

The growing number of robot installations will have an impact on our society in general and the labor force in particular. In addition to robotics, other manufacturing automation technologies (CAD/CAM, flexible automation, computer-integrated production management systems, etc.) will have similar effects. The obvious impact on the labor force will be the displacement of human workers by robots. There will also be other effects that are not as obvious, effects on the organization of the workplace and in the nature of the work performed by humans. Not only will direct labor operators be affected by the introduction of robotics and related technologies; professional and semiprofessional staffs will be affected as well. Indeed, the field of robotics will have an impact that goes beyond the immediate work force. Society in general will be affected by this technology in areas such as productivity, education, and international economic competition. The applications of robotics will gradually extend beyond manufacturing into the service sector. The spectacle of robots at work in banks, hospitals, grocery stores, and fast food restaurants will not be uncommon in the future. When in the future these kinds of applications will begin is difficult to predict, but the impact on society will be significant.

In July 1981, the Office of Technology Assessment (OTA) held a workshop to explore the social impact of robotics. A summary and statement of the issues was published in an OTA report[9] in February 1982. Five categories of social issues are identified in that report:

Productivity and capital formation
Labor
Education and training
International impact
Other applications (beyond current industrial applications)

The purpose of this chapter is to consider the effects that robotics has had and will have in these areas. We will use the above list as an outline for the chapter, exploring each of the five areas (although not all at the same level of detail). In our coverage of the labor issue, we will consider the effects of robotics and automation on direct labor in manufacturing, and we will also consider the likely effects on professional and semiprofessional staffs (engineering, production planning, and the clerical support for these functions). We will not attempt to draw any grand conclusions about these social issues, nor will we make any proposals for national policy to deal with them. Our objective is simply to describe the issues and their impact.

18-1 PRODUCTIVITY AND CAPITAL FORMATION

In our previous discussions of applications and applications engineering, we have described how robots can reduce cost and increase productivity in manufacturing. Productivity improvement in the United States is an important social issue and a major national concern. The conventional definition of productivity is the following:

$$\text{Productivity} = \frac{\text{units of output}}{\text{units of input}}$$

The units of output can be reduced to monetary terms for purposes of comparing the products of different industries. Labor hours have traditionally been used as the units of input in productivity measurement, and the resulting ratio gives an indication of labor productivity only. However, capital (equipment), technical knowledge, and various other inputs are also considered to be ingredients contributing to productivity. When these other inputs are combined, the ratio is called the total factor productivity. Technical knowledge is a factor which should be interpreted to include several facets. One facet is certainly the technological improvements that are incorporated into successive generations of capital equipment. For example, a robot purchased today can be expected to have certain improvements in technology (e.g., intelligence, programming ease, accuracy, input/output interface capabilities, etc.) as compared to a robot purchased 10 years ago for perhaps the same price. Another facet of technical knowledge is the managerial expertise in operating the business. Manufacturing management practice has been improved, for example, by computerized methods such as materials requirements planning (MRP), shop floor control, and other computer-integrated manufacturing techniques.

Robotics is an input to the productivity ratio which represents both capital and technical knowledge. As an input, it is a substitute for human labor in determining productivity. Presumably by making this substitution, productivity is improved. In the bargain, human workers are denied work opportunities with potentially serious financial and emotional consequences to themselves and their families. Whether or not the productivity improvement is worth the sacrifice by the workers constitutes an important social issue.

Robotics technology, as we have described it in Chap. One, is a form of programmable automation. With programmable automation, the production system is designed with the capability to change its sequence of operations so as to process different product configurations. This type of automation is therefore ideally suited to batch manufacturing.

The batch manufacture of medium and small lots of discrete products has traditionally been very dependent on manual labor and is characterized by relatively low productivity and poor utilization of equipment. Table 18-1 shows one of the findings in a study of the machine tool industry by the Lawrence Livermore National Laboratory.[1] As indicated by the last row of data in the tabulation, the proportion of time spent in productive use of the machine tools is very low in the medium-volume (batch production) and low-volume (job shop production) categories. Traditionally, these categories tend to be quite dependent on manual labor. The limitations imposed by the use of manual labor are an important reason why there are nonproductive losses during plant shutdown (weekends and holidays) and second and third shifts.

The opportunity offered by the use of robotics and other forms of programmable automation is to increase the productivity and equipment utilization by significant margins in the batch manufacturing industries. If these systems could be made sufficiently independent of human labor, they could be operated 24 hours per day for 7 days per week. The resulting nonproductive time would be significantly reduced, and the economics of the production process would be substantially improved.

Table 18-1 Time allocations in three basic types of discrete parts manufacturing†

Plant activity	High production, %	Batch production, %	Job shop production, %
Plant shutdown	27	28	34
Losses from second and third shifts	—	40	44
Equipment failures	7	6	—
Inadequate storage	7	—	—
Tool changes	7	7	
Load/unload, noncutting, etc.	14	4	
Setups, gauging, etc.		7	
Setups, gauging, loading, etc.			12
Losses from nonoptimal cutting			2
Other idle time			2
Work standards, allowances, etc.	16		
Productive cutting	22	8	6
	100	100	100

†Compiled from Ashburn et al.[1]

Capital formation is another social issue that is related to productivity. Economists often attribute the capability to create new investment capital to the growth of productivity. The important social questions concerned with capital formation, as developed in the OTA exploratory workshop,[9] seem to be the following:

1. Will there be sufficient capital available to fund the construction of new plants and the modernization of existing plants that will use robotics and other automation technologies?
2. Will there be sufficient capital to build the factories that will produce robots in the quantities needed for the new plants?
3. Will there be enough capital to fund research and development by entrepreneurs who wish to develop new types of robots and related products?

The answers to these questions will depend largely on how robotics is perceived by business managers and the investment community to be a promising technology in which to invest. Investment capital in the United States is generally allocated to those opportunities that seem most attractive compared to the alternatives. If robotics is perceived as a technology that will generate a return on investment for those who want to buy robots, build robots, or develop robots, then the capital will be available.

18-2 ROBOTICS AND LABOR

The Effect of Robotics on Direct Labor

In Chap. One (Sec. 1-4), we speculated about the future growth of the robotics industry and created the projected sales and installation base shown in Fig. 1-8. In the present section, we consider the potential effects of these projections on the direct labor force.

Nearly all robot installations involve the substitution of the robot for one or more human workers. When robots are installed in existing production operations, this involves a direct displacement of the workers. And when robots are installed in new facilities, the direct labor operators who would have performed the work in these facilities are not hired. The rate of substitution would typically be one to three human workers for each robot that is installed. Using the projected values for robot installations shown in Fig. 1-8, and using a substitution rate of three workers per robot installed (greatest impact scenario), the numbers of workers displaced or not hired as a result of the introduction of robotics are presented in Fig. 18-1.

These numbers of robots compare with a present total human work force of about 103 million, according to the Bureau of Labor Statistics (1983). This total is expected to increase into the early 1990s but not with nearly the growth rate of robots. Of this 103 million person work force, approximately 19 million are employed directly or indirectly in manufacturing. This number may

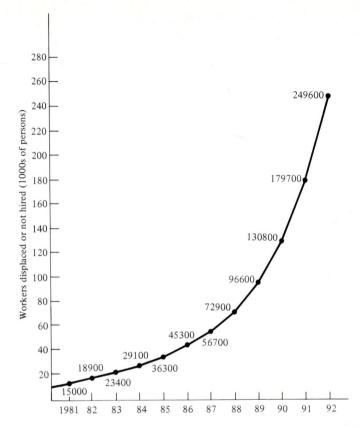

Figure 18-1 Projected number of human workers displaced or not hired as a result of the installation of robots in manufacturing. These projections are determined by multiplying the installed base in Fig. 1-8 times a rate of substitution of three workers per robot. Since three is considered to be a relatively high substitution rate, the projected values are probably high.

decline slightly over the next decade because of the introduction of robotics and other computer-automated production technologies into current industrial operations. On the other hand, some economic forecasts[6] suggest that there will be a net gain in the number of jobs in manufacturing due to the growth of new industry, including the robotics industry itself.

Using the above values, one can see that the number of workers who are replaced by robots constitutes a relatively small proportion of the unemployment problem in the United States. In 1990 the contribution to unemployment by those workers shown in Fig. 18-1 who are displaced or not hired as a result of robot installations stands at about 0.13 percent of the total workforce of 103 million. (The total work force in 1990 will probably be greater than 103 million, and a larger work force would have the effect of reducing the unemployment rate due to robotics.) This percentage is minor compared to the overall unemployment rate which currently stands at around 8 percent. The impact is somewhat more substantial when compared with the 19 million people who are employed in manufacturing. Assuming the number of manufacturing workers in the United States remains at 19 million in the year 1990, the rate of unemployment among these workers resulting from robot installations will be around 0.69 percent.

The nature of the jobs in manufacturing will evolve as a result of the introduction of robotics, computerized procedures, and other automation technologies. The shift will tend to be from direct manual participation in the production operations toward indirect labor jobs. The direct labor jobs tend to be highly structured and routine, with a minimum of irregularities and exceptions to be concerned about. The workers can readily perceive the direct relationship between their efforts and production output. Direct labor workers are often rewarded according to this relationship by means of an incentive system.

Indirect labor jobs are not subjected to the same high levels of routine and structure as direct labor jobs. These jobs usually require higher skill and knowledge levels than direct labor tasks. Examples of these jobs that will tend to grow with the number of robot installations include maintenance personnel, programmers, setup workers, and other technical support specialists. In general, the jobs will tend to possess a higher technological content in such areas as computers and electronics. A category of job called robot technician will likely develop, and the responsibilities of this position will include functions such as testing, programming, installing, troubleshooting, and maintaining industrial robots.[11]

Effect of Robotics on Labor Unions

The effect of these shifts in the structure of factory jobs is a matter of concern to the labor unions. Unless economic growth results in a general increase in employment, the labor displacements and shifts in job structure described above will mean a small population of workers who will be candidates for union membership. The unions acknowledge that these are likely to be the effects of robotics and other automation technologies. Instead of resisting the introduction of these technologies, they are attempting to deal with their membership problems in positive ways. Two approaches being used by the unions are:

1. Trying to recruit among indirect labor and semiprofessional workers in addition to their regular direct labor sources of membership.
2. Trying to participate in the benefits of automation and robotics on behalf of their members.

The first approach is the obvious strategy for increasing membership. If the traditional sources of membership are reduced in size, then it seems reasonable to seek membership growth from new sources that are readily accessible. Although some unions have had success in this area, the success has not generally been sufficient to offset declines in regular union member-ship sources. The problem for the unions has been the tendency on the part of the technically trained indirect personnel to be less interested in union mem-bership than direct labor factory workers. Highly technical and semiprofes-sional personnel are likely to identify and associate their own interests with those of management and professional staff, much more so than direct labor. They are more likely to see a career path in the organization leading to professional promotion and financial reward that cannot be realized from trade union membership.

The second approach reflects a recognition of the economic merits of the new manufacturing technologies and the inevitability of their eventual adop-tion in industry. In this approach the unions are bargaining for contracts aimed at participating in the benefits of automation. The ways of participating include retraining, reemployment, and sharing of the cost savings between management and labor. The examples which follow do not pertain exclusively to robotics, but to the broader field of automation and related new tech-nologies in manufacturing.

The United Auto Workers Union has adopted a position that generally promotes the introduction of new technologies. The union's view is that new technologies are essential in promoting economic progress that will benefit the union membership in the form of safer working conditions, better working hours, and higher wages and fringe benefits. At the same time, the union expects certain benefits in return. Some of these benefits include:

Sufficient notice to the union when the company is planning to introduce new technologies.

New technologies should be introduced in such a way as to cause the least amount of displacement to its members. Normal attribution of workers should be used to reduce the impact of displacement whenever possible.

Retraining of members to perform the jobs required to operate the new technologies.

Another example of a union position is the "Technology Bill of Rights" developed by the International Association of Machinists. Its proposal is to impose a "tax" on automation and related technologies that displace factory workers. The tax would be paid by the company that is introducing the new technology and would represent a portion of the cost savings and productivity improvements that result from the installation. The monies provided would be utilized to finance retraining of the displaced workers for subsequent reemployment.

Quality of Working Environment

Historically, U.S. manufacturers have tended to neglect the importance of the working environment and the quality of work life. However, there is an increasing awareness that this issue is a contributing factor in worker satisfaction and productivity.[10] In this subsection we address this issue by attempting to explore both the positive and negative aspects that relate to robotics.

In the past, robots have been installed in applications that were considered to be unsafe or unpleasant for human workers. If the workers displaced in these applications were able to find better jobs in better environments, then the clear impact of robotics was to improve the quality of work life. As the technology of robots becomes more sophisticated the applications will no longer be limited to environments that are undesirable for humans. In these cases the workers may be reduced to tasks that support the robot operation, such as loading and unloading parts from the automated operation. Some argue that the effect of robotics in such applications will be to reduce the skills of human labor and to degrade the quality of the work environment for those who become a component in a mechanized workcell.

If the operations can be made fully automated (by robotics and other computer-based technologies), human workers will not be required to perform manual tasks which are repetitive and routine. In these situations, management experts believe that the nature of the work involved will be more challenging, will require more knowledge-based skills, and will encompass a greater variety of tasks. The jobs may allow workers to oversee the manufacturing or assembly process from beginning to end rather than being limited to a single specialized step in the process. This would require workers to assume more responsibility for the overall results of the process, leading to job enhancement and greater worker satisfaction.

The opposing viewpoint on this issue is that greater responsibility for the overall process may lead to greater job stress. The complexity of the automated production system, combined with the pressure to minimize downtime because of the high cost of lost production time, will surely mean greater work-related stress for the individuals responsible for the operation of the system.

Impact on Professional Staff

As production operations become more automated, with robotics and other similar technologies, the work of professional and semiprofessional personnel will change. The automated factory will be less oriented around the use of manual labor and more oriented to the use of sophisticated equipment and computer-based systems. Because of this shift away from direct labor and toward machines, more emphasis will be placed on activities relating to project planning, machine maintenance, process optimization, computer systems and software, and systems analysis. The professional and semiprofessional staffs must be technologically proficient to perform these new activities.

The manufacturing planning activities in the factory will become more and more automated by increased use of CAD/CAM (computer-aided design and computer-aided manufacturing) systems. These planning activities are largely clerical and routine and they include process planning (the preparation of route sheets), estimating, purchasing, and many of the tasks associated with production planning and control. The use of integrated systems of computer-aided design and computer-aided manufacturing will allow the repetitive and routine portions of these planning functions to be accomplished automatically. Some of the automated planning software packages which are currently available but not all widely used, include CAPP (computer-aided process planning), MRP (material requirements planning), and automated numerical control part programming.

As direct manual labor is gradually replaced by automated systems, the need for work measurement (direct time study) and the use of piece rate incentive systems will be reduced. It is unclear whether work measurement, as it is traditionally accomplished today, will even be done in the automated factories of tomorrow. Industrial engineers who have typically been responsible for work measurement in production plants, will be faced with the problem of measuring the work and designing the appropriate incentive systems for indirect labor and other personnel who will operate the equipment in the automated factory. New types of work standards related to product quality, production yield, throughput, lead time, and other measures will have to be devised.

Robotics is a technology which is at the same time highly specialized and highly interdisciplinary in nature. Engineering staffs who develop robot systems and applications must reflect this interdisciplinary nature. Robotics is a combination of computer science, machine tool technology, mechanism design, and control systems. Its application requires a mixture of electrical engineering, human factors, engineering economy, workplace layout design, and robot programming. The implementation of robot systems and other forms of automation requires not only engineers who represent the individual disciplines but engineers who are also capable of integrating their own disciplines and those of others.

18-3 EDUCATION AND TRAINING

Several issues related to education and training are raised by the mass introduction of robotics into society. We are not referring to the specific training issues discussed in Chap. Seventeen, but rather the more general social issues that must be confronted. Many of these issues are identified in the two reports published by the Office of Technology Assessment[9,10] listed among the references. They include:

Shortage of trained technical staffs in robotics and other programmable automation technologies

Need for a more technologically literate work force

Shortage of technical instructors and state-of-the-art laboratory equipment in the schools

Need for retraining and job counseling programs designed for the displaced
  workers

As the imperative to introduce automation grows, the number of technical
experts in robotics and other technologies will need to be expanded in order to
keep pace with the growth. The technical expertise needed to implement
robotics includes computer science, engineering (especially electrical, industrial, and mechanical), software programmers, and related technical support specialists (electronics, mechanical and electrical maintenance, robot
programmers, etc.). One of the obstacles cited in both of the *OTA Reports* was
the low level of technological literacy among the general labor force in the
United States. To be capable of comprehending the technologies related to
robotics and other programmable automation areas, basic foundation skills in
science and mathematics are required of the work force. Deficiencies in these
basic skills have already been recognized as a problem in the retraining of
workers for jobs in these areas.

Some of the additional education and training requirements cited by
OTA[10] were the need for the development of multiple skills and "cross-
training" of workers who must perform a variety of different functions in the
factory production environment, and the increasing need for using analytical
and problem-solving skills as opposed to strictly manual skills. At the
engineering level, there is the need for engineering specialists to comprehend
the complete sequence of activities that must take place in the design-to-
manufacturing process and how computers, machines, and people must be
integrated to achieve optimum production systems.

One of the limitations in the growth of robotics and automation may prove
to be the instructional system that is responsible for providing education and
training in these fields. Except in certain "pockets of excellence" in the
country, there exists a general shortage of instructors, state-of-the-art laboratory facilities, and other resources needed to implement modern educational
programs in high-technology production areas. A common feature of the
successful educational programs observed by OTA was the close cooperation
and collaboration that exists among education, industry, labor, and govern-
ment in defining needs, purposes, and curricula.

Finally, there is a need to institute more programs for retraining and job
counseling of displaced workers. The rapid advances that are taking place in
the technology of manufacturing suggest that there will be a continuing need
for these kinds of programs. Past experience with retraining programs has
shown that participation rates are low and dropout rates are high, often
because the programs have failed to build on the existing strengths of the
students and to provide opportunities to increase basic skills applicable to a
wide variety of technical areas. One of the important needs in the job
counseling field is access to dependable up-to-date information on job market
trends and developing vocational opportunities. This information would be
useful not only to displaced workers but also to young people who must make
career decisions.

18-4 INTERNATIONAL IMPACTS

There is a great deal of concern in the United States about international competition in the field of robotics. The principal source of this competition is Japan. European countries, notably West Germany and Sweden, also have established robotics industries. The international competition exists in two ways. First, there is a competition between countries in the development and marketing of robotics technology itself. The current belief in the United States is that the state-of-the-art in the domestic technology is as advanced as in foreign countries. Given that computer science has become an increasingly significant component of robotics technology, and given the traditional strengths in this area that the United States has enjoyed, there is every reason to believe that robots built in this country are at least as sophisticated as robots built in Japan or Europe.

The second way in which international competition exists is in the application of robotics technology to manufacture products more efficiently. In this competition area, the Japanese are generally conceded to be the world leaders. In terms of numbers of robot applications in manufacturing, Japan has roughly five times the number of robot applications as compared to the United States. Table 18-2 provides one assessment of the relative numbers of robot installations for various countries as of the end of 1982. This table not only

Table 18-2 Robot installations operating at end of 1982

Country	Number	Percent of total
Japan	31,900	66
United States	6,301	13
West Germany	4,300	9
Sweden	1,450	3
Italy	1,100	2
France	993	2
United Kingdom	977	2
Belgium	305	<1
Poland	285	<1
Canada	273	<1
Czechoslovakia	154	<1
Finland	98	<1
Switzerland	73	<1
Netherlands	71	<1
Denmark	63	<1
Austria	50	<1
Singapore	25	<1
Korea	10	<1
Total	48,428	

Source: Robot Institute of America, *Worldwide Robotics Survey and Directory*, 1983.

demonstrates the proficiency of the Japanese in the robot applications area, but it also suggests that the capability of Japan to produce robots is far greater than any other country.

18-5 OTHER APPLICATIONS

Nearly all of the present applications of robots are in industrial situations. In the future, robot applications will no doubt extend to fields outside of manufacturing. The possibilities include hazardous work environments, defense applications, space exploration, and undersea operations. There are also opportunities for robots to be used in service industries in restaurants, hospitals, garbage collection, and similar activities. The future technology of robotics, and the potential applications outside of manufacturing which this technology will bring, are the subjects of the last two chapters of the book.

REFERENCES

1. A. Ashburn, R. Hatschek, and G. Schaffer, "Machine Tool Technology," Special Report 726, *American Machinist*, October 1980, pp. 105–128.
2. J. Awerman and D. Cappello, "Positive Employee Relations Paves the Way for Robots," *Robotics Today*, December 1982, pp. 35–36.
3. R. U. Ayres and S. M. Miller, *Robotics: Applications and Social Implications*, Ballinger, Cambridge, MA, 1983.
4. M. Cooley, "The Impact of New Technology on Working People," *Decade of Robotics*, IFS Publications, Bedford, England, 1983, pp. 98–99.
5. M. P. Groover, J. E. Hughes, Jr., and N. G. Odrey, "The Societal Impact of Factory Automation," *Industrial Engineering*, April 1984, pp. 50–59.
6. M. W. Karmin, "High Tech—Blessing or Curse?" *U.S. News & World Report*, January 16, 1984, pp. 38–44.
7. A. Kochan, "Trade Unions: Recognising the Need for New Technology," *Decade of Robotics*, IFS Publications, Bedford, England, 1983, p. 97.
8. L. Kuzela, "IAM Envisions a Tax on Automation," *Industry Week*, May 30, 1983, p. 19.
9. Office of Technology Assessment, "Exploratory Workshop on the Social Impact of Robotics, Summary and Issues," *OTA Report*, Washington, D.C., February 1982.
10. Office of Technology Assessment, "Computerized Manufacturing Automation—Employment, Education, and the Workplace," *OTA Report*, Washington, D.C., April 1984.
11. R. R. Schreiber, "New Perspectives: The Upjohn Report," *Robotics Today*," April 1983, pp. 61–62.
12. T. L. Weekly, "The UAW Speaks Out on Industrial Robots," *Robotics Today*, Winter 1979–80, pp. 25–27.

NINETEEN

ROBOTICS TECHNOLOGY OF THE FUTURE

The approach adopted in the preceding chapters of this book has been to describe the state of the art in the technology, programming, and applications of industrial robots. However, the state of the art in this field is changing rapidly, and it is difficult for us to be satisfied with merely providing a description of the present status. The purpose of this chapter is to describe some of the research and development that is currently taking place and to forecast some of the future advances in robotics technology that will result from these efforts. We include robot programming and related software developments within the scope of robotics technology. The implications of these technological developments in terms of future robot applications will be discussed in Chap. Twenty.

We can hypothesize a likely profile of the future robot based on the various research activities that are currently being reported (see list of references at the end of the chapter). The features and capabilities of the future robot will include the following (it is unlikely that all future robots will possess all of the features listed):

Intelligence. The future robot will be an intelligent robot, capable of making decisions about the tasks it performs based on high-level programming commands and feedback data from its environment.
Sensor capabilities. The robot will have a wide array of sensor capabilities, including vision, tactile sensing, and others.
Telepresence. It will possess a telepresence capability, the ability to communicate information about its environment (which may be unsafe for humans) back to a remote "safe" location where humans will be able to make judgments and decisions about actions that should be taken by the robot.
Mechanical design. The basic design of the robot manipulator will be

mechanically more efficient, more reliable, and with improved power and actuation systems compared to present day robots. Some robots will have multiple arms with advanced control systems to coordinate the actions of the arms working together. The design of the robot is also likely to be modularized, so that robots for different purposes can be constructed out of components that are fairly standard.

Mobility and navigation. Future robots will be mobile, able to move under their own power and navigation systems.

Universal gripper. Robot gripper design will be more sophisticated, and universal hands capable of multiple tasks will be available.

Systems integration and networking. Robots of the future will be "user friendly" and capable of being interfaced and networked with other systems in the factory to achieve a very high level of integration.

The following sections will present a review of these areas of robotics development.

19-1 ROBOT INTELLIGENCE

The concept of an intelligent robot is not a new concept. In Chap. Two, the "intelligent robot" was defined as one of four basic control systems categories for industrial robots. Although this type of robot must be considered relatively primitive at the present time, the level of sophistication of the intelligent robot will evolve in the future as the prerequisite technological components (e.g., sensors, mobility, programming languages, networking, etc.) are developed. The concept of the intelligent robot includes the capability to receive high-level instructions that are expressed as commands to do a general task, and to translate those instructions into a set of actions that must be followed to accomplish the task. The future intelligent robot will also be aware of its environment and will be able to make decisions about its actions based in part on an interpretation of its environment. We discussed these issues in the context of future generation robot languages in Chap. Nine, and we discussed the field of artificial intelligence in Chap. Ten. In this chapter we will present a summary of an alternative approach to robot control developed at the National Bureau of Standards (NBS).

The Industrial Systems Division of NBS[1,2,3,7] has designed the framework for a real-time hierarchical control system that will be capable of responding to goal-oriented instructions and will receive sensor data regarding its environment. The following summary of the NBS system will provide a narrative of the likely way in which the future intelligent robot will operate.

In the NBS Real-Time Control System (RCS), a hierarchical control structure is used. The overall scheme of the system as it would be applied in an automated factory is illustrated in Fig. 19-1. In the operation of the RCS, goals expressed as commands at the highest level are decomposed at each successively lower level into simpler commands. The levels at the top of the

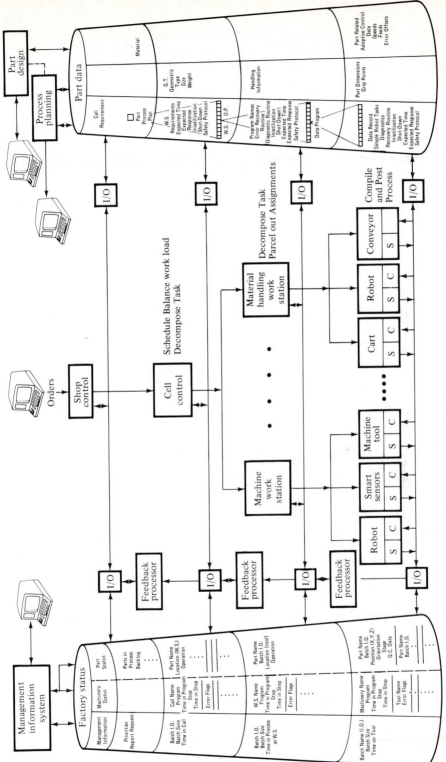

Figure 19-1 Hierarchical control structure used in the Real-Time Control System at the National Bureau of Standards. (*Reprinted by permission from Albus[1]*)

hierarchical pyramid would correspond to factory shop orders, and these orders would be decomposed respectively into workcell instructions, and workstation commands. In turn, these commands would be subdivided into tasks, moves, and primitives. The flow of commands and status information at these lower levels of the RCS are shown in Fig. 19-2. At the lowest level, the signals to drive the robot, gripper, and other equipment are generated.

Each control level in the hierarchical structure has a clearly defined and bounded control function with a limited number of inputs and outputs. The configuration of each control module is the same, regardless of its level in the hierarchical structure. The module configuration is illustrated in Fig. 19-3. The input portion of the module receives status information from feedback sensors and from the control module at the level immediately below. It also receives commands from the next higher level in the control structure. The module preprocesses the input data, combining and converting it into the

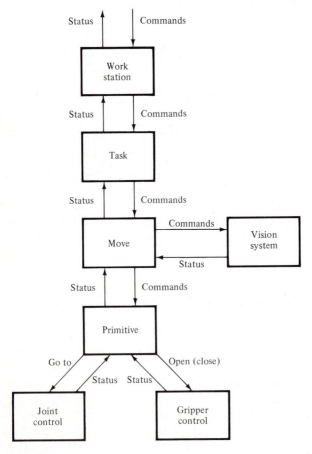

Figure 19-2 The flow of commands and status information in the NBS Real-Time Control System.

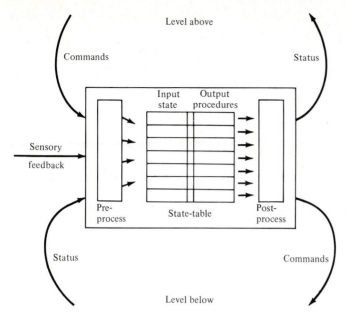

Figure 19-3 Configuration of each control module, irrespective of the level in the control hierarchy.

proper format for the processing which is to be accomplished at that level. The processing is done by means of a state-table, consisting of a list of output procedures which are determined according to the status conditions represented by the input data. A postprocessor in the control module is utilized to convert the output procedures and relevant other data from the state-tables into formats that are required by components external to the immediate control module. These components would consist of other control levels or pieces of equipment that are controlled by the module.

19-2 ADVANCED SENSOR CAPABILITIES

The sensors used in robotics have previously been discussed in Chaps. Six and Seven. It is anticipated that the typical robot of the future will be much more richly endowed with sensor capabilities than today's machines. These sensor capabilities would permit the robot to be more aware of its environment, to communicate with human operators more readily, and to make use of the higher level of intelligence described in the previous section. Included among the advanced sensor capabilities will be three-dimensional vision and tactile sensing. The following paragraphs will address some of the research developments currently in progress in these areas.

Three-Dimensional Machine Vision

Although three-dimensional vision has been demonstrated to be technologically feasible (for example, the development work of Robotic Vision Systems, Inc.), there are a limited number of economically feasible systems being applied in industry today. It is expected that high-resolution three-dimensional vision sensors currently under development will be used in future applications to permit the robot to recognize objects more readily, to measure the distances between itself and objects using range-finding techniques, and to avoid obstacles in its path. Three-dimensional vision sensors would also facilitate safety monitoring by enabling the robot to easily determine the position of a human in the workcell, and to have its manipulator avoid contact with the human. The development of high-speed microprocessors will assist in the development of three-dimensional machine vision. One approach to three-dimensional vision is described in Ref. 15.

Tactile Sensing

Compared to human touch-sensing capabilities, current tactile sensors are quite rudimentary. Sophisticated tactile sensing would permit a variety of characteristics to be determined about the robot's environment or the object it is handling. These characteristics include the roughness of the contacting surface, the elasticity of the material, the weight of the object, its shape, and other physical features. Togai[20] reports that the Japanese are developing three types of tactile sensors in their work on the next generation robot:

1. *Shear sensors.* This is the capability to sense slip between an object and the gripper finger surfaces. Applications of this sensor would be in handling both rigid objects as well as materials that are soft and limp, such as textiles and garments.
2. *Contact sensors.* Contact sensors with multiple contact pads, similar to the Lord Corporation sensors described in Chap. Six, would be useful in recognizing objects and in positioning objects optimally in the gripper.
3. *Force sensors.* These would be applied for detecting an object's elasticity, for holding an object with the proper gripping force, and for following contours with a certain specified force level.

19-3 TELEPRESENCE AND RELATED TECHNOLOGIES

As defined in Chap. One, telecherics is a technology that involves the use of remote control manipulators, or teleoperators, to perform handling tasks. The technology is used for hazardous handling tasks (e.g., handling radioactive materials) inside an unsafe environment. It is anticipated that some robots of the future may be employed as sophisticated teleoperators, combining their own intelligence with the intelligence and judgment of humans located in some remote position. (Chapter Twenty will explore a number of possible applications where the future robot would work in an environment that is hazardous for humans.) For example, the tasks performed may be so complicated that human assistance is required to deal with the complexities. When used in this way, the robot will need a two-way channel of communication with the humans: first, it will need to acquire information (e.g., sensor data) about its task and its environment and to transmit this information back to the humans; and, second, it will need to receive complex instructions and commands from the humans.

Telepresence is concerned with the first portion of this channel of communication. The functions of telepresence are: information acquisition using advanced sensor technologies, feedback of this information to a remote location, and display of this information in a manner that facilitates interpretation by humans.

The objective of information acquisition and feedback is to gather sensor data about the robot's task and its environment in a manner that simulates the human sensory functions. Three-dimensional vision and tactile sensing, as described in the previous section, would be two of the more important sensory capabilities involved. Other capabilities would also be involved, including the possibility that the robot, through its own sensors and intelligence, would be able to make observations and draw logical conclusions about its situation, so that these can be communicated to the remote location.

The objective in the display function is to make the information as complete and realistic as possible so that the humans almost believe that they are in the place of the robot. They will therefore be able to relate to the task situation more realistically. In effect, the humans will project their presence into the situation. The presumption is that this mode of operation will improve the performance of the remote robot by making its actions and reactions more like those of the humans who are guiding it.

The capability for the robot to use speech synthesis to communicate information about its task and environment is also a possible means for the future robot to augment the display function in telepresence. Speech synthesis is a technology that exists today. There are presently two commonly used techniques for speech synthesis by machine. The first relies on the fact that spoken English (or other human language) makes use of a limited number of sounds called "phonemes." Phonemes are the smallest units of speech that distinguish one utterance from another in the language. By stringing together the appropriate phonemes, whole words can be generated. A typical system based on this technique would consist of a microprocessor that selects the appropriate phoneme codes out of memory according to a prescribed set of rules. The codes would then be processed by a speech synthesis device so that the output would emulate the human voice. The problem that is encountered with this first speech synthesis technique is the inability of the system to make speech sound "natural."

The second common technique for speech synthesis uses spoken words that have been recorded by a human. Since the possibility exists for very large vocabularies to be required with the accompanying demands on computer memory, an approach known as linear predictive coding (LPC) is often used to compress the amount of information that must be stored in memory. LPC allows the system to record the human speech, and to compress the data and store it in memory. When used for speech synthesis, the LPC data are expanded and the system sounds almost exactly like the original speaker. Consequently, the LPC approach provides a more natural sounding synthesized speech than the phoneme-based technique. Of course, the limitation on the LPC system is that only those words that were recorded are available for use. In today's industrial environment, a limited vocabulary is usually adequate for most applications.

The use of speech synthesis to enhance the telepresence functions is a likely technology in the future of robotics. What is presently missing is the sophisticated artificial intelligence required to synthesize thoughts and observations into sentences that can then be communicated by means of speech.

The opposite direction in the channel of communication between the robot and remotely located humans is voice programming, the oral communication of instructions to the robot. Speech, as an input to robots, computers, and other machines, has been an area of research and development since the 1960s. The most natural way for a human to communicate is orally.

There are a number of issues and research problems involved with the use of voice programming. First, there is the problem of getting the voice data into a form that the computer can comprehend. The solution usually involves digitizing the amplitude of the speech at different frequencies. The speech recognition system identifies each spoken word by comparing its frequency spectrum against a model stored in computer memory. The problem is complicated by the fact that different persons enunciate their words differently and possess different voice tones and speech patterns.

Another problem area is the separation of spoken words from each other. Most people speak in what is described as continuous speech, running the words together. The difficulty for the speech recognition system is to distinguish one word from the next to make the interpretation of each separate word. This problem is usually solved by speaking discretely, one word at a time.

These problems can be simplified by the use of a limited vocabulary and by restricting the number of operators who use the system. For example, a robot control system might only need to make use of the words, "up," "down," "right," "left," "fast," "slow," "go to ⟨point A⟩," and "stop." In the training of the system, each operator would repeat these words into the system a number of times so that the controller could learn the pattern of the user's voice in enunciating each vocabulary word. In the subsequent use of the system, the operators would be able to command the robot by speaking the words so that the speech recognition system could understand each word in sequence. For each word or set of words, the robot would accomplish some corresponding action or task. The assumption behind voice programming is that the human operator would be able to command and communicate with the robot more quickly and in a way that is more natural to the operator. Uses for voice programming in robotics would include applications in hazardous environments, and one-of-a-kind jobs where programming time is critical. Another area where this technology holds great promise is in the design of robotic aids for handicapped persons.

Both voice programming and speech synthesis are available today for implementation in robotics. However, they have seen very little use outside of the research laboratories. What is missing is a sufficient need to utilize these technologies in today's applications. It is anticipated that certain future applications, those requiring closer communication between the robots and their human masters, will use either or both of these methods.

19-4 MECHANICAL DESIGN FEATURES

Future robots will be designed to be mechanically more efficient, and will use improved power and actuation systems compared to current robots. Some future robots may have more than a single arm and will require advanced control systems to coordinate the actions of the arms working together. The design of the robot is also likely to be modularized, so that robots for different purposes can be constructed out of components that are fairly standard. One of the improvements in the area of mechanical design is the direct-drive robot.

The Direct-Drive Robot

The effects of backlash and compliance on the performance of a robot manipulator were discussed in Chap. Three. One way to reduce these effects is to couple the actuator directly to the joint. This robot design has been called the "direct-drive mechanical arm" (or direct-drive robot, for short).[4,5]. Direct drive involves locating the motor or actuator for a given joint contiguous to that joint, thus eliminating the need for a power transmission mechanism between the joint and the motor. In the typical robot of today, the motor is usually positioned remotely in the robot base or body.

The direct-drive robot design would provide a number of additional benefits beyond the elimination of backlash and other mechanical deficiencies. The robot would be inherently more efficient since there would be no transmission losses. Another benefit is that the joints would be backdrivable, thus allowing joint–space force sensing (see Chap. Six) or selectively applied joint compliance to facilitate assembly (see Chap. Fifteen). Additionally, the direct-drive robot would require less maintenance due to fewer components in its fabrication. Finally, because fewer parts are needed to assemble the robot, this would presumably produce a favorable cost benefit.

Unfortunately, implementing the direct-drive robot has been difficult. A common task for a manipulator is to hold a load steady against gravity. When electric motors are used, this results in high currents running through the motor at stall. In order to support the load without the help of a power transmission to multiply the motor's stall torque, a very high current must be used to generate the necessary holding torque. If a relatively small motor is used, there is a significant risk that the motor will overheat. In a dc motor, the power in the armature is a function of the square of the current. The temperature rise of the armature is proportional to the power. On the other hand, if the motor is relatively large, it becomes too heavy to be practical for locating a motor at each joint in the robot arm. It is difficult to find a suitable compromise between these two problems.

There are a number of ways in which the direct-drive technology is being made practical for use in robots. One approach involves the configuration of the arm itself. If the joint is oriented so that the motor does not have to support the load against gravity, the motor is only required to supply the torque necessary to overcome friction and to accelerate and decelerate the mass of the total payload (arm and load). When the payload is at rest, the motor requires relatively little power, thus allowing the opportunity to dissipate heat. The SCARA design (Selectively Compliant Arm for Robot Assembly) has the joints oriented so that the axes of rotation of the principal joints are vertical, hence the load is carried by the joint frame rather than the motor. Figure 2-8 (Chap. Two) shows the SCARA design.

Another approach that facilitates the direct-drive robot design is in the design of the motors themselves. Developments are being made in the materials being used in motor design. One of the design considerations is that the torque that a motor is capable of producing is a function of the strength of the magnetic fields in the motor. New magnetic materials such as samarium–cobalt and neodynium–iron hold the promise of motors with higher torque capacities for the same current input.

Multiple-Arm Coordination

Many robot tasks could benefit from the use of two or more hands. Assembly tasks are typical examples of this case. In walking machines (described in the following section), there is a need to coordinate the actions of the legs in order to provide balance and propulsion for the machine. At the present time, it is possible to provide only a crude level of coordination of the motion of the two arms through the use of interlocks. No machine is available that is capable of the hand-to-hand coordination required to thread a needle, for example. A much simplified version of this situation is the capability of a robot equipped with machine vision to coordinate its actions with a moving conveyor to pick up a part in motion. While attacks on this type of problem have met with limited success, the degree of difficulty in coordinating two manipulators with each other is much greater. One aspect of the problem which is in need of solution is in the area of arm dynamics. The robot controller must be able to precisely determine the robot's present as well as its future locations quickly enough to achieve the level of control required in hand-to-hand coordination.

19-5 MOBILITY, LOCOMOTION, AND NAVIGATION

Today's industrial robots tend to be planted at one location. By contrast, materials and people in a factory generally move about. Providing robots with the capacity to move under their own power would greatly increase their potential utilization. Robot mobility in the factory environment can be used either to move materials from one place to another, or to perform jobs that require movement significantly beyond the work volume of today's stationary robots. Outside of the factory, there are many possible applications for a mobile robot capable of self-navigation. There are two basic ways in which robots can be made mobile: wheeled vehicles and pedded vehicles (walking machines). The following subsections will examine some of the possibilities within these two categories.

Wheeled Vehicles

The current state-of-the-art in self-propelled machine locomotion is the automated guided vehicle (AGV). AGVs are typically battery-powered, three- or four-wheeled carts that follow a specific path around the factory or warehouse floor. The most common way of defining the path is by means of a guide wire buried just below the surface of the floor. On the underside of the vehicle is a sensor system which detects the wire and guides the vehicle along the defined path. In the more sophisticated systems, instructions to start/stop or change routes are communicated to each vehicle electronically over radio frequencies. Automated guided vehicle systems are typically used in today's applications for moving materials in warehouses and factories.

Some development work is now being done to add manipulators to the AGVs. In this way the guided vehicle would be able to transport materials between workcells and to load and unload the machines in the workcells. In effect, this type of guided vehicle-and-manipulator would represent a possible method of implementing a mobile robot cell, described in Chap. Eleven. Gantry robots, an example of which is pictured in Fig. 2-6 (Chap. Two), represent one means of providing mobility for the manipulator.

In order for robot vehicles to be able to be used in uncontrolled environments, they would need to navigate autonomously without the use of tracks or guide wires in the floor. To accomplish autonomous navigation, the vehicles would need to be capable of several "intelligent" functions. These would include: "scene analysis" (most probably implemented by means of machine vision), "trajectory planning" (the ability to plan alternative routes from starting point to destination point and to select the best route among them), and "obstacle avoidance" (the ability to develop strategies to navigate around obstacles along the planned vehicle path based on the scene analysis). These capabilities would require the use of advanced sensor technologies and artificial intelligence. Prototype vehicles have been demonstrated at Stanford University[11] and at Carnegie-Mellon University which were able to view the scene in front of the vehicle and navigate around obstacles. However, the vehicles required impractically long times to move a few feet. The navigation problem is an active research area.[9,19]

Walking Machines

Wheeled vehicles have limitations; they can only travel over relatively smooth surfaces. Vehicles with tank tracks would be an improvement for rough terrain. Walking machines[12,14,21] offer the greatest versatility for dealing with a variety of surfaces and obstacles. However, walking machines must overcome all of the same technological hurdles as autonomous locomotive wheeled vehicles, with the additional problem of coordinating the motions of the legs. In addition, since it is assumed that such vehicles will be used over rough terrains, they must be highly adaptive to the irregularities of the terrain.

There are a number of factors that must be considered in the design and control of walking machines. These factors include the number of legs, gait selection, balance, and coordination of the legs. Research has been done on one-legged, two-legged, four-legged, and six-legged machines. A one-legged

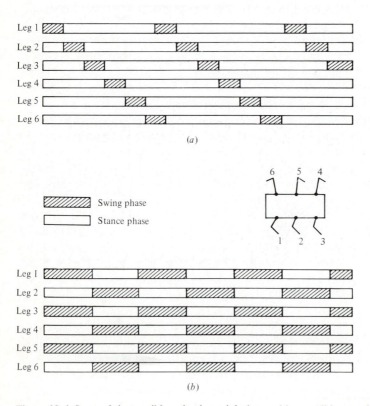

Figure 19-4 Some of the possible gaits that might be used by a walking machine.

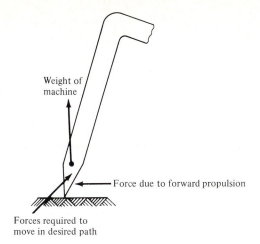

Weight of
machine

Force due to forward propulsion

Forces required to
move in desired path

Figure 19-5 Diagram showing the forces on the end of the leg of a six-legged pedipulator.

Figure 19-6 ODEX 1, a six-legged walking machine. (*Photo courtesy of Odetics, Inc., Anaheim, CA.*)

machine would have no coordination problems, but would be much more difficult to balance than a six-legged machine. The gait of a walking device describes the sequence in which the legs are brought into use during the walking motion. It generally refers to the duration of the stance phase and swing phase of each leg. The stance phase occurs while the leg is on the ground providing support or propulsive force. The swing phase occurs while the leg is in the air preparing for the next stance. Some gaits are more stable than others. Six-legged walkers (such as cockroaches) employ a gait known as the "alternating tripod" gait. In this gait, the pairs of three legs alternate between the stance phase and the swing phase. This enhances the stability of the walking machine by providing that there are three legs on the ground at a time. Another gait, called the "crawl" gait, has only one leg in the air at a time; while a gallop has no legs in the air at certain moments during the gait. Figure 19-4 illustrates some of the possible gaits that might be used by a walking machine.

During the stance phase, a leg is subject not only to the forces required to support the machine, but also the forces generated by the motion of the body of the walking machine. Figure 19-5 illustrates the forces on the leg of a six-legged pedipulator. The leg must be able to control the motion of the foot to ensure that the body maintains motion along the desired path (usually a straight line path).

Walking machines present a significant challenge to the robot designer since the problems in their design include force sensing (in the foot), multileg coordination, balance, navigation, obstacle avoidance, and others. Figure 19-6 shows one commercial design of a six-legged walking robot, called the ODEX 1.[17]

19-6 THE UNIVERSAL HAND

The design of the end effector is a critical consideration in the application of robotics to industrial operations. The end effector must typically be designed for the specific application. Our discussion in this section will deal exclusively with gripper design. By comparison to the human hand, a robot's gripper is very limited in terms of its mechanical complexity, practical utility, and general versatility. In order to realize the full potential of future robotics technology, grippers must be designed more like the human hand, both in their sensory and control capabilities as well as their anatomical configuration.

The servoed gripper represents one step in the direction toward increased versatility and utility. It is a technology which exists today. The servoed gripper is typically manifested as a device with two parallel jaws or fingers that are equipped with sensors and controlled by a servomechanism such as a dc servomotor. By definition the device has position sensing and it is often equipped with some type of tactile and proximity sensing. An example of a gripper equipped with tactile-sensing capabilities is shown in Fig. 6-3 (Chap. Six). This gripper is capable of sensing forces being applied to a part or to some other object.

Another example of a servoed gripper is a prototype hand equipped with tactile and position sensing that was developed at the Massachusetts Institute of Technology. This prototype was demonstrated in an application in which the gripper would close on the rim of a paper cup and store the force versus deflection data. In this way the robot was able to determine whether it had picked up one cup or two cups, since the force for two cups would be greater for the same deflection. A servoed gripper such as this would be capable of handling a large assortment of parts in assembly applications and other tasks.

The anthropomorphic hand is another approach in the pursuit of the universal gripper. Development work is proceeding at a number of research centers to develop an articulated hand with attributes similar to those of the human hand.[10,13,18] The human hand configuration, with its four articulated fingers opposed by the thumb, is a most universal tool. It can do much more than simply hold an object. For example, the human hand is sufficiently dexterous that people can rotate an object in their hands without putting the object down. Most of the work in articulated hand design for robotics involves three fingers rather than five, with two or three joints per finger. The complexity of the design problem is roughly equivalent to the problem of designing a three-armed robot with tactile and force sensing and a control system that is sophisticated enough to coordinate the actions of the three arms. Aside from the control problem, the designer must be able to package about nine motors or actuators into a very small volume (perhaps slightly larger than the human hand). One approach to solve this packaging problem has been to use tendons (small cords and pulley systems) and to locate the motors in the base of the hand or in the arm of the robot. An anthropomorphic hand design utilizing this general configuration is shown in Fig. 5-17 (Chap. Five).

19-7 SYSTEMS INTEGRATION AND NETWORKING

A technological area which goes beyond robotics is in systems integration and computer networking. Robotics is only one of many computer-oriented technologies that will be used in the future automated factory. These technologies include computer-aided design, computer-aided process planning, manufacturing resource planning, manufacturing information systems, expert systems (and other applications of artificial intelligence), computer numerical control, flexible manufacturing systems, and other examples of production automation. The problem facing designers of the products in these technologies is to make them compatible with each other so that they can be integrated into a single factory control system. A related problem is to design the robots and other systems so that they can be easily installed and connected to the existing factory network.

In the factory of the future, all of the production equipment will share a common data base, perhaps in the general arrangement diagrammed in Fig. 19-1. In order for robots to participate in the factory network system, they must be able to communicate with the network. A good deal of effort is currently being exerted in the United States to define standard interfaces and protocols for the networks. At present, robots are able to share some of the data with host computers, such as motion programs or machine status information. Eventually, more data will need to be communicated between the various stations in the network in order to realize the full capabilities of the factory-wide real-time control system described in Sec. 19-1.

REFERENCES

1. J. S. Albus, *Brains, Behavior, and Robotics*, BYTE Books, Peterborough, NH, 1981, chaps. 9–11.
2. J. S. Albus, A. J. Barbera, and R. N. Nagel, "Theory and Practice of Hierarchical Control," *Proceedings*, Twenty-Third IEEE Computer Society International Conference, September 1981, pp. 18–39.
3. J. S. Albus, C. R. McLean, A. J. Barbera, and M. L. Fitzgerald, "Hierarchical Control for Robots in an Automated Factory," *Conference Proceedings*, 13th International Symposium on Industrial Robots and Robots 7, Chicago, IL, April 1983, pp. 13-29–13-43.
4. H. Asada and T. Kanade, "Design of Direct-Drive Mechanical Arms," *Technical Report CMU-RI-81-1*, Carnegie-Mellon University, April 1981.
5. H. Asada, K. Youcef-Toumi, and R. Ramirez, "M.I.T. Direct-Drive Arm Project," *Conference Proceedings*, Robots 8, Detroit, MI, June 1984, pp. 16-10–16-21.
6. R. Ayres and S. Miller, "Industrial Robots on the Line," *Technology Review*, May/June 1982, pp. 34–47.
7. A. J. Barbera, M. L. Fitzgerald, J. S. Albus, and L. S. Haynes, "RCS: The NBS Real-Time Control System," National Bureau of Standards, Working Paper (undated).
8. J. F. Engelberger, *Robotics in Practice*, AMACOM, New York, 1980, chap. 9.
9. M. Julliere, L. Marce, and H. Place, "A Guidance System for a Mobile Robot," *Conference Proceedings*, 13th International Symposium on Industrial Robots and Robots 7, Chicago, IL, April 1983, pp. 13-58–13-68.
10. D. Lian, S. Peterson, and M. Donath, "A Three-Fingered, Articulated, Robotics Hand," *Conference Proceedings*, 13th International Symposium on Industrial Robots and Robots 7, Chicago, IL, April 1983, pp. 18-91–18-101.
11. H. P. Moravec, "Obstacle Avoidance and Navigation in the Real World of the Seeing Robot Rover," PhD Thesis, Stanford University, Palo Alto, CA, 1981.
12. K. Pearson, "The Control of Walking," *Scientific American* **235**, 72–86 (1976).
13. D. A. Petersen, "Development of a 16-Axis End Effector," *Conference Proceedings*, Robots 8, Detroit, MI, June 1984, pp. 17-1–17-8.
14. M. H. Raibert and I. E. Sutherland, "Machines that Walk," *Scientific American* **248**(1), 44–53 (1983).
15. M. Rioux, "3-D Camera Based on Synchronized Scanning," *Conference Proceedings*, 13th International Symposium on Industrial Robots and Robots 7, Chicago, IL, April 1983, pp. 17-48–17-60.
16. *Robotics Research and Advanced Applications*, American Society of Mechanical Engineers, New York, 1982.
17. M. Russell, "ODEX 1, The First Functionoid," *Unmanned Systems*, Fall 1983.
18. J. Salisbury and J. Craig, "Articulated Hands: Force Control and Kinematic Issues," *The International Journal of Robotics Research* **I**(1), (1982).
19. M. Takano and G. Odawara, "Development of New Type of Mobile Robot TO-ROVER," *Conference Proceedings*, 13th International Symposium on Industrial Robots and Robots 7, Chicago, IL, April 1983, pp. 20-81–20-89.
20. M. Togai, "Japan's Next Generation of Robots," *Computer*, March 1984, pp. 19–25.
21. M. Weiss, *Bachelor's Thesis*, Massachusetts Institute of Technology, Cambridge, MA, 1977.

TWENTY

FUTURE APPLICATIONS

The kinds of technological advances described in Chap. Nineteen will permit robots to be used in new applications for the future. It is not possible to predict all of the future opportunities for the use of robots or to forecast the order in which the applications will occur. A combination of economic and technical factors will determine how these applications will be introduced. It seems clear that future robot applications will include not only manufacturing operations but also nonmanufacturing operations as well.

Ayres and Miller[2] present a list of tasks that robots can accomplish today and may be able to accomplish in the future. The list is both amusing and revealing, and we present it with permission of the publisher in Table 20-1. Engelberger[5] also provides a listing of potential robot applications for the future. His list of near term applications includes batch assembly, order picking, wire harness manufacturing, packaging, textiles processing, and medical laboratory handling tasks. "Farther out" robot applications include garbage collection, fast food preparation and delivery, gasoline dispensing, and nuclear maintenance and cleanup.

This final chapter will explore some of the possible applications, with greater emphasis on the nonmanufacturing uses of robots in the future. Considering that employment in manufacturing constitutes only about 18 percent of the total work force (19 million workers out of a total employment of 103 million as indicated in Chap. Eighteen), greater opportunities for robot applications may ultimately be found in the nonmanufacturing sector of the economy. We will divide our presentation of the future applications into three areas: (1) manufacturing, (2) hazardous and inaccessible environments, and (3) service industries. Before discussing these three application areas, let us examine the kinds of common chatacteristics that these applications will possess, and the capabilities of the robots that will perform them.

Table 20-1 Tasks that robots can do today and may be able to do in the future†

Things present (or past) robots can do	Things next generation robots will be able to do	Things a very sophisticated future robot may be able to do	Things no robot will ever be able to do (probably)
Play the piano	Vacuum a rug (avoiding obstructions)	Set a table	Cut a diamond
Load/unload CNC machine tools	Load/unload a glass blowing or cutting machine	Clear a table	Polish an opal
Load/unload die casting machines, hammer forging machines, molding machines, etc.	Assemble large and/or complex parts, TV's, refrigerators, air conditioners, microwave ovens, toasters, automobiles	Juggle balls	Peel a grape
		Load a dishwasher	Repair a broken chair or dish
		Unload a dishwasher	Darn a hole in a sock/sweater
Spray paint on an assembly line		Weld a cracked casting/forging	Play tennis or pingpong at championship level
Cut cloth with a laser	Operate woodworking machines	Make a bed	
Make molds		Locate and repair leaks inside a tank or pipe	Catch a football or a frisbee at championship level
Deburr sand castings	Walk on two legs	Pick a lock	
Manipulate tools such as welding guns, drills, etc.	Sheer sheep	Knit a sweater	Pole vault
	Wash windows	Make needlepoint design	Dance a ballet
Assemble simple mechanical and electrical parts: small electric motors, pumps, transformers, radios, tape recorders	Scrape barnacles from a ship's hull	Make lace	Ride a bicycle in traffic
	Sandblast a wall	Grease a continuous mining machine or similar piece of equipment	Drive a car in traffic
		Tune up a car	Tree surgery
		Make a forging die from metal powder	Repair a damaged picture
		Load, operate, and unload a sewing machine	Assemble the skeleton of a dinosaur
		Lay bricks in a straight line	Cut hair stylishly
		Change a tire	Apply makeup artistically
		Operate a tractor, plow or harvester over a flat field	Set a multiple fracture
		Pump gasoline	Remove an appendix
		Repair a simple puncture	Play the violin
		Pick fruit	Carve wood or marble
		Do somersaults	Build a stone wall
		Walk a tightrope	Paint a picture with a brush
		Dance in a chorus line	Sandblast a cathedral Make/repair leaded glass windows
		Cook hamburgers in a fast food restaurant	Deliver a baby
			Cut and trim meat
			Kiss sensuously

† Reprinted with permission from Ayres and Miller's *Robotics: Applications and Social Implications*, Copyright 1983, Ballinger Publishing Company.

20-1 CHARACTERISTICS OF FUTURE ROBOT TASKS

The most prominent characteristic about present-day robot applications is that they require the robot to perform a repetitive motion pattern. Although the motion pattern is sometimes complicated, variations in the motion pattern are minimum. Also, the level of sensor technology required in the application is fairly low.

The enhancements in the technological capabilities of future robots will permit the applications to evolve in new directions. Some of the important characteristics of future robot tasks that will distinguish them from typical present-day applications are the following:

The tasks will be increasingly complicated. In addition to repetitive tasks, robots will perform semirepetitive and even nonrepetitive operations.

The tasks will require higher levels of intelligence and decision-making capabilities from the robot. Advances in the field of artificial intelligence will be incorporated into the design of robot controllers.

Some of the tasks will require "robust mobility," the capability to move about the work area without relying on rails or moving platforms to execute the move.

The robot tasks of the future will commonly make use of a variety of sensor capabilities, including vision, tactile sensing, and voice communication.

Many of the future tasks performed by robots will require a higher level of end effector technology. The requirements for hand articulation and tactile sensing capabilities will be far in advance of today's gripper devices. The concept of the universal hand will be much closer to reality.

The greater variety of robot applications will require that robot anatomy become more specialized and differentiated according to the applications. The physical configuration of the robot will be designed for the specific purpose that the robot is supposed to serve. The economics of this specialization will be improved by the use of techniques such as flexible automation, modularized construction, and standardization of components.

Tasks that are performed in inaccessible environments will require significant improvements in robot reliability because of the difficulty in servicing, maintaining, and repairing the machine. The reliability improvements will also be incorporated into robot designs that are not used in these kinds of environments.

The inaccessible environments may require the use of a telepresence capability, so that humans can instruct the robot during the task.

Many of these applications characteristics correspond to the technological capabilities of the future robot profile described in Chap. Nineteen.

20-2 FUTURE MANUFACTURING APPLICATIONS OF ROBOTS

We indicated in Chap. One that at the end of 1984 there were an estimated 9700 robots installed in the United States, nearly all of them working in manufacturing. By 1990, the installed base is projected to grow to 43,600 robots. And by the year 2000, if the projected trends continue, the installed base will increase to approximately one million units. Not all of these units will be employed in manufacturing operations, but many of them will. The question is: what will they be doing? Will they perform the same kinds of tasks that robots perform today, or will future robot applications in manufacturing be different?

Let us attempt to estimate the application trends to see how robots are likely to be used in manufacturing operations in the future. We will also consider the various problem areas that are presented by these operations and how future robots might be capable of dealing with the problems.

Table 20-2 presents a listing of current robot applications with an estimate of the share of robots in each of the categories. Also presented in the table are our estimates of the percentages in each application category in 1990. The present biggest application areas for industrial robots are in the spot-welding and the materials handling and machine loading categories. The handling of materials and machine tending are expected to continue to represent important applications for robots, but the relative importance of spot welding is expected to decline significantly by the time all (or nearly all) of the automobile body lines are converted over to robot spot welding before 1990. The most significant growth in shares of manufacturing applications are expected

Table 20-2 Estimated distribution of robot applications in 1984 and in 1990†

	1984 percent	1990 percent
Materials handling, machine loading	30–35	20–30
Processing operations:		
Spot welding	32–40	5–10
Arc welding	5–8	15–20
Spray coating	8–12	4–8
Other processing operations	5	3–7
Assembly and inspection	7–10	25–35
Other manufacturing operations, not covered by the above categories	nil	2–8
Nonmanufacturing	1–2	2–6

† Based on estimates and projections of the authors, and several sources listed in the references.

to be in assembly and inspection and in arc welding. It is anticipated that new uses for robots will be found in manufacturing and that these will constitute a growing share of the applications.

Nonmanufacturing uses of robots are very limited in numbers and dollar value at the time of this writing (July 1984). Examples of these current uses include research and development applications and teaching robots in colleges and universities. Teaching robots constitute a growing share of the market in terms of numbers of units; however, the price per unit for these robots is low compared to the price of an industrial grade robot. By 1990, the non-manufacturing applications will still constitute a relatively small proportion of the total robot installations. Sometime after the turn of the century, this proportion is projected to grow into a majority.

Assembly Applications

The assembly process represents an important future application for robots, and we have discussed this application in Chap. Fifteen. The area of assembly in which robots are expected to be used is in batch production operations. In the mass production of relatively simple products (e.g., flashlights, pens, and other mechanical products with fewer than 10 components), robots will probably never be able to compete with fixed automation in terms of speed and throughput rates. Even with lower-cost robots in the future, the economics will favor the use of high-speed specialized machines to accomplish the assembly tasks for these products. It is in the batch assembly of medium and small lots (e.g., electric motors, pumps, and many other industrial products) and in the high production of more complex assembled products (e.g., automobiles, televisions, radios, clocks), that robots are most likely to be utilized. However, these kinds of operations are currently the domain of human workers who possess the intelligence, dexterity, and adaptability needed for the tasks that go far beyond the capabilities of present-day robots.

This general area of assembly automation is sometimes referred to by the name programmable assembly. The present state of the art in programmable assembly is such that relatively few robots are employed in this technology. An estimated 5 percent of the current systems use robotics technology, but this proportion is expected to grow to 30 percent in 1990.[12] This suggests that advances in robot technology directed at the assembly process, combined with a better understanding of programmable assembly techniques, will occur during the period between 1985 and 1990. Some of the technological improvements needed to introduce robots in greater numbers into the assembly process include:

Improvements in sensor technology (especially machine vision)
Higher accuracy and repeatability
Higher speeds
Changes in design concepts and fastening methods for products to permit easier assembly by robots

More versatile grippers

Improved off-line programming methods that will permit complex robot programs to be developed from design data with the aid of advanced CAD/CAM software and to be downloaded directly to the assembly workstation for the required assembly tasks

A large field of applications for programmable assembly systems using robot technology is electronic assembly. The tremendous growth potential in the electronics industry over the next two decades provides a substantial impetus for the robotics industry to develop new robotic assembly systems.

Arc-Welding Applications

Another application expected to grow in importance is arc welding. Most continuous arc-welding operations are accomplished manually today. Present state-of-the-art robot arc-welding installations almost invariably involve the production of medium or high quantities of items. In this situation the robot must be programmed to perform the required welding cycle and the parts to be welded must be placed in fixed locations. Programming the robot to do the welding cycle takes considerably longer than the actual welding task. The location requirement is satisfied by means of a special welding fixture to hold the parts and a human fitter who works in the robot cell as explained in Chap. Fourteen. The productivity of these semiautomated welding cells can be two or three times as high as the corresponding manual cell in which a fitter and a welder work together. This is because of the low arc-on times usually encountered in manual welding operations. The economics of the application require that the production quantities must be sufficient so that the productivity gains per unit of product can overcome the initial cost of the programming time and the special fixture.

One of the technical problems described in Chap. Fourteen that arises in using robots for arc welding is the variation in the part edges that are to be welded. Human welders are capable of compensating for these variations during the welding operation, but the conventional playback robot cannot. This inability of playback robots to follow the variations in the welding gap has inhibited their use in the arc-welding process. Several sensor technologies are being developed to deal with the problem of part edge variations. It is anticipated that the widespread adoption of these sensor technologies will be an important factor in the expanded use of robots for arc welding in 1990 and beyond.

Parts Handling and New Robot Applications

Parts handling and machine loading represent a third large area of future robot applications, although the proportion of these applications compared to others will probably decline modestly. Perhaps the biggest limitation in using robots for these functions is the problem of locating and orienting the part so that the robot can fetch it at the beginning of the work cycle. In the past, there has been only one solution to this problem and that is to present the parts to the robot in a known position and orientation. This solution requires the parts to be prepositioned and preoriented for the robot application by means of some form of parts-handling device. Additional expense is involved to engineer the parts-handling capability and sometimes an extra manual operation is required to load workparts into the device. The problem of part position and orientation in robot applications provides reinforcement to a general argument among factory automation specialists that part orientation must be established when processing of the part initially begins and should never be lost during subsequent manufacturing and assembly operations. This is certainly not the common practice in today's factories geared largely toward manual operations in which parts are usually stored in random arrangements in tote pans, bins, and boxes.

Running counter to this part orientation argument are the recently introduced commercial systems capable of retrieving workparts that are all mixed together in a tote pan. These systems are called "bin-picking" systems. They are based on the use of machine vision and we have previously described their operation in Chap. Seven. It is anticipated that this bin-picking capability will be an important factor which will allow robots to be used in an increasing number of parts-handling, machine-loading/unloading, and many other factory operations. According to the Delphi study by the Society of Manufacturing Engineers and the University of Michigan[12] the proportion of robots sold in the United States with bin-picking capability will be 5 percent of the total market in 1990 and 10 percent in 1995. Robots equipped with machine vision to accomplish the bin-picking problem might simultaneously perform visual inspection operations on the parts that are being retrieved.

In Table 20-2, there is an entry listed as "other manufacturing operations, not covered by the above categories." We expect that several new robot applications in manufacturing will be developed that are not discussed in Chaps. Thirteen through Fifteen, and that these new applications might very well grow to become a significant portion of the total. The advances that are being made in robotics technology and related computer software will allow new uses of industrial robots that are today only found in the laboratory or not yet even seriously considered. These possibilities include wire harness assembly, garment manufacturing, shoemaking, product packaging, food-processing operations, dipping cycles in electrochemical plating, and a host of other unanticipated operations.

Flexible Manufacturing Systems

Finally, another technology that will spur future robot applications in manufacturing is computer-integrated flexible manufacturing systems (FMS). A flexible manufacturing system can be defined as a group of automated machine tools (usually numerically controlled machines) that are interconnected by means of a materials handling and storage system, and which

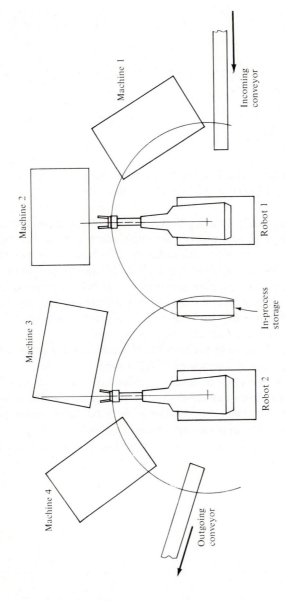

Figure 20-1 Layout of a machining cell that uses two robots to handle parts between machines. Note the in-process storage for transferring parts between robots. Parts are delivered into and out of the cell by conveyors.

operates as an integrated system under computer control. Although these systems started appearing approximately 15 years ago, the application of FMS technology is only now beginning to grow significantly. It is anticipated that flexible manufacturing systems will become less expensive to install and operate as improvements are made in the technology and as we learn more about them. The improving cost advantage compared to other forms of production will result in an increase in the share of U.S. manufacturing activity that is performed by FMS technology.[2] Robots are being used increasingly in flexible manufacturing systems and machining cells to perform the materials handling function. Industrial robots with properly designed grippers have been found to be ideal for handling rotational workparts in this type of application. Figure 20-1 shows a possible layout for a machining cell that uses a combination of robots and conveyors to handle parts. The conveyors are used to bring parts into and out of the cell and the robots are used to handle parts between machines in the cell.

20-3 HAZARDOUS AND INACCESSIBLE NONMANUFACTURING ENVIRONMENTS

Manual operations in manufacturing that are characterized as unsafe, hazardous, uncomfortable, or unpleasant for the human workers who perform them have traditionally been ideal candidates for robot applications. Examples include die casting, hot forging, spray painting, and arc welding. The workers who are displaced by robots in these operations are usually relieved to be out of the workplace as long as they are given alternative jobs that are better. It is anticipated that the same reaction will apply to hazardous manual tasks that are performed in nonmanufacturing situations.

The desire to remove a human worker from an unsafe environment is a worthy ambition and will undoubtedly lead to the development of new applications for robots. Additional robot applications will be developed for environments that are either inaccessible or altogether inhospitable for human beings. Examples of potential nonmanufacturing robot applications that are in hazardous or inaccessible environments include the following:

Construction trades
Coal mining
Hazardous utility company operations
Military applications
Fire fighting
Undersea operations
Robots in space

We will explore some of these possibilities, but not in great detail. For the interested reader, several of the references provide more comprehensive discussions of the topics.[1,2,6,8,13]

Robots in the Construction Trades

As indicated in Table 20-2, nearly all of the present applications of robots are in manufacturing, where they substitute for human workers in operations that are manual labor intensive. The construction industry represents an interesting opportunity for applying robotics technology because it is also based largely on the use of manual labor. There are three features about the manufacturing operations that have made robots relatively easy to apply. First, many (though certainly not all) of the operations where robots are substituted for human labor are hazardous. This feature has promoted the acceptance of robots in manufacturing by the workers. Second, the production operations can be performed at a single work location. And third, the tasks are highly repetitive. These three features apply only in varying degrees in the construction trades. First, some construction work must be considered hazardous since it is performed at high elevations. In spite of safety percautions that can be taken, there is no doubt that fabricating the steel frame of tall buildings or constructing high bridges is dangerous for the construction workers. Second, in the building trades, robots would typically be required to move about the construction site rather than remain at a single location. Providing the robot with mobility is difficult enough by itself. What makes it especially difficult in this application is that the construction site usually consists of dirt piles, ruts, ditches, and debris, and the robot would not only have to negotiate around these obstacles, but it would also have to climb stairs and squeeze through doorways as well. The third feature of many manufacturing jobs that make them ideal for robots to be used is the repetitive nature of the work. In high production and even in batch manufacturing, the robot performs either the same or a limited number of motion patterns over and over again. Some construction jobs can almost be considered repetitive. Examples of these jobs would include ditchdigging, bricklaying, applying roof shingles, painting, laying ceramic tile, installing insulation, and other similar tasks. These jobs require a repetition of motion cycles that are very similar, but the location of the work is always changing and this requires a translation of the motion pattern in space with each new cycle.

Another reason why construction work is an interesting possibility for the application of robotics is that many of the common construction machines use mechanisms similar to the mechanisms used in robotics. Backhoes, front-end loaders, ditchdiggers, and certain construction cranes reflect these similarities with their hydraulically operated manipulators and digging claws. All of these machines currently work under the control of human operators who direct the machines to accomplish their digging or lifting functions.

Solving the robot mobility problem might be done by borrowing from the construction machines identified above. For robots that perform digging operations, large wheels could be used to move about the construction grounds, with hydraulically operated legs to stabilize the robot at each location. For digging long trenches, the robot would move through a series of locations forming the line of the trench, with a similar motion cycle repeated at each location to dig the section of the trench. Getting the robot to follow the correct trench line and to accomplish the proper amount of excavation at each section, without any human assistance, would require a combination of advanced programming, sensor technology, and artificial intelligence that goes slightly beyond the current state-of-the-art in robotics.

For performing work internal to a building during construction, an entirely different anatomy would be required in the robot design. Perhaps in this case, the optimum method of solving the mobility problem would be to use a pedipulator. A walking robot might be designed which could climb a staircase as well as move across a flat floor. Another problem in the design of an internal construction robot would be to provide it with the ability to adapt to the variety of different tools (drills, hammers, chisels, paint brushes, sanders, caulkers, etc.) that are used in the building trades. For a technologically sophisticated construction robot, this might involve a simple change of end effectors. The ability to effectively utilize all of these tools is again a problem in advanced robot programming, sensor technology, and artificial intelligence.

These technical problems will be solved probably within the coming decade. When this happens, the development of construction robots will be reduced to a problem of economics. The prospect of being able to keep a robot working nearly 24 hours per day without having to pay time-and-a-half or double-time will probably be appealing enough to motivate the introduction of robotics in some form into the construction trades before the next century.

Underground Coal Mining

Among industrial occupations, underground coal mining is one of the most dangerous and unhealthy jobs that humans can do. The sources of the dangers include fires, explosions, poisonous gases, cave-ins, and underground floods. Although safety conditions have dramatically improved since the early 1900s, deaths in mining accidents still run at a much higher rate than the average of other work accidents. Around 1900, fatalities in mining accidents were running around 450 per 100,000 workers, based on figures presented in Ref. 2. This has been reduced, especially since 1969 with the enactment by Congress of more stringent mine safety laws, to around 80 deaths per 100,000 per year. This compares with an average rate of about 10 fatalities for all industrial work accidents. Even if the mine worker survives the accident statistics illustrated in these numbers, there is still the problem of "black lung" disease that is so common among those who have been exposed to the environment in underground mines for many years. Yet coal is perceived as a major source of energy in the United States for many years into the future. The problem of extracting the coal from the earth in a manner that does not place humans at undue risk for their lives is one that might be addressed through the use of robotics technology.

Highly mechanized systems are currently being used in the mining industry but their operation requires the attention of human workers for guidance and control. Examples of these systems include power undercutters and rotating head machines for digging and excavating at the mine face, conveyors for bringing the coal up from the excavation site to the surface, and loading machines for transferring the freshly dug coal to the conveyors. Full automation in mining operations would involve the implementation of machinery which could accomplish one or more of these functions without humans in attendance.

One possible "robotic" machine for doing the tunneling and retrieval of the coal from the cut face of the mine is pictured in Fig. 20-2. The mechanical complexity of the tasks to be accomplished are suited to robotics technology; however, once again the programming, sensory, and intelligence requirements go somewhat beyond the available state-of-the art. To be fully automatic, the tunneling robot would have to be capable of moving forward into the mine as the excavation proceeds. Accordingly, mobility would have to be designed into the machine, probably by means of tank treads. The profile of the tunnel (both in terms of the direction into the forward mine surface and the shape of the opening to be tunnelled) could be placed in the robot's memory. A vision

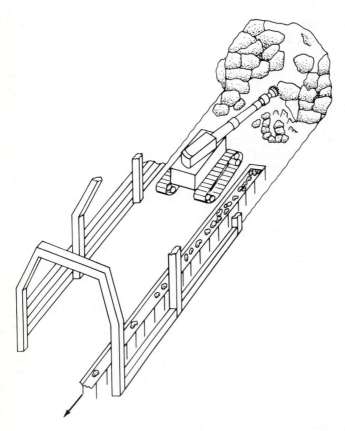

Figure 20-2 Robotic coal mining machine for tunneling and scooping coal from the mine face.

system would be the likely sensor system to determine how the cutting is progressing and to guide the robot in its digging operations. The dirty, dusty environment in the mine as a result of the excavation process presents a problem in mechanical reliability, and the design of the machine would have to take this problem into account. Also, the large forces encountered during the drilling process would add to the wear and tear on the machine. Periodically, as the tunnel is carved out, roof support would have to be constructed and the conveyor systems would have to be extended forward into the mine opening. These tasks might still be accomplished by human workers, or alternatively by sophisticated construction robots equipped for such work.

The technical complexity of the coal mining operation described above poses a difficult problem for the designer of a coal mining robot which must work independently of human attendance. However, given that there will be a commercial demand for coal well into the next century and that there are approximately 150,000 workers in underground coal mines in the United States alone, the need for such a machine is great.

Utilities, Military, and Fire Fighting Operations

There are many other nonmanufacturing work situations that present hazards or potential hazards to those employed in the work. These include certain utility company operations, military operations, and fire fighting. In each case, robotics technology might be utilized to reduce some of the risks to humans.

Some of the activities associated with utility company operations are unsafe for humans and represent possible applications for robots. The most prominent examples are maintenance and repair operations in radioactive boilers and the handling of nuclear fuels and other radioactive substances. Other examples include utility pole construction and repair and other high wire activities, coal pile grooming and compaction, boiler and condenser tube inspection, and inspection of flue gas stack liners and ducts during planned boiler outages. Additional utility company activities might include construction work (e.g., construction robots) for power stations and security monitoring of plant and surrounding areas.

Contemplating the various possible uses for advanced humanoid robots in military operations, the manning of naval vessels, fire fighting, and even some police work represents a fascinating mental exercise that approaches science fiction. The prospect of sending robots on a suicide mission deep into enemy territory without risking the lives of friendly soldiers must surely be a source of great interest to military planners. Many of the more routine tasks that are accomplished in military or naval operations and the logistics support for these operations might be performed by robots. Examples include refueling of vehicles out in the field, driving trucks in a convoy following a lead truck which is driven by a human, loading of field artillery or ship cannons, working in the engine room on board ship, and construction work for pontoon bridges and other temporary structures.

The use of robots for certain fire fighting, police, and disaster control functions is also a possibility. In fire fighting, robots could be used to extinguish fires in smoke filled atmospheres, guide people to safety in disaster situations, and operate the fire fighting equipment. Future police robots might augment municipal public safety departments in duties ranging from routine traffic control to dangerous functions such as explosive mine disposal.

Undersea Robots

The ocean represents a rather hostile environment for human beings due principally to extreme pressures and currents. Even when humans venture into the deep, they are limited in terms of mobility and the length of time they can remain underwater. It seems much safer and more comfortable to assign aquatic robots to perform whatever tasks must be done underwater. Among the possible uses of undersea robots are the following: exploring for minerals, gathering geological samples, underwater mining and drilling operations, retrieving lost objects, underwater construction, and undersea farming and fishing.

Underwater robots, properly protected against corrosion, could operate for extended periods of time at virtually any depth. They would be largely self-contained, equipped with on-board power supply, sensors, computer, and several manipulators for various functions. The power supply would probably consist of batteries which would not be nearly so heavy underwater as they are for vehicles on land. Other forms of power using fuel cells might also be adapted for underwater operation over extended periods of time. Hydraulic actuators and motors could operate using water as the hydraulic fluid. Hydraulics would be used to operate the robot arms and to move the robot in the water. Mobility would be provided by means of propellers or water jets combined with controlled buoyancy. The robots could be designed to be virtually weightless underwater, achieving the appropriate buoyancy by means of air chambers filled with various volumes of air and water. The ratios of air and water would be changed to permit the robot to regulate its underwater depth. The on-board sensors and computer would permit the robot to perform many of its normal functions independent of any human interference. However, radio communication would allow instructions to be given from surface ships and to provide sensor readings from the robot. Video cameras mounted on the robot would permit those on the surface to see underwater and perhaps to give commands accordingly.

The U.S. Navy has been using underwater vehicles since the mid-1960s with features similar to some of the ones described above. The term used by the Navy for these "robots" is remotely operated vehicle or ROV. Instead of being self-contained, these vehicles usually have an umbilical cord to the surface for power and control. Applications for these undersea vehicles have consisted mainly of recovering military ordnance lost near the coast, using a gripper mounted at the end of a manipulator arm to grasp the items. Figure 20-3 illustrates the Navy's Cable-controlled Underwater Recovery Vehicle.

Figure 20-3 U.S. Navy's CURV III (Cable-controlled Underwater Recovery Vehicle). Note gripper arm in front of vehicle in lower left of picture. (*Official Photograph, U.S. Navy*)

Robots in Space

Space is another inhospitable environment for humans, in some respects the opposite of the ocean. Instead of extremely high pressures in deep waters, there is virtually no pressure in outer space. In order to permit humans to survive the extreme conditions, they must be contained in some form of life-support system that provides pressure, air, and other requirements. In future space travel to faraway planets, the sheer enormity of the distances involved compared with the limitations on rocket velocity means that humans would be required to spend, perhaps, years away from earth in order to accomplish a space voyage within our own solar system. (Travel outside of the solar system would require more time than humans have available.) The safety issues involved in space travel would be considerable. Reliability of the equipment over extended time periods would pose a significant risk for human space travelers.

Robots would not need the elaborate support systems required for humans, and the time factor in space travel would have no emotional or psychological effects on robots. Equipment reliability would still be a problem but it would be only a reliability problem. There would be no threat to human life from equipment that fails in space travel if no humans were on board. These considerations have surely been on the minds of the engineers, scientists, and managers involved in the space program.

Technologies related to robotics have been used in the space program in several instances. In the U.S. Lunar Surveyor program during the 1960s, a remote controlled manipulator was used to dig a trench on the moon's surface and to perform other similar functions. The Lunar Exploration program by the U.S.S.R., in the late 1960s, also utilized a manipulator to capture soil deposits for return to the earth. In the U.S. Mars Viking program in 1976, a computer-controlled manipulator arm was used in a number of scientific experiments on the Martian surface. Finally, in 1982 the U.S. Space Shuttle started to use a 48-ft-long manipulator arm to remove payloads from the cargo bay of the shuttle and to handle various items in space. Figure 20-4 shows a picture of the shuttle arm.

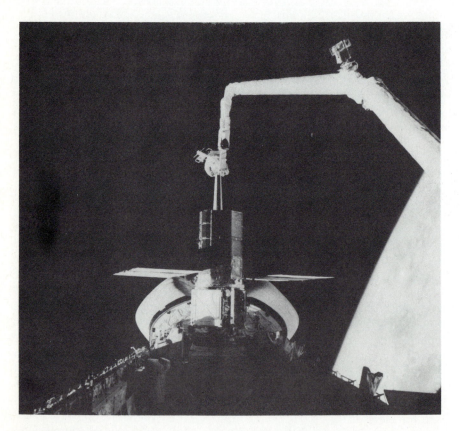

Figure 20-4 Remote manipulator arm on-board the U.S. Space Shuttle for handling cargo and other chores in space. Note radius of earth between shuttle bay and manipulator arm. (*Photo courtesy of National Aeronautics and Space Administration, Kennedy Space Center, Florida*)

The functions that would be performed by future robots and manipulators in space include exploration, construction in space, rescue missions, maintenance and repair, space transportation, materials processing, and other industrial operations in space. Space exploration by robots could add tremendously to our scientific knowledge of neighboring and more distant planets without risking human life. Mobile robots could be programmed to roam the surface of the planet, gather samples, take measurements, perform experiments, analyze the data, and send the results back to earth. On-board computers executing artificial intelligence and other advanced software would be able to make decisions on where to explore, what samples to gather, and which samples to bring back to earth if a return trip is contemplated.

Robots could also be used in the construction of space stations, factories, and large cargo vehicles that are built in outer space. The robots could be used to move materials, help in docking maneuvers for sections of the construction, and perform other functions that would assist the human workers who are supervising the project. These applications would allow the number of humans required to accomplish the project to be reduced, thereby reducing the need for more life-support systems in space. Rescue missions for astronauts or construction workers stranded in space could be carried out by robots. Other uses of space robots would include maintenance and repair operations on the equipment, and space travel involving the transportation of humans and/or cargo through space. In each of these applications, humans would control the robots using high-level commands and the robots would have adequate intelligence to carry out the instructions.

Certain materials-processing operations could be profitably carried out in outer space. Examples of these operations include containerless processing of liquid materials, diffusion processes in liquids and vapors, and solidification of molten metals without convection or sedimentation. Some biotechnology processes could also be performed beneficially in space. The distinctive environmental conditions offered by space which are advantageous in these processes are zero gravity and zero atmospheric pressure (close to a perfect vacuum). In addition to other sophisticated forms of automation, the use of robots to accomplish these manufacturing processes in space would reduce the need for human attendants and their associated life support systems, and would probably lead to lower production costs for the resulting materials.

20-4 SERVICE INDUSTRY AND SIMILAR APPLICATIONS

In addition to nonmanufacturing robot applications that are considered hazardous, there are also opportunities for applying robots to the so-called service industries. The possibilities cover a wide spectrum of jobs that are generally nonhazardous. We present the following subsections to illustrate the potential applications.

Teaching Robots

The concept of "teaching robots" may extend beyond the use of small safe machines in college classrooms and laboratories. Such robots are widely used today for teaching the principles of programming (as well as limited applications) to undergraduates and two-year technical school students. In the future, teaching robots might be useful in elementary school systems. Children would be likely to consider a small robot (close to the size of a child) to be a friendly machine and would be willing to "play" with the machine in an interactive mode to learn basic skills and concepts, much in the same way that personal computers are used today in many elementary schools. Robotic "teachers' helpers" would multiply the capabilities of human teachers, perhaps increasing the permissible student–teacher ratio.

Retail Robots

Intelligent robots might be used in certain repetitive functions in retail establishments, such as cleaning, straightening the merchandise, checkout at cash registers, and merchandise restocking.

Fast Food Restaurants

Engelberger[4] indicates that Unimation, Inc. was once asked by an officer of MacDonald's chain if a robot could accomplish some of the routine food preparation tasks required in a typical fast food restaurant. Fast food store operations are very labor intensive, especially in stores that stay open 24 hours per day. The skill levels required of the employees are very modest and many of the tasks are quite repetitive. With certain changes in the organization of the work in these restaurants, it is not difficult to imagine that robots could accomplish some of the tasks, such as cooking the food, dispensing beverages and ice cream, and making up orders based on instructions from a human order-taker.

Bank Tellers

Automatic tellers are used today for simple transactions such as deposits and withdrawals. Telephone checking is just beginning to be used as this chapter is being written. There will no doubt be a continuation of the trends in banking automation into the future, with the possibility that friendly teller robots may some day perform nearly all of the common customer-related transactions in a bank. Such a robot would have to be able to communicate in a manner which is unintimidating and convenient to the customer (voice recognition and speech synthesis technologies would have to be advanced beyond today's state-of-the-art). It would also have to add, subtract, count money, and obtain quick access to the central computer file to determine a customer's account status.

Garbage Collection and Waste Disposal Operations

Collecting garbage is another operation performed by humans today which is mostly routine. There have been a number of attempts to mechanize garbage collection operations involving the use of large fabricated steel containers that could be readily picked up and hauled away by specially equipped trucks. Most house collection operations today still rely on one truck driver and one or two workers who must collect the garbage cans and empty them into the back of the garbage truck. These latter functions could surely be perfomed in the future by mobile robots specially designed for lifting the garbage cans.

Cargo Handling, Loading, and Distribution Operations

Making up orders in a distribution warehouse, and loading them onto trucks or railroad boxcars, typically require a combination of clerical and physical labor that is routine and prone to mistakes when done manually. For large distribution centers, automated storage and retrieval systems (AS/RS) are used to computerize and mechanize these clerical and manual functions. Installation of an AS/RS facility is usually a multimillion dollar investment. For the smaller warehouse that either cannot afford to install an automated storage and retrieval system or whose volume of operations does not warrant a large system, robots or robotic-type devices may become useful for some of the order picking and loading functions. As these functions are currently organized around the use of manual labor, the robots would require mobility and the capacity to handle variations in the shape and size of the items and containers used in warehouse operations. Although order picking is repetitive in a general sense, the locations of the items to be collected are different, and the robot would require sufficient intelligence to deal with these nonroutine portions of the typical order-picking cycle. The robots would also need to be able to receive ordering instructions from the warehouse computer files.

Security Guards

Security guards lead a lonely existence, periodically roaming through the building to check for intruders and other irregularities. The duties also include sitting in front of closed-circuit TV monitors whose cameras are trained on entrances, exits, and other areas of the building and surrounding grounds.

Robot technology could be employed to perform some of these duties. Mobile robots, equipped with sensors to detect the presence of human intruders, could wander through the building on a random schedule designed to foil the intentions of burglars who might rely on a regular timetable to carry out their sinister activities. Sensing the presence of humans in unauthorized building space, the robot would communicate its observations to a central station manned by human security guards who are prepared to take appropriate action.

Medical Care and Hospital Duties

Much of the work that is done in hospitals by staff nurses, practical nurses, nurse's aides, orderlies, and technicians is clerical and routine. Robots are likely to perform some of this work in the future. Some of the hospital functions that might be automated include delivering linens, making beds, clerical duties such as entering patient records into computer file, delivering medicines and supplies from the hospital pharmacy and central supply, and transporting patients for different services in the building. Some of the duties might even include aspects of patient care such as monitoring vital signs, and passing water and food to the patients.

A related medical care activity that might be performed by robots or robotic devices at some point in the future is assistance for paraplegics and other physically handicapped persons. Providing handicapped persons with full-time robot servants is a meritorious social objective that might eventually be realized.

Agricultural Robots

Although the labor content required to operator a farm has been drastically reduced over the last 60 years by mechanized equipment, there still remain opportunities for further automation. The Japanese[10] have identified a variety of tasks that might be accomplished with the help of future robots in the agricultural and related industries. These tasks include harvesting, soil cultivation, fertilizer spreading, and application of insecticides. Related areas of potential robot applications might be found in forestry and livestock care and management. The possibility of using robotic devices to shear sheep in Australia has been explored, and some of this work is illustrated in Fig. 20-5.

Household Robots

The prospect of a domestic robot in nearly every home provides a tremendous market potential and a tremendous commercial opportunity for the company that captures that market. Chores that might be accomplished by a household robot include dishwashing, rug vacuuming, making beds, furniture dusting, window washing, and certain food preparation tasks. Many of the technical problems that must be solved in the design of a construction robot (discussed in Sec. 20-3) also arise in the case of a household robot. The robot would need to be capable of mobility and obstacle avoidance in order to find its way around the house or apartment. A truly versatile robot would be capable of movement on more than a single floor, and this would add a degree of difficulty to the mobility problem. The robot would also need to receive high-level oral commands (e.g., "wash the dishes," "clean the rug," "make the beds," etc.) and to reduce those commands to a detailed set of actions that must be carried out one by one in order to perform the given chore. In addition to its regular duties during the day, the household robot could be on duty at night, per-

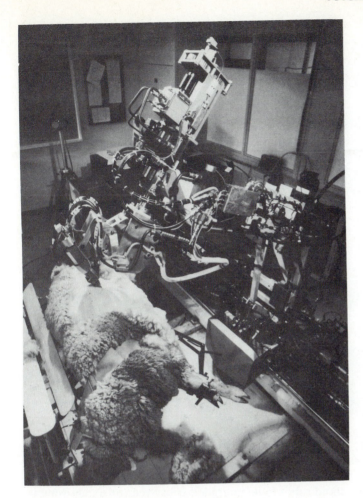

Figure 20-5 Research on sheep-shearing robots is of interest in New Zealand and Australia. (*Photo courtesy of Cary Wolinsky/Stock Boston*)

forming monitoring functions with its sensors to make sure the house is secure against burglars, and to act as a smoke detector and fire alarm.

Figure 20-6 illustrates one of the first commercially available "household robots." The cost of a highly functional household robot would be limited not by the intelligence requirements for the machine, but by its mechanical and sensor requirements. It is anticipated that advances in microprocessor technology will permit powerful computers (relative to today's standards) to be mounted on-board future robots and that the cost of these computers will be a minor portion of the total robot price. The development costs for software used in the household robot will be spread over many thousands (perhaps millions) of units, thus allowing the software portion of the price to be minimized. It is probable that various software packages will be commercially

Figure 20-6 The RB5X Robot, "the world's first mass-produced, general-purpose robot designed specifically for home use". (*Photo courtesy of RB Robot Corp.*)

available for the household robot, just as different software is available for today's personal computers. New software introduced to the market would permit an existing household robot to be upgraded every year or so, allowing it to accomplish increasingly complex tasks.

The mechanical structure of the robot and its sensor systems would probably establish a lower limit on the price of a household robot. Even if manufacturing costs have been significantly reduced by the economies of mass production, the material costs of a robot large enough to perform useful household chores would be substantial. Albus has estimated that the price of such a robot would be in the range $4000 to $6000 (in 1980 dollars). The choice for an average household might be between buying a new car or a new household robot. And if the decision is based on how much of the family's time is affected by each of the two alternatives, it would probably turn out that the robot would have a bigger impact on the family's lifestyle.

Specially designed robots might be capable of performing lawn and garden work. The possibilities include mowing the lawn, spreading fertilizer and other chemicals, grass trimming, and clipping a hedge or bush. These robots could be powered by gasoline engines, similar to today's tractor-type mowers. A simple instruction, such as "mow the lawn" would engage the robot to accomplish several hours work, requiring it to reduce that macro-level command into a complex sequence of travel motions to finish the job.

The use of domestic robots in hotels for cleaning and making over the guest rooms would add an extra dimension to the market for this class of robot. These machines could be kept busy a high proportion of the day and their worth to the hotel would be measured in terms of the work they could accomplish compared to a human maid employed by the hotel. The investment criteria for the hotel would be similar to that used for current industrial robots in manufacturing applications.

20-5 SUMMARY

In the preceding chapters of the book, we have discussed the technology, programming, and applications of robots: how they work, how to work them, and what work they can do. In the present chapter, we have examined the prospects for smarter, mobile robots in the future to manufacture products more cheaply, build bridges more safely, explore outer space, search under the sea, help doctors in patient care, and assist homemakers with domestic chores.

A substantial opportunity exists in the technology of robotics to relieve people from the boring, repetitive, hazardous, and unpleasant work in all forms of human labor. There is a social value as well as a commercial value in pursuing this opportunity. The commercial value of robotics is obvious. Properly applied, robots can accomplish routine, undesirable work better than humans and at lower cost. As the technology advances, and more people learn how to use robots, the robotics market will grow at a rate that will approach the growth of the computer market over the past 30 years. One might even consider robotics to be a mechanical extension of computer technology.

The social value of robotics is that these wonderfully subservient machines will permit humans more time to do work that is more challenging, creative, conceptual, constructive, and cooperative than at present. There is every reason to believe that the automation of work through robotics will lead to substantial increases in productivity, and that these productivity increases year by year will permit humans to engage in activities that are more cultural and recreational. Not only will robotics improve our standard of living; it will also improve our standard of life.

In the first chapter of the book, we mentioned Carel Capek's science fiction play about sinister robots which ultimately brought great harm to humans. It seems appropriate in this final chapter to express our belief that the field of robotics, in contrast to Capek's play, offers the promise of great commercial and social benefit to humankind.

REFERENCES

1. J. S. Albus, *Brains, Behavior, and Robotics*, BYTE Books (McGraw-Hill), 1981, chap. 11.
2. R. U. Ayres and S. M. Miller, *Robotics, Applications and Social Implications*, Ballinger, Cambridge, MA, 1983.

3. L. Conigliari, "Trends in the Robotics Industry," *Technical Paper MS82-122*, Society of Manufacturing Engineers, Dearborn, MI, 1982.
4. J. F. Engelberger, *Robotics in Practice*, AMACOM (Division of the American Management Association), 1980, chap. 9.
5. J. F. Engelberger, "The Household Robot: by 1993," *Decade of Robotics*, IFS Publications, Bedford, England, 1983, pp. 12–13.
6. P. Foster, "Can the Robot be the Miner's Friend," *Decade of Robotics*, IFS Publications, Bedford, England, 1983, pp. 102–103.
7. W. B. Gevarter, "Robotics: An Overview," *Computers in Mechanical Engineering*, August 1982, pp. 43–49.
8. E. Heer, "Robots in Space," *Decade of Robotics*, IFS Publications, Bedford, England, 1983, pp. 104–107.
9. V. D. Hunt, *Industrial Robotics Handbook*, Industrial Press Inc., New York, 1983, chap. 14.
10. Japan Industrial Robot Association, *The Robotics Industry of Japan*, *Today and Tomorrow*, Fuji Corporation, Tokyo, Japan, 1982.
11. Office of Technology Assessment, *Exploratory Workshop on the Social Impacts of Robotics*, Washington, D.C., February 1982.
12. D. N. Smith and R. C. Wilson, *Industrial Robots*, *A Delphi Forecast of Markets and Technology*, Society of Manufacturing Engineers, Dearborn, MI, 1982.
13. Wernli, R., "The Silent World of the Undersea Robot," *Decade of Robotics*, IFS Publications, Bedford, England, 1983, pp. 100–101.

INDEX